Statistics and Decisions

An Introduction to Foundations

Statistics and Decisions

An Introduction to Foundations

Steven H. Kim
Massachusetts Institute of Technology

CRC Press
Taylor & Francis Group
Boca Raton London New York

CRC Press is an imprint of the
Taylor & Francis Group, an **informa** business

A CHAPMAN & HALL BOOK

First published 1992 by Van Nostrand Reinhold

Published 2019 by CRC Press
Taylor & Francis Group
6000 Broken Sound Parkway NW, Suite 300
Boca Raton, FL 33487-2742

© 1992 by Taylor & Francis Group, LLC
CRC Press is an imprint of Taylor & Francis Group, an Informa business

First issued in paperback 2019

No claim to original U.S. Government works

ISBN 13: 978-0-367-45033-5 (pbk)

**Visit the Taylor & Francis Web site at
http://www.taylorandfrancis.com**

**and the CRC Press Web site at
http://www.crcpress.com**

Library of Congress Cataloging-in-Publication Data

Kim, Steven H.
    Statistics and decisions : an introduction to foundations / Steven
H. Kim.
        p.   cm.
    Includes bibliographical references and index.

    1. Probabilities.   2. Statistics.   3. Statistical decision.
I. Title.
QA273.K49115   1992
B19.2—dc20   91-36316
                                                 CIP

Library of Congress Catalog Card Number 91-36316

To my sister Eun

Contents

Part II.   STATISTICS

Part III.   DECISIONS

APPENDICES

Preface

Our lives involve decision making in an uncertain world. The scientist designs an experiment without knowing what types of results will be obtained, even if the work succeeds; an engineer designs a prototype without full knowledge of the artifact's performance; the marketer promotes a new service without fully understanding the predilections of the consumer; the executive sets a strategy with only dim knowledge of the evolving economy and society. At a more personal level, we commit to engagements while unaware of our future inclinations; we enter into relationships without fully understanding the consequences; and we choose careers with imperfect knowledge of the implications.

Topics such as these are the intended domain of decision theory. However, despite the pervasiveness of probabilistic phenomena and decision making under uncertainty in our lives, most of us are largely unaware of the tools, techniques, and results from decision theory.

Courses in decision theory are often taught in schools of management, but seldom in departments of science and engineering. This situation may be due, in part, to the constrained nature of texts or references in the field. A number of books on the market introduce the field in an informal way, and others provide an excellent in-depth view. These books serve their respective objectives well. From a pedagogical perspective, however, each type of book has its own limitation. The informal books, almost by definition, lack rigorous mathematical treatment; on the other hand, the technical texts assume prior knowledge of background subjects, predominantly in the form of probability and statistics.

The need exists for a book that provides the necessary prerequisites in probability and statistics as well as the key ideas in decision theory. This book is intended to fill that role.

The book should appeal primarily to students and practitioners who desire to apply decision-theoretic thinking to their own work, rather than to specialists whose goal is to contribute to the field of decision theory. Even so, my hope is that the book will serve as a readable and systematic introduction to the field. As alluded to above, the book may serve as a background to more advanced texts in engineering as well as the physical and social sciences.

The only mathematical prerequisite for this book is calculus, including the manipulation of several variables. The modular structure of this book will facilitate its use as a supplement to lectures or as a self-study guide. For example, an instructor wishing to focus on decision theory and its applications may rely solely on Part III; this approach presupposes some background in probability and statistics among the students.

From a different perspective, the book also facilitates differential levels of rigor in instruction. For instance, Part I presents probability both casually and formally. In particular, Chapter 2, which deals with the conceptual issues in probability, is an informal treatment of the subject, while the mathematical foundations and proofs are relegated to the more formal presentation given in Chapter 3.

In a similar way, Part II on statistics partitions the presentation into casual and formal treatments. As a result, the instructor may provide for an informal introduction to decision theory by assigning only the conceptual chapters.

In summary, this book is intended to be distinguished from other introductory texts in mathematical statistics and decision theory in a number of ways:

- *Appeal to intuition.* Concepts are first introduced in intuitive ways, then explained in rigorous fashion while striving to maintain readability.
- *No prior knowledge of probability.* A background in probability theory is not necessary, as the relevant topics are introduced in Part I.
- *Critical concepts in statistics.* The conceptual foundations of statistics are featured, including estimation and hypothesis testing. Other techniques such as regression and experimental design, although of great utility in many areas of applied statistics, have been bypassed to conserve space.
- *Foundations of decision theory.* The foundations of decision theory are emphasized, including their basis in statistical methods.
- *Informational perspective.* Issues bearing on information processing concerns are elaborated. These topics—such as the sufficiency of data, the invariance of procedures, and sequential techniques—relate to both theoretical and practical consequences.

These features of the book render it especially attractive as an introduction to statistical methods and their application to diverse arenas.

The book should be of interest to students and researchers in a wide range of fields, from engineering and the natural sciences to management, economics and psychology.

I am indebted to Elias Towe for thoughtful comments on the manuscript. The word processing and graphics were cheerfully accomplished by Maureen Kelly and Linda Cohen. Any errors which remain, however, can be credited to my account.

S.K.
Cambridge, Massachusetts

Statistics and Decisions

An Introduction to Foundations

1

Introduction

Decisions are our constant companions. They accompany us throughout the day, whether at home, school, office, or leisure.

Often these decisions are easy to make because they are fully comprehensible and simple to resolve. This happens, for instance, in choosing among several job offers, one of which is better than the others along every dimension: salary, environment, opportunity for growth, and so on.

Other decisions are easy to make because the perceived consequences are roughly similar. Should I wear my blue slacks or gray slacks today? Should we go swimming or jogging?

All too often, however, the problems are more difficult. The difficulty arises when the consequences of alternative decisions are worlds apart. It also occurs when our understanding of the problem is imperfect, whether due to incomplete data, poor problem specification, probabilistic phenomena, or cognitive limitations.

Examples of nontrivial problems are the following:

- A zoologist is investigating a species of bird which may be on the verge of extinction. Part of the study involves an estimation of the average number of offspring and their survival rate. Unfortunately, observations are difficult and the data sporadic. What can the investigator conclude on the basis of a handful of data?

- A production manager is concerned about the unusually high rate of products that fail at the inspection stage. He must determine whether each piece of manufacturing equipment is calibrated properly–that is, ensure that the operating performance lies within the optimal range for each device.

- A pharmaceutical company has developed a new drug to eradicate the common cold. In a 3-year test involving 100 subjects, 62 have developed no symptoms of the affliction. Is the drug effective, or is the result due merely to chance?

- A start-up company has created a household robot to serve as a sentry, vacuum carpets, and perform other minor chores. Should the company market a simple model priced at $3,000 to appeal to middle-income families? Or should it develop a sophisticated $27,000 version that would appeal to a luxury market? Or, as a safeguard against competitors a year or two down the road, should it promote both models to capitalize on its current monopoly position?
- An investor has her financial assets tied up in stocks, bonds, and money market funds. It appears that the economy is poised for a period of healthy expansion. Should she change her portfolio, moving funds from, say, bonds to stocks? If so, should she do it today or next month? And how much should she transfer? Which stocks should she purchase?
- An artist has been commissioned to create a painting for a major exposition. He goes through one preliminary sketch after another, each improving on the last, yet none of them perfectly satisfactory. How long should he continue? At what point does he say "Enough!"?

Problems such as these can be addressed through the techniques of decision theory. The scenario with the zoologist is an example of *estimation*: determination of the true value of a parameter based on partial information. The task of the production manager is one of *localization*: establishing the interval that contains the unknown value of a parameter, or ensuring that the parameter lies within a given interval. The problem faced by the pharmaceutical company is one of *hypothesis testing*: deciding whether a proposition is true or false. The problem with the robotic company falls under the category of *multiple actions*: settling on one among several courses of action. The portfolio and painting scenarios illustrate the problem of *sequential procedures*.

These categories are listed separately for convenience in discussion. However, many problems involve features of two or more categories. For instance, the robotic marketing case involves a multiple action problem; but in order to make an informed decision, the company may perform a marketing survey to determine the percentage of prospective customers in each price range—a task of estimation. In a similar way, the financial investment example involves a *sequential* task of deciding *when* to modify the portfolio of assets; but the decision must be made in conjunction with the *multiple actions* available to the investor.

BACKGROUND TO DECISION THEORY

The origins of decision theory, as it is known and practiced today, lie primarily in the fields of statistics and game theory. Statistics focuses on drawing inferences from data. The data may describe a phenomenon completely, as in the case of the grades in a particular class; but often the information is incomplete, as in data on the masses of stars in the galaxy. Statistical analysis, then, deals with rational procedures for drawing conclusions from data.

The logical foundations of statistics lie in probability theory, a field dealing with uncertainty. Probabilistic analysis deals with the modeling of uncertain phenomena and inferences concerning their properties, such as long-run tendencies.

The fields of probability and statistics are the mathematical cornerstones of decision theory. As mentioned earlier, however, the study of decision making owes an intellectual debt to game theory as well. The field of games deals with decision making among multiple agents. These agents may interact cooperatively, as with partners in a game of cards; they may compete, as often happens among rivals in the business environment or in athletic events; or they may interfere destructively, as in a chess game or on a battlefield. In particular, decision theory may be viewed as a special case of game theory in which an agent seeks to maximize his gains in the face of actions taken by nature. Unlike most games, however, nature is a neutral player and takes no deliberate action to help or hinder the decision maker.

PLAN OF THE BOOK

As the title suggests, this book provides a systematic introduction to the foundations and key results in decision theory. Part I deals with concepts and techniques in probability. Chapter 2 introduces the basic concepts of probability, whereas Chapter 3 offers a more formal view of the field.

Part II presents the field of statistics. Chapter 4 highlights the conceptual issues, whereas Chapter 5 outlines the field in a rigorous way. The next chapter presents estimation and localization, whereas the last chapter in this part deals with hypothesis testing in greater detail.

Part III delves into decision theory. Chapter 8 is a conceptual introduction, whereas Chapter 9 presents a formal background in terms of the theory of theory. The last two chapters explore the extension of decision theory to more sophisticated contexts. In particular, these relate to the problems of multiple decisions and sequential decision making.

The material in this book is designed to prepare the reader to tackle more advanced applications of probability, statistics, and decision theory. An example lies in certain stochastic problems in economics and finance. Another example is the analysis of knowledge on the performance of intelligent systems. References to these and several other topics are presented under the heading "Further Reading and Selected Bibliography."

I
Probability

2

Concepts of Probability

We live in the midst of uncertainty. We get up in the morning without knowing whether to expect rain or shine; we utter words without knowing how they will be perceived; we embark on projects without knowing how well they will fare; we enter races without assurance of completing the course—let alone winning the event.

Despite the vagaries of everyday life, we can better understand where we stand by turning to probabilistic methods. The field of probability deals with the modeling of uncertain events and the analysis of their properties.

Some probabilistic arguments are exemplars of common sense. If the likelihood of getting a head on each toss of a coin is one-half, what is the probability of observing two heads in a row? The answer, one-quarter, seems so obvious as to obviate further discussion.

On the other hand, many probabilistic situations are not so obvious. What is the chance of observing an odd number of heads over three tosses of a fair coin? We need to pause for a moment to deduce the answer of one-half.

How about the likelihood of obtaining the sum of 7 or above on 2 tosses of a die? What if the likelihood of observing an odd number on the second toss is triple that of an even number when the first toss yields a 5?

It is clear that even for restricted domains such as coins and dice, the scenarios can quickly escalate into problems of discouraging proportions. These domains are trivial compared to most tasks we encounter as a matter of course at home, work, or play.

For such heavy-duty problems, the concepts and techniques of probability can provide relief by reducing formidable problems to manageable scale. This chapter introduces the basic ideas of probability, whereas Chapter 3 presents a more rigorous approach.

The next section explores the notion of an event, which is a prelude to probabilistic arguments. This is followed by the concept of probability and its relationship to different types of events. The idea of a random variable is then introduced. The behavior of a random variable is specified by its distribution pattern, which determines characteristics such as the average value and degree of dispersion.

2.1   EVENTS

A *trial* refers to the occurrence of an experiment or natural phenomenon. The trial is defined by a condition or an activity; examples are found in the throw of a die, the turn of the weather, or the course of the economy.

An *event* refers to the result of a trial that might, but need not, transpire. The event might be characterized by the roll of a "6" on a die, the onset of a hurricane this evening, or a plunge in the stock market next week.

The limiting cases of events are also considered to be events. Let Ω denote the *total* or *certain* event characterized by every possible outcome of a trial, and $\varnothing$ the *null* or *impossible* event containing no outcomes. Both Ω and $\varnothing$ are viewed as events. In the case of a die, the total event Ω is "1, 2, 3, 4, 5, or 6 occurs," whereas an example of the null event is "7 occurs."

A *complete* or *exhaustive* set of events is a collection whose union is the total event Ω. In rolling a die, let E_o denote the event "An odd number occurs" and E_e the event "An even number occurs." Then we say that the set $\{E_o, E_e\}$ is complete with respect to Ω.

A set of events is *mutually exclusive* or *incompatible* if no more than one member of the set can occur. In tossing a coin, the events "head turns up" and "tail turns up" are mutually exclusive.

A set of events may, of course, be exhaustive as well as mutually exclusive. For the die example, the two events "an odd number occurs" and "an even number occurs" have both these properties.

The *sum* of a set of events is defined by their union. If E_1, E_2, ..., E_n represent n events, then their sum is written as $E_1 \cup E_2 \cdots \cup E_n$.

Let E_i denote the event in which the number i turns up on a die. Suppose D is the sum of E_1 and E_2—that is, $D = E_1 \cup E_2$. Then D is the event in which a "1" or a "2" turns up.

The *product* of a set of events is an event in which all the underlying events arise. The product of n events E_1, E_2, ..., E_n is written as $E_1 \cap E_2 \cap ... \cap E_n$ or E_1 & E_2 & $\cdots$ & E_n. For the die example, let E_o be the event "an odd number occurs" and $E_{4,5,6}$ the event "4, 5, or 6 occurs." Then $C = E_o$ & $E_{4,5,6}$ is the joint event where the outcome is odd and is a "4," "5," or "6." The event C is equivalent to the event E_5, which denotes the occurrence of a "5."

2.2 PROBABILITY

The notion of probability corresponds to the chance that some event of interest will occur. An intuitive concept of probability is based on the idea of *objective assessment*. For instance, the probability of observing a "3" on the roll of a die is 1/6 since the event of interest is simply 1 out of 6 contenders of equal likelihood.

The idea of a rational characterization of events leads us to the classical definition of probability.

- *Probability as Objective Assessment*. Suppose that a trial can result in N outcomes, each being equally likely. If M of these outcomes correspond to the event E, then the probability of E is given by $P(E) = M/N$.

In other words, the probability of an event is given by the ratio of favorable outcomes to the total number of equally likely outcomes.

The value of a probability P will range from 0 to 1, inclusively. The probability of the null event is 0 and that of the total event is 1; that is, $P(\varnothing) = 0$ and $P(\Omega) = 1$.

Regarding probability as the likelihood of an individual trial is a comfortable view in simple contexts. In rolling a die, it is a simple matter to define all six outcomes and to regard them as being equally likely.

In most contexts, however, the outcomes are *not* equally likely to occur. Some experiments may be designed to investigate the very proposition that alternate outcomes are equally likely. This occurs in a demographic context in which the question is "Is a newborn baby as likely to be male as female?"; or in an industrial setting such as "How often does the *morning* shift operate without breaking down in our sedan factory?"; or in an economic context such as "What is the chance that economic recovery is initiated in the automotive sector as opposed to other industries?"

For such applications, a second view of probability is more appealing. This is based on the idea of probability as *long-term frequency*: the probability of an event E is defined by the likelihood that E occurs in a long sequence of observations. For instance, if 512 out of 1000 babies sampled at random are male, then the probability that a newborn will be a boy is 0.512.

From the frequentist perspective, our level of belief in an estimated probability increases with the number of trials. The "real" probability is viewed as the limit obtained through observations of increasingly larger size. We define these notions more precisely as follows.

- *Probability as Long-Term Frequency*. Let $E_1, E_2, E_3, \ldots$ be a set of observations of an event E. The indicator function I_j monitors E_j; that is, $I_j = 1$ if E occurs on observation E_j and $I_j = 0$ otherwise. Then the probability of E

is given by

$$P(E) = \lim_{n \to \infty} \frac{\sum_{j=1}^{n} I_j}{n}$$

In some sense, this definition embodies an empirical view of probability: it relies on actual observations of some event E.

On the other hand, there are at least two practical difficulties with the frequentist view of probability. The most obvious is that we cannot afford the luxury of an infinite number of observations. A second difficulty is that of convergence. Consider a coin-tossing scenario, whether through manual tosses or computer simulation. Our task is to estimate the probability p of obtaining heads. Even if the coin were "truly" fair, having an underlying parameter of $p = 1/2$, we are not likely to observe *precisely* 50 heads out of 100 tosses, or 5000 heads out of 10,000 trials. In fact, the observed value of p will most likely move into the neighborhood of .5, then fluctuate around this central value, assuming both higher and lower numbers. However, it has little chance of converging to the true value of $1/2$ with indefinite accuracy.

Despite these drawbacks, however, the view of probability as long-term frequency is intuitively appealing and will serve in many practical contexts.

The frequentist view of probability has another limitation in relation to the scope of its applications. By definition, it relies on multiple observations of a phenomenon and is, therefore, of little use when an event may occur only once, or perhaps not at all. Examples arise in contexts such as the following:

- "What is the probability that it will rain during my sister's first ski trip?"
- "What is the probability that robots will outnumber humans by the year 3000?"
- "How likely is it that the unidentified objects on the radar are enemy missiles?"
- "What is the probability that a meteor caused the extinction of the dinosaurs?"

In the first two examples, the event of interest belongs to the future; the third case involves the present; the last scenario involves the past. In each case, the event relates to a single trial. What does it mean for the weatherman to say "The probability of rain tomorrow is .5"? In this context there is only *one* tomorrow, the one following *this* day. Since the event is binary—it will occur or not—the posterior estimate of rain *after* tomorrow is strictly 0 or 1.

One way to interpret probabilities in these contexts is to regard the numbers as subjective or personal estimates. According to the *subjective* view, a probability is a personal estimate of likelihood that makes sense only in relation to other trials having similar estimates.

- *Probability as Subjective Assessment.* The probability of an event E is a personal estimation of the chance of the occurrence of E. It varies from a low of 0 to a high of 1, depending on the strength of the belief in the event.

For instance, to say that the probability of nuclear war is .4 is to express a stronger conviction in the event than to offer a value of .1.

The subjective view of probability can be related to empirical phenomena. This is achieved by comparing among multiple trials of similar probability. If a value of p is given as a subjective probability, then the trials associated with that value should result in an observed proportion close to p. To illustrate, suppose that the probability of rain on Sunday is estimated as .3. Then about 30% of the cases in which that proclamation was made should result in rain.

Of the three definitions of probability discussed in this section, the notion of probability as a subjective assessment is perhaps the most general. According to the subjective perspective, there is no explicit attempt to relate probabilities to empirical phenomena. It, therefore, avoids the conceptual difficulties associated with the first two definitions. The alternative perspectives of probability, however, are more of philosophical than of practical consequence.

In fact, for many applications, alternate viewpoints of probability are equivalent in operational terms. For instance, my evaluation of the probability of rolling a "4" on a die is 1/6 according to the view of probability as objective assessment: a "4" is 1 out of 6 equally likely outcomes. By my own experience and those of others, a "4" turns up about 1 out of 6 times in throwing dice, so I infer that the long-run frequency of the event is 1/6. Either or both of these rationales move me to assess the probability of rolling a "4" on a die as 1/6, according to the subjective perspective; I suspect that among events to which I assign a chance of 1/6, I will be correct about 1 out of 6 times. In this context, the three views of probability converge to the same result in operational terms.

The modern definition of probability avoids ties to empiricism altogether. The definition is based on an axiomatic foundation and is purely a mathematical construct. We will examine this subject further in Chapter 3.

In the meantime, the subjective view of probability is sufficiently general for our purposes. It will serve as a comfortable intuitive foundation for the remainder of this chapter.

2.3 CONDITIONAL VERSUS UNCONDITIONAL EVENTS

If an event consists of a set of outcomes that are mutually exclusive, then the probability of the event is given by the sum of the individual probabilities. For instance, let the event E be the compound event $E_1 \cup E_5$, where E_i is the probability of obtaining i on the roll of a die. Then the probability of E is

$$P(E) = P(E_1) + P(E_5) = 1/6 + 1/6 = 1/3$$

In general, if an event E consists of mutually exclusive subevents E_1, $E_2, \ldots, E_n$, then the probability of the compound event is

$$P(E) = \sum_{i=1}^{n} P(E_i)$$

Most of our discussion so far has focused on *simple* or *unconditional* probabilities in which the outcome of a trial depends only on the initial conditions. For instance, the probability of obtaining E_i is $1/6$ assuming that a die is ideal and each of the 6 faces is equally likely to turn up.

A *conditional* probability is one that depends on some other event or stipulation. For instance, the probability of drawing a king from a stack of cards during a game depends on the identity of the other cards withdrawn prior to this trial. The notation $P(E|D)$ is used to denote the probability that event E occurs under the assumption that event D has already occurred.

Example

An urn contains billiard balls, each of which is green or red. The urn is opaque, and the experimenter is as likely to draw one ball as any other. If there are g green balls and r red ones, what is the probability of drawing a green ball? In addition, what is the conditional probability of drawing a green ball given that a single green one has already been withdrawn from the urn? What is the conditional probability of drawing a red one in the same situation?

Let n denote the initial number of balls in the urn; that is, $n = g + r$. Further, G represents the event of drawing a green ball and R the event of a red ball.

Then the (unconditional) probability $P(G)$ of picking a green ball at the first drawing is given by the ratio of green balls to the total number: $P(G) = g/n$. In a similar way, $P(R) = r/n$. Since the trial must result in either a green or a red ball, the corresponding probabilities must sum to 1:

$$P(G) + P(R) = g/n + r/n = 1$$

Let G' and R' be the conditional probabilities of drawing a green and red ball, respectively, on the second trial assuming that a green one has already been removed. In that case we have $g - 1$ green ones among a total of $n - 1$ balls. Hence, the conditional probability of drawing a green ball, given that a green has already been removed, is

$$P(G'|G) = \frac{g-1}{n-1}$$

By a similar argument, the conditional probability for picking a red ball at the second step is given by

$$P(R'|G) = r/(n-1) \qquad \blacklozenge$$

2.4   INDEPENDENT EVENTS

A set of events is *independent* if the events have no effect on each other. The joint probability of a collection of independent events is given by the product of their respective probabilities.

To illustrate, let L be the event that I receive a letter from my father today, and S the event of snowfall tomorrow. Since L and S are independent, the probability of the joint event—denoted $P(L \,\&\, S)$ or $P(L, S)$—is given by the product of the individual probabilities: $P(L, S) = P(L) \cdot P(S)$. In particular, if $P(L) = 1/30$ and $P(S) = 1/2$, then $P(L, S) = 1/60$.

We may state the general formula for the joint probability of independent events as follows:

Proposition. Let E be the joint event consisting of independent subevents E_1, $E_2, \ldots, E_n$. The probability that E occurs is given by the product of the respective probabilities

$$P(E) \equiv P(E_1, E_2, \ldots, E_n) = \prod_{i=1}^{n} P(E_i)$$

In the above formula, the symbol "$\equiv$" means "is defined as."

2.5   DEPENDENT EVENTS

A set of events is called *dependent* if it is not independent. In other words, at least one event affects at least one other event in the set.

The joint probability of a collection of dependent events must take into account the conditional probabilities that characterize their dependence. To illustrate, let E_o denote the event of obtaining an odd number on the roll of a die, and E_h the event of a high number (4, 5, or 6). These events are not independent: knowledge of one event affects the likelihood of the other.

The unconditional probabilities are each one-half: $P(E_o) = 1/2 = P(E_h)$. However, knowing that E_o has occurred affects the chance that E_h has also occurred, and vice versa. In particular, the probability that the outcome is odd given that it is high is $P(E_o|E_h) = 1/3$: of the three high outcomes (4, 5, or 6), exactly one is odd (5). Conversely, the probability that a number is high

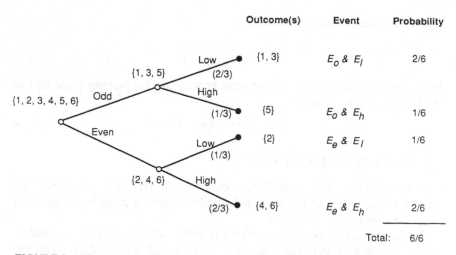

	Outcome(s)	Event	Probability

$\{1, 2, 3, 4, 5, 6\}$ — Odd — $\{1, 3, 5\}$

Low (2/3) → $\{1, 3\}$ — E_o & E_l — 2/6

High (1/3) → $\{5\}$ — E_o & E_h — 1/6

Even — $\{2, 4, 6\}$

Low (1/3) → $\{2\}$ — E_e & E_l — 1/6

High (2/3) → $\{4, 6\}$ — E_e & E_h — 2/6

Total: 6/6

FIGURE 2–1. Event tree for die-tossing example, in terms of parity (odd or even) followed by magnitude (high or low). E_o denotes the event "Odd number"; E_e denotes "Even number"; E_l denotes "Low number"; E_h denotes "High number." The braces indicate possible outcomes and parentheses represent probabilities. For instance, the initial set of possibilities is $\{1, 2, 3, 4, 5, 6\}$. The event "Odd" may occur with probability 1/2, and the corresponding outcomes are $\{1, 3, 5\}$. At this juncture the conditional probability of the event "Low" given that "Odd" occurred is 1/3.

given that it is odd is $P(E_h|E_o) = 1/3$: of the three odd outcomes (1, 3, or 5), exactly one is high (5).

The joint probability must, therefore, take into account these contingent probabilities. The probability that both E_o and E_h occur is the probability that E_o occurs *and* the probability that E_h occurs after adjusting for the fact that E_o occurs as well. More specifically, the joint probability is given by the product of the unconditional and conditional probabilities:

$$P(E_o, E_h) = P(E_o) \cdot P(E_h|E_o) \qquad (2\text{--}1)$$

Since $P(E_o) = 1/2$ and $P(E_h|E_o) = 1/3$, the joint probability is $(1/2)(1/3) = 1/6$. This line of argument is depicted in Figure 2–1 as a tree of events.

The situation is symmetric between the two events. In other words, the joint event can also be specified in terms of the occurrence of E_h and the conditional validity of E_o given that E_h occurs. The formula is

$$P(E_h, E_o) = P(E_h) \cdot P(E_o|E_h)$$

The corresponding tree of events is presented in Figure 2–2.

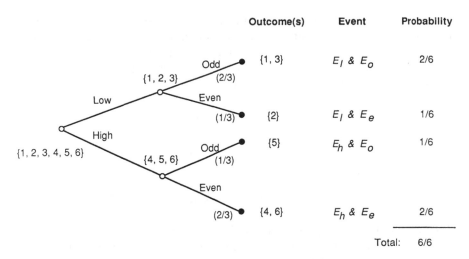

FIGURE 2-2. Event tree for die-tossing example, in terms of magnitude (low or high) followed by parity (odd or even).

We state the general formula for two events as follows:

Proposition. Let D and E be two events. Their joint probability is given by either of the following formulas:

$$P(D, E) = P(D) \cdot P(E|D)$$
$$P(D, E) = P(E) \cdot P(D|E)$$

In the proposition, the conditional probability can be viewed as a way to subtract out the influence of the unconditioned event. For instance, $P(E|D)$ is the probability that E occurs, after adjusting for the fact that D is known to occur as well. In this way, the probability of the conditional event "$E|D$" can be viewed as being independent of D.

As a result, $P(D, E) = P(D, "E|D")$. Since $E|D$ is the component of E that is independent of the occurrence of D, the adjusted event $E|D$ occurs independently of D. Therefore, their joint probability is given by the product of the respective events: $P(D, "E|D") = P(D) \cdot P(E|D)$.

Total Probability

Consider a set of mutually exclusive and collectively exhaustive events E_1, E_2, ..., E_n. Since the E_i are exhaustive, their union is the certain event: $E_1 \cup E_2 \cup \cdots \cup E_n = \Omega$.

Let D be another event; then D & E_i is the joint event where D and E_i both occur. Now the joint probabilities are pairwise independent. In other words, the joint event $D \& E_i$ is mutually exclusive in relation to any other joint event $D \& E_j$. Therefore, saying "D occurs, and at least one of E_i or E_j occurs" is equivalent to saying "at least one of these compound events occur: D and E_i, or D and E_j." In other words, we can write

$$D \& (E_i \cup E_j) = (D \& E_i) \cup (D \& E_j)$$

This relation holds for any number of subevents E_i, as well as the entire collection:

$$D \& (E_1 \cup E_2 \cup \cdots \cup E_n) = (D \& E_1) \cup \cdots \cup (D \& E_n)$$

The left-hand side is equal to the event "D occurs, as well as *any* of the E_i." But the E_i are collectively exhaustive and sum up to the certain event Ω. The equivalent statement is "D occurs, as well as the certain event Ω." Since the certain event must always occur by definition, we conclude "D occurs." In algebraic terms, we have shown the following:

$$D = D \& \Omega = D \& (E_1 \cup \cdots \cup E_n)$$
$$= (D \& E_1) \cup \cdots \cup (D \& E_n)$$

As stated previously, each compound event $(D \& E_i)$ is disjoint from any of its siblings $(D \& E_j)$. Therefore, the probability of their union is the sum of their respective probabilities. From the above equation, we infer

$$P(D) = \sum_{i=1}^{n} P(D \& E_i)$$

Since $P(D \& E_i)$ is equivalent to $P(E_i)P(D|E_i)$, we can also write this formula as

$$P(D) = \sum_{i=1}^{n} P(E_i)P(D|E_i) \tag{2-2}$$

This relation is called the formula for *total probability*.

Example

Let D be the event of obtaining a defective product among the items sampled in a factory. The factory has two production lines. The event E_1 refers to sampling from one production line, and E_2 the other line. The events E_1 and

E_2 are mutually exclusive and exhaustive in terms of obtaining samples. The probability of observing a defective item is

$$P(D) = P(D, E_1) + P(D, E_2)$$
$$= P(E_1)P(D|E_1) + P(E_2)P(D|E_2)$$

Hence, the total probability of observing a defective item can be determined from the chances of sampling from each line and the corresponding conditional probability of having a defective item.

In particular, suppose the two lines are equally likely to be selected. Then $P(E_1) = P(E_2) = .5$. Now assume that the probabilities of defective items from the two lines are $P(D|E_1) = .01$, and $P(D|E_2) = .03$. Then the overall probability of observing a defective item is

$$P(B) = .5(.01) + .5(.03) = .02 \qquad \blacklozenge$$

Bayes' Rule

A conditional probability may be expressed in terms of joint and unconditional probabilities. Consider two events, A and B. The chance that B occurs, once it is known that A has occurred, is given by their joint probability normalized by the probability of the first event A. In symbolic form, we have

$$P(B|A) = \frac{P(A, B)}{P(A)} \qquad (2\text{--}3)$$

This formula is called *Bayes' Rule*, in honor of a minister who proposed it in the eighteenth century (Bayes, 1763). This result can also be obtained through a rearrangement of Eq. (2–1).

The intuition behind Bayes' Rule is as follows. The denominator on the right-hand side of the equation is a normalizing factor for the joint probability. Suppose that the probability of A is high, say .9. Then $P(B|A)$ is only slightly higher than that of the joint probability $P(B, A)$. In the extreme, when A is certain to occur, $P(A) = 1$ and the conditional probability $P(B|A)$ simply matches the joint probability $P(A, B)$.

On the other hand, suppose that the probability of A occurring is low. Since the chance of A occurring is small, the chance of their joint occurrence—namely, $P(A, B)$—is also small. But B could be closely associated with A so that whenever A occurs, B is highly likely to follow suit. Then $P(B|A)$ should be high. This is achieved by having $P(A)$ in the denominator of Bayes' Rule. For instance, if $P(A)$ is .1, then $P(B|A)$ is greater than the straightforward joint probability $P(A, B)$. In fact, the adjusted value is larger by a factor of $1/0.1 = 10$.

Let us explore some alternative expression for Bayes' Rule. We know that the joint probability of A and B can be written as $P(A, B) = P(B)P(A|B)$. As a result, an equivalent form of Bayes' Rule is

$$P(B|A) = \frac{P(B)P(A|B)}{P(A)} \tag{2-4}$$

The numerator in Eq. (2–3) has been expanded in terms of the unconditional probability of B and the conditional probability of A given B.

Bayes' Rule provides a useful relationship when the conditional probability of one sequence of events can be determined more easily than the converse. These concepts, as well as the notion of total probability, are clarified in the next example.

Example (Bayes' Rule)

In his childhood, Morton used to be very fond of marbles. Yesterday he found in his attic several containers of marbles: one box and two cylinders. The box contains 10 green marbles and 20 red ones. Each of the cylinders, however, contains 15 green marbles and the same number of red ones. Suppose Morton picks one of the three containers with equal probability, then a marble at random from that container. If the marble thus selected is green, what is the probability that it came from the box?

Since there are two cylinders but only one box, the probability of selecting the box is $P(\text{Box}) = 1/3$. Conversely, the chance of selecting a cylinder is the complement: $P(\text{Cylinder}) = 2/3$.

The conditional probabilities for the color once the container type is selected are simple to deduce. For instance, once the box is chosen, the chance of selecting a green marble is $P(\text{Green}|\text{Box}) = 10/30 = 1/3$. In a similar way, given that the container is a cylinder, the chance of selecting a green marble is $P(\text{Green}|\text{Cylinder}) = 15/30 = 1/2$. These numbers are shown in the event tree of Figure 2–3.

The *path probability* at each terminal node in the event tree is the chance of reaching that spot. For instance, the probability of reaching the bottom node is the chance of selecting a cylinder, followed by the chance of selecting a red marble:

$$P(\text{Cylinder, Red}) = P(\text{Cylinder})P(\text{Red}|\text{Cylinder}) = (2/3)(1/2) = 1/3$$

To return to the original question, we must determine the probability that

Path Probability

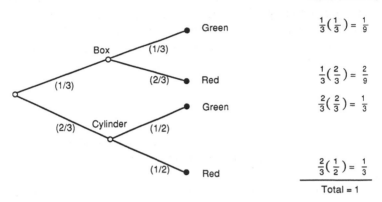

$$\frac{1}{3}\left(\frac{1}{3}\right) = \frac{1}{9}$$

$$\frac{1}{3}\left(\frac{2}{3}\right) = \frac{2}{9}$$

$$\frac{2}{3}\left(\frac{2}{3}\right) = \frac{1}{3}$$

$$\frac{2}{3}\left(\frac{1}{2}\right) = \frac{1}{3}$$

Total = 1

$$P(\text{Box}|\text{Green}) = \frac{P(\text{Box, Green})}{P(\text{Green})} = \frac{P(\text{Box})\,P(\text{Green}|\text{Box})}{P(\text{Box})\,P(\text{Green}|\text{Box}) + P(\text{Cyl.})\,P(\text{Green}|\text{Cyl.})} = \frac{(1/3)(1/3)}{(1/3)(1/3) + (2/3)(1/2)} = 1/4$$

FIGURE 2–3. Illustration of Bayes' Rule. The probability of selecting a box is 1/3, whereas the chance of a cylinder is 2/3. The chance of selecting a green marble, once a box is selected, is 1/3. The *path probability* is the chance of reaching the corresponding endpoint. For instance, the path probability of reaching the box node, followed by the red node, is first 1/3, then 2/3; the overall value is $P(\text{Box})P(\text{Red}/\text{Box}) = (1/3)(2/3) = 2/9$.

the container is a box, given that the marble is green. We can write this as

$$P(\text{Box}|\text{Green}) = \frac{P(\text{Box, Green})}{P(\text{Green})} = \frac{P(\text{Box})P(\text{Green}|\text{Box})}{P(\text{Green})} \qquad (2\text{--}5)$$

The first equality reiterates Eq. (2–3) and the second relies on Eq. (2–4).

The denominator can be expanded in terms of the total probability of obtaining a green marble. From Eq. (2–2) we have

$$P(\text{Green}) = P(\text{Box})P(\text{Green}|\text{Box}) + P(\text{Cylinder})P(\text{Green}|\text{Cylinder})$$

$$= (1/3)(1/3) + (2/3)(1/2) = 4/9$$

We can now return to Eq. (2–5) to obtain

$$P(\text{Box}|\text{Green}) = \frac{(1/3)(1/3)}{4/9} = \frac{1}{4}$$

which is the desired result. ◆

This example illustrates the convenience of Bayes' Rule. The conditional probability of selecting a box, once the marble is known to be green, is not obvious. But it can be computed from other quantities that are easier to determine, such as the conditional probability of picking a green marble once a box is selected.

2.6 RANDOM VARIABLES

A *random variable* is a variable that might assume any one of several values in trials conducted under identical conditions. Each of the alternative values corresponds to a different outcome. A set of outcomes is also called a *random event*.

Random variables may be classified broadly into two categories on the basis of their granularity. A *discrete* random variable is one that takes quantized values. Examples are found in the number of strokes played in a game of golf, the tubes of toothpaste purchased in a store, or the number of planes arriving at an airport in a single day.

In similar fashion, a *continuous* random variable is one that can assume any value over a finite or infinite interval. This occurs in the case of the temperature in a room, the amount of rainfall during a month, or the life span of a person.

A random variable is characterized completely by its *density* function, which defines the relative likelihood of assuming one value over the others. When the outcomes for a random variable can be ordered along a numerical dimension, then its density function may be integrated to yield the cumulative *distribution* function for the variable. These concepts are examined in greater detail below.

Discrete Random Variables

The *density* function for a discrete random variable defines the probability of obtaining each outcome of a trial. The *density* function for a discrete variable is also called its *mass* function. More specifically, let $x_1, x_2, \ldots, x_n$ be the values that a random variable X can assume.[1] Then the probability that X takes on value x_i is written in any of the following ways: $P(X = x_i)$, or $p_X(x_i)$, or simply $p(x_i)$ when it is clear that the underlying variable is X.

---

[1] Practitioners in probability and statistics tend to use capital letters to denote variables and lowercase letters for constants. This may be disconcerting at first to readers in the natural sciences, engineering, and most other fields, who are used to the opposite convention. The concerned reader will be relieved to hear, however, that a few weeks of regular exercise with problem sets will eliminate the discomfiture; what at first seems idiosyncratic will appear entirely natural.

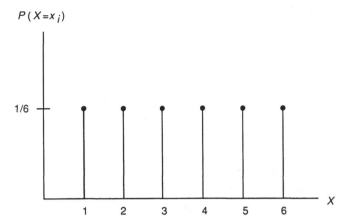

FIGURE 2–4. Density (or mass) function for the die-rolling example. Random variable X can take any of six values: $x_1 = 1$ through $x_6 = 6$.

In a coin-tossing context, we could specify the following quantities: $x_1 \equiv 0$ if the outcome is a tail, and $x_2 \equiv 1$ for a head. Here we have used the symbol "$\equiv$" to denote "defined as." For a fair coin, the corresponding probabilities are each one-half: that is, $p(x_1) = p(x_2) = 1/2$.

For the case of a die, the probability for each of the six possible values is $1/6$. The probability density function for this example is shown in Figure 2–4.

We have seen that the probabilities for all the outcomes must sum to 1. Suppose $x_1, x_2, \ldots, x_n$ are the outcomes for a random variable X. Then $\sum_i P(x_i) = 1$.

The *distribution* function of a random variable is defined by the cumulative sum of the density for all values up to and including the argument. Let $x_1, x_2, \ldots, x_n$ be a set of values for random variable X, and $p(x_i)$ the corresponding probabilities. The distribution function $F_X(x_i)$ is the probability that random variable X takes on values up to, and including, x_i:

$$F_X(x_i) = \int P(X \leq x_i) = \sum_{j=1}^{n} P(x_j)$$

The distribution function F_X may also be written as $F(X)$ or simply as F when the underlying variable is understood.

The distribution function for the die example is given in Figure 2–5. For instance, the value $F(2)$ is $1/3$ since

$$F(2) = P(X \leq 2) = P(X = 1) + P(X = 2) = 1/6 + 1/6 = 1/3$$

The distribution function $F(X)$ is 0 for $X < 1$, and 1 for $X \geq 1$.

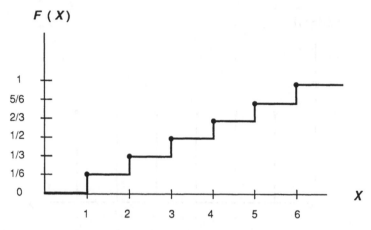

FIGURE 2–5. Distribution function for the die rolling example. The distribution is the cumulative sum of the density given in Figure 2–4. That is, $F(x_i) = \sum_{j=1}^{i} P(x_j)$.

Regardless of the values of the random variable X, its distribution $F(X)$ is the cumulative sum of non-negative values. It is, therefore, monotonically increasing. The function assumes the value 0 for $X = -\infty$, and reaches 1 for $X = \infty$. In symbolic form, we can state these three properties as follows:

- $F(X = x) \leq F(X = y)$ for $x \leq y$
- $\lim\limits_{x \to -\infty} F(X) \equiv F(-\infty) = 0$
- $\lim\limits_{x \to \infty} F(X) \equiv F(\infty) = 1$

Continuous Random Variables

A continuous random variable is one that takes values in an interval, such as the weight of a kangaroo. The probability that a continuous random variable X takes on a specific value x in an interval A is 0. For instance, there is no chance of meeting a kangaroo whose weight is *exactly* 5 kilograms.

On the other hand, it is possible to assign a nonzero probability to the event that X assumes a value within some infinitesimal strip dx on the interval A. This value defines the relative likelihood of observing a value for X in one differential slice $(x, x + dx)$ versus another slice $(y, y + dx)$. The *density* function $f_X(x)$ of the continuous random variable X defines the relative likelihood of X, assuming a value in a differential strip containing x (as opposed to some other strip). In Figure 2–6, the fact that $f(X)$ is greater at x than at y, implies that the variable X is more likely to fall in an infinitesimal band around x than around y.

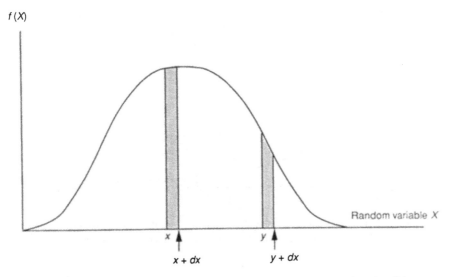

FIGURE 2–6. Density function for a continuous random variable X. The function $f(x)$ defines the relative likelihood of observing X at value x. More specifically, if $f(x)$ is larger than $f(y)$, then the likelihood of observing X in interval (x, dx) exceeds that for (y, dy). The integral of F over the interval (x, y) gives the probability of X falling in that interval.

The integral of $f(X)$ over some interval A defines the probability of observing X within A. In Figure 2–6, the chance of observing X within the interval $(x, y]$ is given by

$$P(x < X \leq y) = \int_x^y f(X)\,dX$$

Since the density $f(x)$ is 0 for any particular value x, the interval of integration may be open or closed, in part or in whole. In other words, the probability of observing X in some interval A is identical whether A is taken to be open, closed, half-open, or half-closed: (x, y), $[x, y]$, $(x, y]$, or $[x, y)$.

An example of a density function is the exponential form

$$f(x) = \begin{cases} \lambda e^{-\lambda x} & \text{if } x \geq 0 \\ 0 & \text{otherwise} \end{cases}$$

This function of X is characterized by the quantity λ, called a *parameter* of the density. The value of λ defines the steepness of the curve in Figure 2–7: the higher the value of λ, the steeper the initial slope as X increases from 0.

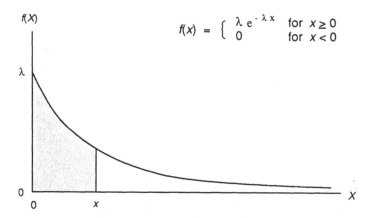

FIGURE 2–7. Example of a density function. Random variable X has an exponential density function. The shaded area defines the probability that X assumes any value up to x.

The shaded area in the figure corresponds to the probability that the variable X takes on a value of x or less.

The exponential density arises in many applications. One classs of applications relates to the time taken for an event to occur. Suppose that the chance of an event occurring is constant over time, being small but fixed from one moment to the next. Then the duration until the next occurrence of the event is an exponential function.

This situation often occurs in the longevity of light bulbs, the time to decay for a radioactive isotope, and the duration between two customers arriving at a bank. In this case, the parameter λ defines the rate of occurrence of the event. If $\lambda = 5$/hour, then on average five events occur each hour. As the value of λ increases, the slope in Figure 2–7 begins at a higher point and drops more quickly as a function of X. Because the area under the curve defines the probability of X occurring in the corresponding interval, a large value of λ gives more weight to small values of X. In other words, the event is more likely to occur sooner than later.

The distribution function for a continuous variable is defined by the cumulative integral of its density. More specifically, let X be a continuous random variable. Then its *distribution* function F_X is given by

$$F_X(x) \equiv P(X \le x) = \int_{-\infty}^{x} f(X)\,dX$$

Since the density f is non-negative, the distribution F for a continuous variable X has a number of properties which correspond to that for a discrete variable. First, F is monotonically increasing. Second, it vanishes for negative

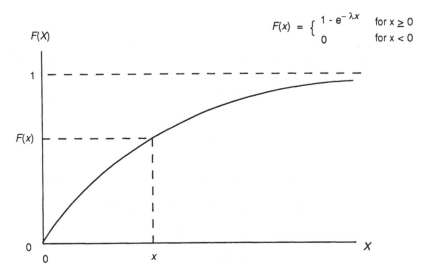

$$F(x) = \begin{cases} 1 - e^{-\lambda x} & \text{for } x \geq 0 \\ 0 & \text{for } x < 0 \end{cases}$$

FIGURE 2–8. Distribution function for the exponential form. $F(x)$ is the probability that X takes on any value up to x; in other words, it corresponds to the shaded area in Figure 2–7.

infinity: $F(-\infty) = 0$. Third, it attains its maximum value of 1 at positive infinity: $F(\infty) = 1$.

The distribution function for the exponential density with parameter λ is

$$F(x) = \begin{cases} 1 - e^{-\lambda x} & \text{if } X \geq 0 \\ 0 & \text{otherwise} \end{cases}$$

As indicated in Figure 2–8, the value of $F(X)$ vanishes for negative values of X and approaches 1 in the limit as X extends to infinity.

The probability that a random variable falls in any interval A can be expressed in terms of the distribution function. Suppose x and y are the endpoints of interval A. Then the chance of X falling in A is given by

$$P(x < X \leq y) = \int_x^y f(X) \, dX$$

$$= \int_{-\infty}^y f(X) \, dX - \int_{-\infty}^x f(X) \, dX$$

$$= F(y) - F(x)$$

In Figure 2–6, for instance, the lightly shaded area corresponds to the probability that X will fall within x and y. Its value is given by $F(x) - F(y)$.

Example

Let X be the duration of a message in a computer network. Suppose X has an exponential density with parameter 3/second. What is the probability of observing a long message in excess of 2 seconds?

The density function is given by $f(x) = 3e^{-3x}$. The corresponding distribution is $F(x) = 1 - e^{-3x}$. The probability that X is 2 or less is

$$P(X \leq 2) = F(2) = 1 - e^{-3(2)}$$

The chance of X exceeding 2 is given by the complement with respect to unity:

$$P(X > 2) = 1 - P(X \leq 2) = 1 - F(2)$$
$$= 1 - (1 - e^{-6}) = e^{-6}$$
$$\cong .0025$$

In other words, the chance of encountering a message exceeding 2 seconds is about 0.25%.
◆

2.7   EXPECTATION OF A VARIABLE

As discussed earlier, a random variable is completely defined by its density or distribution function. Knowledge of the density function f is tantamount to knowledge of the distribution function F because F can be obtained from f through summation or integration. Conversely, the density f can be obtained as the marginal value or derivative of F.

The density can be used to infer a number of characteristics of the underlying variable. The two most important attributes are measures of location and dispersion. We examine the first of these attributes in this section and relegate the second attribute to the next section.

The *mean* or *expectation* of a random variable is a composite of its values weighted by the corresponding probabilities. The mean is, therefore, a measure of central tendency: the value that the random variable takes "on average."

Let $x_1, x_2, \ldots, x_n$ be the outcomes for a discrete random variable X. The *mean* or *expectation* of X, indicated by the notation μ_X or $E(X)$, is defined as

$$E(X) \equiv \sum_{i=1}^{n} x_i p(x_i)$$

The symbol E is also called the *expectation operator*.

The mean value need not assume one of the outcomes of the random variable. Suppose X is the outcome when a fair die is rolled. For each value of X from 1 to 6, the probability is 1/6. The mean of X is given by

$$E(X) = \sum_{i=1}^{6} x_i p(x_i)$$

$$= 1(1/6) + 2(1/6) + 3(1/6) + 4(1/6) + 5(1/6) + 6(1/6) = 7/2$$

The result, 3.5, is a measure of the average location of X but is distinct from any of the actual values of the random variable.

The mean of a continuous random variable is obtained by weighting it against its density and integrating over all possible values of the variable. More specifically, let $f(X)$ be the density function for a continuous random variable X. Then the mean of X is

$$E(X) \equiv \int_{-\infty}^{\infty} xf(x)\,dx$$

To illustrate, consider the simple case of a density function X having a uniform value on the interval $[a, b]$. In more explicit terms,

$$f(X) = \begin{cases} \dfrac{1}{b-a} & \text{for } a \le X \le b \\ 0 & \text{otherwise} \end{cases}$$

The function is depicted in Figure 2–9.

The mean of X for this density function is obtained as follows:

$$E(X) = \int_{-\infty}^{\infty} xf(x)\,dx = \int_{a}^{b} x\left(\frac{1}{b-a}\right)dx = \frac{1}{b-a}\left(\frac{b^2 - a^2}{2}\right)$$

$$= \frac{a+b}{2}$$

For the uniform density, the mean lies midway between the two endpoints.

As with discrete variables, the value of the mean need not coincide with a possible value for x. To illustrate, consider the "split uniform" density having value 0.5 in the intervals $[0, 1]$ and $[2, 3]$, but 0 elsewhere. Then the mean will lie in the middle at 1.5, outside both of the two intervals having positive probability.

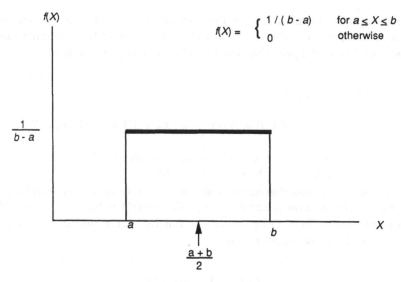

FIGURE 2–9. Uniform density function for random variable X. The mean $E(X)$ is at $(a + b)/2$.

Expectation of Multiple Variables

The sum of random variables is also a random variable. If X and Y represent outcomes of a die, then $Z = X + Y$ represents the sum of the individual outcomes. The quantity Z is characterized by its own density function.

The possible values of X and Y are depicted in Figure 2–10. If the dice are fair, then each of the 36 points in the figure is equally likely to occur. The probability that Z takes value z is directly proportional to the number of points in the corresponding rectangle. The density function for Z is also shown in the figure.

From this density function we can calculate the mean of Z as

$$E(Z) = \sum_{i=1}^{12} ip(z_i) = 2(1/36) + 3(2/36) + \cdots + 12(1/36)$$

$$= 7$$

This result is also the sum of the individual means of X and Y, namely, $3.5 + 3.5$.

More generally, the *mean of a sum* of random variables is the *sum of the means*. Let X and Y be two random variables. Then the mean is given by

$$E(X + Y) = E(X) + E(Y)$$

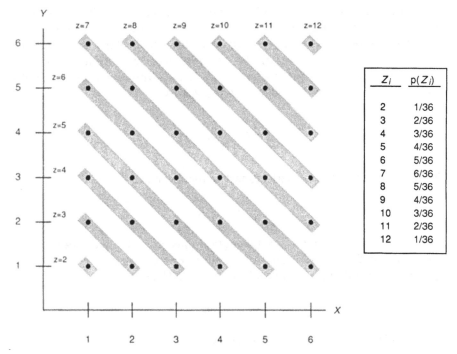

FIGURE 2–10. The space of outcomes for rolling two dice. Each of random variables X and Y takes values between 1 and 6, inclusively, with equal probability. Their sum $Z = X + Y$ varies from 2 to 12; the probability of $Z = z$ is given by the number of points in the corresponding rectangle.

In other words, the expectation operator E has the property of additive linearity. This result applies to three or more variables, so that the mean of the sum of multiple variables is the sum of their means.

2.8   VARIANCE OF RANDOM VARIABLES

Often, we are interested not only in the location of a random variable, as embodied in its mean value, but also the dispersion. An instructor might wish to know not only that the average grade on a quiz is 75, but also how tightly the scores cluster around this value. An executive in charge of airline operations would want to know not only that the mean flight time between Boston and Los Angeles is 6.5 hours, but also how scattered the flight times are.

In the grading example, suppose that one lab section of a biology class consists of three students whose grades are $\{75, 75, 75\}$. A second section consists of three students whose scores are $\{70, 75, 80\}$. Both sections have the same mean score of 75. On the other hand, their dispersions differ: the second set of grades is more scattered than the first, which shows no variation at all.

One way to obtain a measure of dispersion for a particular score is in terms of its deviation from the mean. To be more specific, consider a random variable X which can assume values $x_1, x_2, \ldots, x_n$, with respective probabilities $p_1, p_2, \ldots, p_n$. Then the mean of X is given by

$$\mu \equiv \sum_{i=1}^{n} p_i x_i$$

The *deviation* of a value x_i with respect to the mean is

$$\delta_i = x_i - \mu$$

To illustrate, assume that a genetic engineer has created a new strain of hen that lays 1 or 5 eggs with equal probability. For the outcomes $x_1 = 1$ and $x_2 = 5$, the probabilities p_i are each equal to .5. The mean is

$$\mu = .5(1) + .5(5) = 3$$

The deviations for x_1 and x_2 are given by

$$\delta_1 = x_1 - \mu = 1 - 3 = -2$$
$$\delta_2 = x_2 - \mu = 5 - 3 = +2$$

In other words, x_1 is two units below the mean and x_2 is two units above. The deviation δ_i is a meaningful measure of dispersion for each x_i.

On the other hand, the discrepancy δ_i is of little use as a measure of scatter for the *collective* set of outcomes. In the egg-laying example above, we see that the mean deviation is 0:

$$E(\delta) = .5(-2) + .5(+2) = 0$$

In fact, the expected deviation is 0 for *any* density function. This follows from the fact that the average deviation above the mean exactly matches that below the mean.

Not all is lost, however. We can nullify the destructive interference of positive and negative deviations by squaring their values. In other words, the squared deviation for the ith outcome is

$$\delta_i^2 = (x_i - \mu)^2$$

We can obtain a measure of deviation for the collective set of outcomes by taking the mean of the squared deviations:

$$\sum_i p_i(x_i - \mu)^2$$

The mean squared deviation of a random variable X is more commonly called the *variance* of X. It is denoted by var(X) or σ_X^2, and often simply as σ^2 when the underlying variable X is understood.

To summarize, let $x_1, x_2, \ldots, x_n$ be a set of values for random variable X, with respective probabilities $p_1, p_2, \ldots, p_n$. If μ is the mean of X, then the variance of X is given by

$$\text{Var}(X) \equiv \sum_{i=1}^{n} p_i(x_i - \mu)^2$$

Example

Let us return to the egg-laying context. The variance of the number of eggs is given by

$$\text{Var}(X) = p_1(x_1 - \mu)^2 + p_2(x_2 - \mu)^2 = .5(1 - 3)^2 + .5(5 - 3)^2 = 4$$

If a second strain of hen were to lay 2 or 4 eggs with equal probability, then the mean is still 3. On the other hand, the variance would be smaller. Let Y denote the number of eggs corresponding to the new strain. Then its variance is

$$\text{Var}(Y) = .5(2 - 3)^2 + .5(4 - 3)^2 = 1$$

Suppose that a third strain of hen generates 1, 3, or 5 eggs with equal probability. If U is the corresponding random variable, its mean is still 3. However, the variance is

$$\text{Var}(U) = (1/3)(1 - 3)^2 + (1/3)(3 - 3)^2 + (1/3)(5 - 3)^2 \approx 2.67$$

According to the criterion of the mean squared deviation, the variable U is more tightly clustered than X but less than Y.

Assume that a fourth strain lays 1, 3, or 5 eggs with probabilities .1, .8, and .1, respectively. If V is the random variable associated with this third strain, then the mean is still 3. However, its variance is

$$\text{Var}(V) = .1(1 - 3)^2 + .8(3 - 3)^2 + .1(5 - 3)^2 = 0.8$$

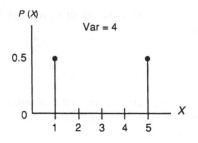

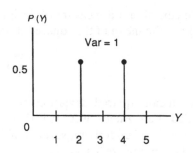

(a) Values {1, 5} each with probability 1/2.

(b) Values {2, 4}, each with probability 1/2.

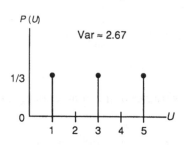

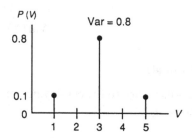

(c) Values {1, 3, 5}, each with probability 1/3.

(d) Values {1, 3, 5} with respective probabilities {0.1, 0.8, 0.1}.

FIGURE 2–11. Variance as a measure of dispersion. Each of the four charts depicts a random variable with mean 3. Chart (d) has the lowest variance of 0.8 because its values are most tightly clustered around the mean.

This value is smaller than all the previous variances. Hence, of the four strains of hen, the fourth lays eggs whose number is most consistent on average. The densities for all four random variables are shown in Figure 2–11. ◆

The variance for a continuous random variable is similar to that for discrete variables. Let X be a random variable with density f and mean μ. The deviation of any particular value $X = x$ from the mean is given by

$$\delta(x) = x - \mu$$

The expected value of the squared deviation is once again an appropriate measure of dispersion. More specifically, the *variance* of a continuous random variable X is given by

$$\text{Var}(X) \equiv \int_{-\infty}^{+\infty} f(x)(x - \mu)^2 \, dx$$

To illustrate, let X be a random variable with the exponential density function

$$f(x) = \lambda e^{\lambda x} I_{[0, \infty)}(x)$$

Here, $I_{[0, \infty)}(x)$ is the indicator function on the interval $[0, \infty)$: when x lies in the interval, the indicator function has value 1, and takes value 0 otherwise. In other words, $f(x) = \lambda e^{-\lambda x}$ for $x \geq 0$ and $f(x) = 0$ when x is negative. The mean of x is given by

$$\mu \equiv E(X) = \int_0^\infty (\lambda e^{-\lambda x}) x \, dx = \frac{1}{\lambda}$$

The variance is

$$\sigma^2 = \mathrm{Var}(X) = \int_0^\infty (\lambda e^{-\lambda x})\left(X - \frac{1}{\lambda}\right)^2 dx = \frac{1}{\lambda^2}$$

Variance in Terms of Expectation

The variance may be written in terms of the expectation operator. The basic component of the variance is the deviation from the mean: $\delta(X) = X - EX$. The variance is obtained by squaring the deviation and taking the expectation of the result. In other words,

$$\mathrm{Var}(X) = E\{[X - E(X)]\}^2$$

where the expectation is taken with respect to variable X. This formula holds whether the random variable is discrete or continuous.

The preceding formula can also be used to derive a useful result for computing the variance. We begin with the relation

$$\mathrm{Var}(X) = E\{[X - E(X)]^2\} = E\{(X - \mu)^2\}$$

The second equation follows from the fact that the expectation of X, namely, $E(X)$, is fixed at value μ. After carrying out the squaring operation, we have

$$\mathrm{Var}(X) = E(X^2 - 2\mu X + \mu^2)$$
$$= E(X^2) - 2\mu E(X) + E(\mu^2)$$

The second equality depends on the linearity property of the expectation operator. For the last two terms, we know that $E(X) \equiv \mu$. Further, the expectation of a constant is the constant itself; that is, $E(\mu^2) = \mu^2$.

The result so far is

$$\text{Var}(X) = E(X^2) - \mu^2$$

Switching back to $E(X)$ for the mean μ, we obtain

$$\text{Var}(X) = E(X^2) - (EX)^2$$

In other words, the variance is the expectation of the square less the square of the expectation.

The preceding formula can be used to calculate the variance of a random variable without bothering to compute each of the deviations. We will also see in later chapters that the formula is useful for obtaining theoretical results in probability and statistics.

Example

Two contenders meet on a competitive field. It might be a pair of gladiators in an arena, two fighter pilots in the sky, a pair of companies in a marketplace, or a couple in playful argument. Let X be the three-valued indicator of success: 1 if the first contender wins, -1 if he loses, and 0 for a draw. If the chances of all three outcomes are equal, what is the variance of X?

The expectation of X and of X^2 are given by

$$E(X) = \tfrac{1}{3}(-1) + \tfrac{1}{3}(0) + \tfrac{1}{3}(1) = 0$$
$$E(X^2) = \tfrac{1}{3}(-1)^2 + \tfrac{1}{3}(0)^2 + \tfrac{1}{3}(1)^2 = \tfrac{2}{3}$$

We can readily determine the variance as

$$\text{Var}(X) = EX^2 - E^2X = \tfrac{2}{3} - 0 = \tfrac{2}{3} \qquad \blacklozenge$$

Standard Deviation

The variance is an effective measure of dispersion. In practical applications, however, the variance can yield quantities which have little physical meaning. Consider the hen example where 1 or 5 eggs are laid with equal probability. The outcomes are $x_1 = 1$ egg, $x_2 = 5$ eggs. The mean value is $\mu = 3$ eggs. The variance is the mean *squared* deviation of these values, or 4 eggs^2. What physical interpretation can be assigned to the units of *eggs squared*? The answer is less than obvious. In general, the variance assumes values whose units are squares of those of the underlying variable.

For convenience in interpretation, we often work with the square root of the variance, called the *standard deviation*. The standard deviation of a random variable, written δ_X or simply δ, is therefore given by

$$\delta_X = \sqrt{\text{Var}(X)}$$

For the above example, the standard deviation of random variable X taking values of 1 egg and 5 eggs with equal probability is 2 eggs.

2.9 INDEPENDENCE OF VARIABLES

In Section 2.4, we explored the nature of independence among two or more events. A similar notion can be defined for random variables.

A set of random variables is said to be *independent* if the joint density of the set can be expressed as the product of the individual densities. Let X and Y be two random variables with individual densities f_X and f_Y, respectively, and a joint density function f_{XY}. Then X and Y are independent if the following condition holds:

$$f_{XY}(x, y) = f_X(x)f_Y(y)$$

The same idea applies when the set contains three or more variables. The joint density of a collection of independent random variables is equal to the product of their individual densities.

When two random variables X and Y are independent, their distribution function $F_{XY}(x, y)$ can also be expressed as the product of their individual distribution functions F_X and F_Y. We can show this as follows:

$$F_{XY}(x, y) = P(X \leq x, Y \leq x)$$

$$= \int_{-\infty}^{x} \int_{-\infty}^{y} f_{XY}(x, y)\, dy\, dx$$

The independence of X and Y implies that $f_{XY} = f_X f_Y$. Thus,

$$F_{XY}(x, y) = \int_{-\infty}^{x} f_X(x)\, dx \int_{-\infty}^{y} f_Y(y)\, dy = F_X(x)F_Y(y)$$

In other words, the distribution function for two independent random variables is the product of their individual distributions.

When two random variables are independent, the expectation of their product is the product of their expectations. We can show this as follows:

$$E(XY) = \int\int xy f_{XY}(x, y)\, dx\, dy$$

$$= \int\int xy f_X(x) f_Y(y)\, dx\, dy$$

$$= \int x f_X(x)\, dx \int y f_Y(y)\, dy = E(X)E(Y)$$

The integrals are to be taken over the entire real line, from $-\infty$ to $+\infty$. The first equality relies on the definition of expectation. The second equality springs from the independence of the random variables, which is tantamount to factoring the joint density into the product of individual densities.

The above result easily generalizes to three or more variables. In particular, the expectation of a product of independent variables is the product of their expectations.

Example

To illustrate the factorization property of the expectation operator for independent variables, we return once more to our trusty dice. Let X and Y be the independent random variables for the outcomes. The mean of their product is given by

$$E(XY) = E(X)E(Y) = (3.5)(3.5) = 12.25$$

since the expectation of each variable is 3.5. ◆

Independence of Events versus Variables

In Section 2.4, we defined a set of events as independent if the events have no effect on each other. The occurrence or nonoccurrence of one event does not affect the chance of any other event in the set.

In particular, if A and B are two events, then they are independent if the conditional probability reduces to the unconditional probability: $P(B|A) = P(B)$. This is equivalent to saying $P(A|B) = P(A)$, as can be shown from Bayes' Rule:

$$P(A|B) = \frac{P(A)P(B|A)}{P(B)} = P(A)$$

The second equality relies on our stipulation that $P(B|A) = P(B)$ if A and B are independent events.

With this orientation, let us examine the relationship between the independence of events and of random variables. To this end, let X and Y be two random variables corresponding respectively to events A and B.

As an illustration, let A be the event "obtaining an odd number on a die." Suppose variable X corresponds to the occurrence of A. Then,

$$P(A) = P(X = 1) + P(X = 3) + P(X = 5) = 1/6 + 1/6 + 1/6 = 1/2$$

The conditional probability of some event B given A is

$$P(B|A) = \frac{P(A, B)}{P(A)} \qquad (2\text{-}6)$$

by Bayes' Rule. The joint probability of A and B is given by

$$P(A, B) = \int_A \int_B f_{XY}(x, y)\, dx\, dy = \int_A \int_B f_X(x) f_Y(y)\, dx\, dy$$

$$= \int_A f_X(x)\, dx \int_B f_Y(y)\, dy = P(A)P(B)$$

The subscript below each integral denotes the region of integration. For instance, the first pair of integrals are to be taken over the values of X and Y corresponding to the occurrence of A and B, respectively. The second equality depends on the factorization of the joint density into individual densities.

By inserting this result into Eq. (2–6), we obtain $P(B|A) = P(B)$. We have just shown that the independence of random variables leads directly to the independence of their corresponding events.

Variance of a Sum of Variables
As we have seen earlier, the sum of random variables is another random variable characterized by its own density function.

In general, the density function for the sum will differ from that of any of the underlying variables. When the underlying variables are independent, however, the variance of the sum is merely the sum of the respective variances.

More specifically, let X and Y be two independent random variables. Then,

$$\text{Var}(X + Y) = \text{Var}(X) + \text{Var}(Y)$$

The result can easily be generalized to three or more independent random variables.

2.10   SOME KEY DISTRIBUTIONS

A random variable is completely defined by its distribution function. Specifying the distribution function is also equivalent to defining its derivative—namely, the mass function for a discrete variable or the density function for a continuous variable. In this section, we explore a number of distribution functions that often arise in the applications of probability theory.

Binomial Distribution

The *binomial* distribution, also called the *Bernoulli* distribution, applies to trials having precisely two outcomes. The random variable X refers to the number of times that one of the two outcomes occurs in a sequence of n observations. An example is found in the number of heads resulting from a series of coin tosses, or the number of wins in a sequence of tennis games, or the number of items failing inspection in a batch of sampled products.

To illustrate, consider an urn containing four white balls and six black balls. If the balls are replaced in the urn after each turn, the probability p of drawing a white ball is 4 out of 10, or .4. If the trial is repeated three times, what is the probability of observing a white ball?

Let W denote the event "Draw a white ball," and B the event "Draw a black ball." Further, we let X be the random variable denoting the number of times event W occurs in n trials.

If $n = 3$, the event "Observe one white ball in the sequence" occurs for a sequence such as $\langle W, B, B \rangle$; in other words, a white ball is drawn followed by two black balls in tandem. The probability of this sequence of events is

$$P(\langle W, B, B \rangle) = P(W)P(B)P(B) = p(1 - p)(1 - p)$$
$$= .4(.6).6 = .144$$

However, precisely one white ball is also observed with the sequences $\langle B, W, B \rangle$ and $\langle B, B, W \rangle$. The probabilities for these events are

$$P(\langle B, W, B \rangle) = P(B)P(W)P(B) = (1 - p)p(1 - p)$$
$$P(\langle B, B, W \rangle) = P(B)P(B)P(W) = (1 - p)(1 - p)p$$

In summary, there are three independent ways to observe exactly one white ball in a sequence of three draws. Since each of these ways occurs with probability $p(1 - p)^2$, the probability of the event "$X = 1$" is given by $3p(1 - p)^2$.

More generally, consider an event X that may occur with probability p, and may fail to occur with probability $1 - p$. Then the probability of observing X precisely k times in a binomial sequence of n trials is

$$P(X = k) = \binom{n}{k} p^k (1 - p)^{n-k}$$

The notation $\binom{n}{k}$ denotes the *binomial coefficient*, giving the number of permutations or patterns in which an event can occur k times in n trials—in other words, how many ways there are to arrange k events in a sequence of n slots. The value of the binomial coefficient is

$$\binom{n}{k} \equiv \frac{n!}{k!(n-k)!}$$

Here, $m!$ denotes the factorial function; that is, $m!$ is the product of all positive integers up to and including m.

Combinations
The nature of the binomial coefficient can be justified as follows. We begin with a simple example relating to all possible combinations of 2 white balls and 3 black balls. This task is equivalent to the query "In how many ways can 2 white balls and 3 black balls be placed in a sequence of 5 boxes?"

Consider the first white ball. It can be placed in any of the 5 boxes; that is, there are 5 possibilities. Consider the second white ball: it can be placed in any of the remaining 4 boxes. Now take the first black ball; it can fit into any of the remaining 3 boxes. The second black ball has 2 possibilities left, and the third black ball has only 1 place to go. Thus far, we have $5 \cdot 4 \cdot 3 \cdot 2 \cdot 1$ possibilities.

However, not all these arrangements are distinct. The reduction in the number of arrangements results from the lack of individuality among balls of a particular color: one white ball is like any other white ball, and a black one is indistinguishable from any other black. Suppose, in particular, that 2 white balls are assigned to boxes 1 and 3. How many ways are there to achieve this? The first white ball can go into box 1 or 3; it has 2 choices. The second ball has only 1 remaining place to go. As a result, the number of potential arrangements must be reduced by 2! to account for the uniformity among white balls. In a similar way, suppose that the 3 black balls are assigned to boxes 2, 4, and 5. The first black ball can go into any of these latter 3 boxes; it has 3 choices. The second black ball can go into either of the remaining 2 boxes; and the last ball has only a unique destiny. The number of possible arrangements must, therefore, be reduced by 3! to account for the indistinguishability among the black balls. The total number of ways to arrange 2 white balls and 3 black balls is, therefore,

$$\frac{5!}{2!3!}$$

More generally, the number of combinations for observing k events out of n trials is

$$\binom{n}{k} \equiv \frac{n!}{k!(n-k)!}$$

For binomial trials that have only two possible outcomes, the occurrence of an outcome is equivalent to the nonoccurrence of the other outcome; and vice versa. Therefore the event "Observe k events out of n trials" is equivalent to the event "Observe $(n-k)$ alternate events out of n trials." This symmetry is reflected in the binomial coefficient, for which

$$\binom{n}{k} = \binom{n}{n-k} = \frac{n!}{k!(n-k)!}$$

A limiting case for the binomial coefficient occurs when $k = 0$. This corresponds, for instance, to "In how many ways can no white balls be selected among n white and black balls?" The situation arises in the single arrangement where *all* the balls are black. This result is matched by the formula

$$\binom{n}{k} = \frac{n!}{0!(n-0)!} = 1$$

where we rely on the convention of defining 0! as 1. The converse limiting case arises for $k = n$ corresponding to the arrangement where all n balls are white. Here the binomial coefficient again equals 1.

Example

A new interstellar spacecraft has redundant computing systems to safeguard the mission against failure. During its 500-year voyage among the stars, each of the 6 computing systems has a .1 probability of failure. Suppose that each system is individually capable of supporting all the mission requirements, and its chance of failure is independent of the other systems. What is the probability that exactly half the computing systems will fail during the voyage? What is the chance that the spacecraft's computational needs are successfully met over the course of the voyage?

Let X be the number of systems that fail during the voyage. Then, X has a binomial distribution with parameter $p = .1$. The probability that precisely 3 systems fail is

$$P(X = 3) = \binom{6}{3} p^3 (1 - p)^3 = \frac{6!}{3!3!}(.1)^3(1 - .1)^3 = .01458$$

The chance that half the systems will fail is 1.458%. The computing requirements of the mission are fulfilled if at least one system survives the trip. This event is equivalent to the event that X is 5 or less, whose probability is the complement of that for X equal to 6. In algebraic terms,

$$P\{\text{survival} \geq 1\} = P(X \leq 5) = 1 - P(X = 6)$$

$$= 1 - \binom{6}{6}(.1)^6(1 - .1)^0 = .999999$$

The chance of failure is one out of a million, and the chance of success is a near certainty.   ◆

The mass function for the binomial random variable tends to increase, then decline. The concave shape of the function reflects the fact that moderate values of the variable tend to be more likely than extreme values.

The binomial function is depicted in Figure 2–12 for the random variable with parameters $p = .4$ and $n = 5$. The mass function $p(X)$ shown in the top half of the figure is skewed toward the left because the parameter p is less than .5. For larger values of p, the event of interest will tend to occur more often. As a result, the probability values $p(X)$ will be higher for large values of X, thereby shifting the pattern of mass toward the right-hand side of the chart. The distribution function $F(X)$ in the bottom half of the figure is simply the cumulative sum of the probability mass $p(X)$.

The mean value of the binomial distribution is obtained as follows. From the definition of the mean,

$$\mu = \sum_{k=0}^{n} k p(k) = \sum_{k=0}^{n} k \binom{n}{k} p^k (1 - p)^{n-k}$$

Since the first term reflected in the sum equals 0 for $k = 0$, the index of the summation can run from 1 to n.

We now observe that the first two factors under the summation can be

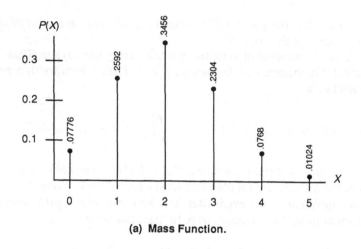

(a) Mass Function.

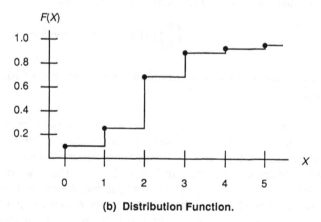

(b) Distribution Function.

FIGURE 2–12. Binomial random variable with parameters $n = 5$ and $p = .4$.

transformed into the expression

$$k\binom{n}{k} = k\frac{n!}{k!(n-k)!} = \frac{n(n-1)!}{(k-1)!(n-k)!} = n\binom{n-1}{k-1}$$

As a result,

$$\mu = np \sum_{k=1}^{n} \binom{n-1}{k-1} p^{k-1}(1-p)^{n-k}$$

where one of the p factors has been extracted outside the summation. The summation is, in fact, simply the expansion of $[(1 - p) + p]^{n-1}$ which equals 1. We conclude that the mean of the binomial random variable is $\mu = np$.

The variance of the binomial density is given by

$$\sigma^2 = \sum_{k=0}^{n} (k - np)^2 p(k)$$

By completing the square and drawing on the same type of argument as before, it is possible to show that

$$\sigma^2 = np(1 - p)$$

Another way to obtain this result is indicated in one of the problems.

To illustrate the computation of the variance for a binomial variable, consider the previous situation with parameters $n = 5$ and $p = .4$. The mean is $np = 5(.4) = 2$, and the variance equals $np(1 - p) = 5(.4)(.6) = 1.2$.

The binomial function is convenient enough to calculate when the values of n and k are small. When those parameters are large, however, the underlying phenomena can often be modeled as Poisson or normal densities. We now turn to these latter distributions.

Poisson Distribution

The Poisson distribution describes an experiment similar to that of the binomial function. The difference is that the number of trials is large, but the event of interest occurs only rarely.

More specifically, let p be the probability of an event occurring in each of n independent trials. Then the product $np \equiv \lambda$ is a measure of the likely number of occurrences of the event in the sequence of trials. For a fixed value of the parameter λ, the value of p vanishes as n grows arbitrarily large. In other words, the Poisson function is given by $\lim_{n \to \infty} \beta(n, k)$, where β is the binomial density function. The form of the Poisson mass function is

$$p(k) = \frac{e^{-\lambda} \lambda^k}{k!} \quad \text{for } k = 0, 1, 2, \ldots$$

The Poisson distribution is parameterized by the single quantity λ.

The density function is shown in Figure 2–13 for different values of the parameter. The top chart depicts the case for $\lambda = 0.5$: since the event of interest is unlikely to occur, the mass is concentrated at 0 and neighboring values. For $\lambda = 1$, the mass moves a little further to the right. The bottom

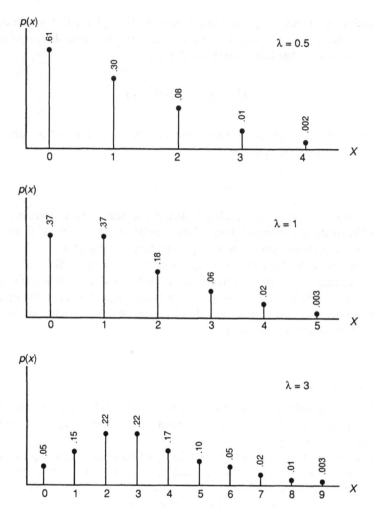

FIGURE 2–13. Poisson mass function for differing values of λ. As λ increases, the center of weight moves to the right and the function becomes more symmetric.

chart corresponds to the case where $\lambda = 3$: the mass shifts even further to the right, and, in fact, the values $X = 2$ and $X = 3$ are the most likely numbers to arise. The pattern is clear: as λ increases, the center of mass shifts to larger values of X. Moreover, the pattern of weights becomes more symmetric around the peak representing the most likely value. The distribution function for the Poisson probability is obtained, as usual, by the cumulative sum of the individual probabilities.

Example

The probability of observing at least one accident at a particular intersection during the day is .01. What are the chances of encountering accidents on exactly 5 days during the month of April? What about no accidents at all during the month?

The probability is $p = .01$ and the number of observations is $n = 30$. The parameter for the Poisson density is $\lambda = np = 30(.01) = 0.3$. The probability of encountering 5 accidents is

$$p(5) = \frac{e^{-.3}(0.3)^5}{5!} = .00001500$$

The probability of observing no accidents at all in April is

$$p(0) = \frac{e^{-.3}(0.3)^0}{0!} = .7408$$

How do these numbers compare with the exact values obtained through the binomial function? The corresponding values for 0 and 5 days of accidents are

$$B(30, 0) = \binom{30}{0}(.01)^0(.99)^3 = .7397$$

$$B(30, 5) = \binom{30}{0}(.01)^5(.99)^{25} = .00001108$$

The results correspond fairly well to those from the Poisson density. As alluded to earlier, the accuracy of the Poisson function increases as n increases—or equivalently, as p decreases—for a fixed value of λ. ◆

The mean and variance of the Poisson distribution can be obtained from those of the binomial function. In particular, λ is substituted for the product np and the probability p pushed toward 0. The result is λ for both the mean and variance. In other words, the mean is $E(X) = np = \lambda$ for the Poisson density function, whereas the variance is $\text{Var}(X) = np(1 - p) = \lambda$ since p drops toward 0.

Normal Distribution

The normal density function plays a central role in the realm of probability and statistics. It arises in many applications where little or no information is available on the underlying random variable, as well as in situations where numerous independent variables impose their aggregate effects.

The normal density can also be viewed as a limiting case of the binomial function. As with the Poisson function, the number n of trials goes to infinity. Unlike the Poisson case, however, the probability p from the binomial distribution can lie anywhere between 0 and 1, rather than vanishing to 0.

The convergence of the binomial function toward the normal density is depicted in Figure 2–14. The top chart in the figure shows the binomial function for $n = 5$ and probability $p = .2$. The second chart shows the binomial function where n has increased to 30, but p is still fixed at .2. For large n, more instances of the underlying phenomena are likely to occur in the sequence of trials. As a result, the center of weight shifts to the right. Furthermore, the function becomes more symmetric around the most probable value, which happens to be $X = 12$ when $n = 60$.

As the number of trials n increases to infinity, and x takes continuous values rather than simply whole numbers, the binomial function approaches the normal density depicted in the bottom chart of Figure 2–14. In other words,

$$\lim_{n \to \infty} B(n, k) = f(x)$$

where $f(x)$ is the normal density given by

$$f(x) \equiv \frac{1}{\sigma\sqrt{2\pi}} e^{-(x-a)^2/2b^2}$$

Here, the parameters a and b can be any real numbers, as can the value of x.

Suppose X is a random variable having the normal density with parameters a and b. Then it is possible to show that the mean is $\mu = a$, and the variance is $\sigma^2 = b^2$. For this reason, the normal density is usually written as

$$f(x) = \frac{1}{\sigma\sqrt{2\pi}} e^{-(x-\mu)^2/2\sigma^2}$$

The normal function is also known as the *Gaussian* density. The normal density $f(x)$ with parameters μ and σ^2 is also indicated at times by the notation $\mathcal{N}(\mu, \sigma^2)$.

Characteristics of the Normal Function
The normal density is shown in Figure 2–15. The maximum value, called the *mode*, of the function occurs at the mean value of $x = \mu$. The function is

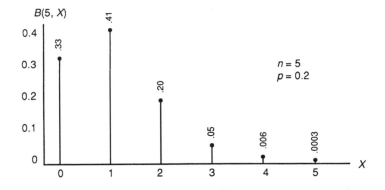

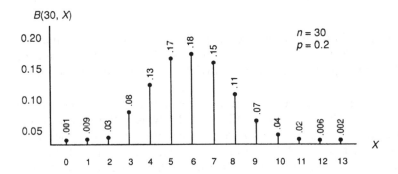

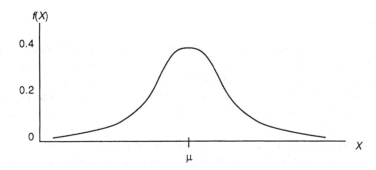

FIGURE 2–14. Normal density as a limiting case of the binomial function for fixed p and increasing n. The top chart shows the binomial function for $n = 5$ and $p = .2$; the middle chart for $n = 30$ and $p = .2$. As n increases, the binomial function approaches the normal density shown in the bottom chart, where $\mu = np = .2n$ and $\sigma = np(1 - p) = .16n$.

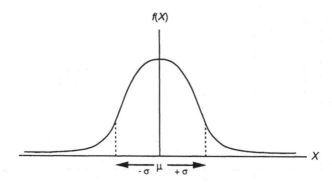

FIGURE 2–15. Characteristics of the normal density. The mean and mode are at μ. The inflection points are at $\mu \pm \sigma$.

symmetric about the mean and decreases monotonically to either side. The points of inflection—where the slope of $f(X)$ changes from becoming negative to becoming positive as a function of X—occurs at the points $(\mu - \sigma)$ and $(\mu + \sigma)$.

The standard deviation σ is a measure of dispersion for the random variable X. The shaded region in Figure 2–16, lying within one σ of the mean, represents 68.26% of the total area under the curve. In other words, the probability that X will fall within the interval $(\mu - \sigma, \mu + \sigma)$ is 68.26%. The areas corresponding to several other intervals are shown in the table accompanying Figure 2–16.

Example

A construction manager has received a shipment of 100 steel beams from her supplier. The length of the beams is distributed normally with a mean of 5 m and standard deviation of 2 mm. If the tolerance on the beams is 5 m ± 4 mm, what fraction of the shipment meets specifications? If the current project requires at most 68 beams, what is the poorest fit that the manager will likely have to tolerate?

The beams have a mean length of $\mu = 5$ m and standard deviation of $\sigma = 2$ mm. The tolerance is ±4 mm, which corresponds to 2 standard deviations. We see from the table in Figure 2–16 that the interval $(\mu - 2\sigma, \mu + 2\sigma)$ covers 95.44% of the total area. In other words, the probability that a beam is within (5 m, 5 m + 4 mm) is 95.44%. The manager can reasonably expect 95 or 96 out of 100 beams to be within tolerance.

For the same table, we see that 68.26% of the total area falls within plus or minus 1 standard deviation of the mean. Since the standard deviation in our case is $\sigma = 2$ mm, we would expect about 68 out of 100 beams to lie in the

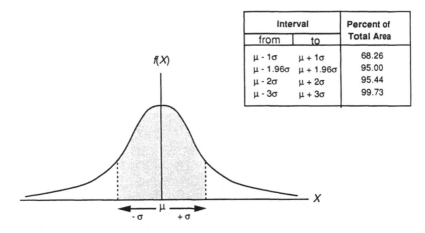

Interval		Percent of Total Area
from	to	
$\mu - 1\sigma$	$\mu + 1\sigma$	68.26
$\mu - 1.96\sigma$	$\mu + 1.96\sigma$	95.00
$\mu - 2\sigma$	$\mu + 2\sigma$	95.44
$\mu - 3\sigma$	$\mu + 3\sigma$	99.73

FIGURE 2–16. Areas under the normal density. The shaded region covers 68.26% of the total area under the curve. The areas corresponding to other intervals are shown in the table.

interval (5 m − 2 mm, 5 m + 2 mm). Therefore, if 68 or fewer of the best beams are selected, the worst among them should lie within 2 mm of the specified length of 5 m.   ◆

Standard Normal Density
The curve of the normal density function depends on both its mean μ and standard deviation σ. The mean determines the location of its peak: as μ increases, the curve shifts to the right, and conversely for decreasing μ.

On the other hand, the standard deviation σ determines the looseness of the clustering around the peak at μ. A large value of σ indicates great dispersion in the underlying variable. Consequently, the density function becomes thicker farther away from the mean, and flatter at the center. This tendency is shown in Figure 2–17, which portrays three curves having the same mean of 0, but differing values of σ. The curve with $\sigma = 2$ is flatter and stouter than the curve for $\sigma = 1$, which, in turn, is broader and squatter than the skinny curve for $\sigma = 0.5$.

The prevalence of the normal density in diverse realms of application suggests the desirability of tabulating its values, thereby presenting the values for ready reference. But this task of tabulation is clearly infeasible for all possible values of μ and σ.

As a practical matter, the normal function is tabulated for precisely one set of values: a mean of 0 and standard deviation of 1. We will soon see that this is not a major handicap; it can be easily adapted to normal functions having arbitrary values of μ and σ.

The normal density having zero mean and unit standard deviation is called

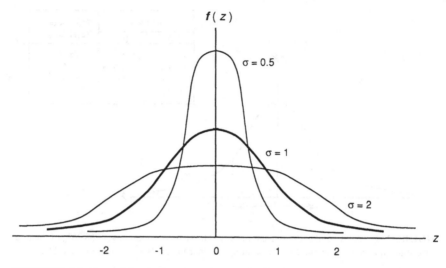

FIGURE 2–17. Normal density function parameterized by its standard deviation. The mean is at 0; as the standard deviation σ increases, the curve becomes flatter at the center and thicker at the tails.

the *standard normal function.* Some values of the standard normal density are shown in Table 2–1. The maximum occurs at the mean of 0, for which the ordinate is $f(0) = 0.3989$ to the accuracy of four significant digits. The function is symmetric about the mean. For instance, the value at $z = -1$, corresponding to the ordinate at 1 standard deviation to the left of the mean, is identical to that at $z = +1$; in other words, $f(-1) = f(+1) = 0.2420$.

Since the normal density describes a continuous random variable X, the probability that X takes on any particular value x is 0. Of greater use in a practical context is the probability that X lies in a particular interval, or set of such. For this reason, the distribution function is more relevant in the applications of probability theory.

The standard normal distribution is the cumulative integral of the standard normal density. More specifically, let Z be a random variable from a normal density with zero mean and unit standard deviation. Then the *standard normal distribution* is given by

$$\Phi(Z = z) = \int_{-\infty}^{z} \frac{1}{\sqrt{2\pi}} e^{-z^2/2} \, dz$$

Some values for the function are given in Table 2–2. For instance, $\Phi(1) = .8415$ indicates that the area under the curve to the left of $z = 1$ is 84.15% of

TABLE 2–1. Values of the standard normal density $f(z)$, for which $\mu = 0$ and $\sigma = 1$. The density is symmetric about the mean; that is, $f(-z) = f(z)$.

z	$f(z)$
0.0	0.3989
0.1	0.3970
0.2	0.3910
0.3	0.3814
0.4	0.3683
0.5	0.3521
0.6	0.3332
0.7	0.3123
0.8	0.2897
0.9	0.2661
1.0	0.2420
2.0	0.0540
3.0	0.0044
4.0	0.0001

TABLE 2–2. Values of the standard normal distribution $\Phi(z)$, for which $\mu = 0$ and $\sigma = 1$. Due to symmetry about the mean of 0, $\Phi(-z) = 1 - \Phi(z)$. A more extensive table is available in Appendix B.

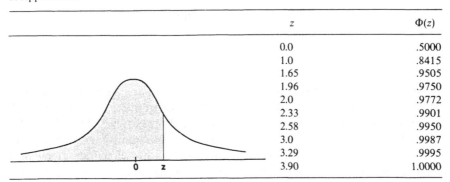

z	$\Phi(z)$
0.0	.5000
1.0	.8415
1.65	.9505
1.96	.9750
2.0	.9772
2.33	.9901
2.58	.9950
3.0	.9987
3.29	.9995
3.90	1.0000

the total. Conversely, the complement given by $1 - \Phi(z)$ gives the area under the curve to the right of the value z.

Since the normal density is symmetric about the mean, the distribution $\Phi(z)$ for negative values of z can be easily obtained from those for positive z. For instance, $\Phi(-2)$ is the area under the curve to the left of $z = -2$. By symmetry, this area equals that under the curve to the right of $z = +2$. This latter value is given by the complement of the area to the left of $z = +2$. In more compact form, the reasoning is as follows:

$$\Phi(-2) = P(Z \le -2) = P(Z \ge +2)$$
$$= 1 - P(Z \le +2) = 1 - \Phi(+2)$$
$$= 1 - .9772 = .0228$$

In other words, Z will fall below 2 standard deviations to the left of the mean with probability 2.28%.

We turn now to the task of converting any old normal variable into standard form. This is easily achieved by subtracting the mean and normalizing by the standard deviation. More specifically, let X be a random variable from a normal density with mean μ and standard deviation σ. Then the transformed variable

$$Z \equiv \frac{X - \mu}{\sigma}$$

is a standard normal variable.

Intuitively, Z may be viewed as the quantity which addresses the query "How many standard deviations from the mean does X lie?" For instance, let X be a normal variable with $\mu = 60$ m and $\sigma = 3$ m. Then the value $X = 54$ m lies 2 standard deviations below the mean:

$$Z = \frac{X - \mu}{\sigma} = \frac{54 - 60}{3} = -2$$

Example

Scores on tests of intelligence tend to be distributed normally. The Stanford–Binet test yields a score called the "intelligence quotient," or IQ, which is calibrated against a mean of 100 and standard deviation of 15. What is the probability that a person chosen at random will have an IQ between 115 and 145? In addition, a psychologist is interested in studying individuals with extreme IQ scores. If the potential subject pool consists of 2.5% of the individuals having the highest scores, and another 2.5% with the lowest, what are the cutoff scores for each category?

The mean and standard deviations are $\mu = 100$ and $\sigma = 15$. For the scores $x_1 = 115$ and $x_2 = 145$, standardized values are

$$z_1 = \frac{x_1 - \mu}{\sigma} = \frac{115 - 100}{15} = 1$$

$$z_2 = \frac{x_2 - \mu}{\sigma} = \frac{145 - 100}{15} = 3$$

The probability that the IQ score falls within 115 and 145 is given by the area within $z = 1$ and $z = 3$ of the standard distribution. That is,

$$P(115 < X \leq 145) = P(1 < Z \leq 3) = \Phi(3) - \Phi(1)$$

$$= .9987 - .8415 = .1572$$

In other words, there is a 15.72% chance that a person chosen at random would have an IQ score between 115 and 145. In the above equations, the equalities may be replaced with inequalities, and vice versa: since X—and therefore Z—are viewed as continuous variables, the probability of attaining any particular value is 0.

For the psychological study, we first need the values of z such that 2.5% of the area under the standard normal density lies above it, and 2.5% below it. The value z_2 for which 2.5% of the probability lies above is the same as that for which 97.5% lies below. From Table 2–2, we see that this is $z_2 = 1.96$. By the symmetry of the normal density, 2.5% of the area lies below $z_1 = -z_2 = -1.96$. The corresponding IQ scores are

$$x_1 = \mu + z_1\sigma = 100 + (-1.96)15 = 70.6$$

$$x_2 = \mu + z_2\sigma = 100 + (1.96)15 = 129.4$$

The psychologist should select subjects with scores of 71 or lower, and those with scores of 129 or higher. ◆

PROBLEMS

The problems below are numbered according to the sections to which they correspond. For instance, Problem 2.8A is the first ("A") exercise pertaining to Section 2.8.

2.1A   A fortune teller relies on a regular deck of cards to make a prediction. She draws a single card from a regular deck and assesses the future. (a) What constitutes a trial as far as the cards are concerned? (b) What is an example of an outcome? (c) What is an example of an event? (d) What is the certain event Ω? (e) What is the null event $\varnothing$? (f) Suppose two cards are drawn at random; what is an example of an event?

2.1B   An investor monitors the stock market every day and pursues the following strategy for selling shares of any stock he may happen to own. If the price of the stock at the end of the day falls 30% below his original purchase price, he sells all his shares the following day (at a loss). If the price at the close of the day exceeds his purchase cost by

50%, he sells the next day (at a profit). Otherwise he holds on to his shares. (a) What constitutes a trial? (b) What are the outcomes? (c) What is an example of an event? (d) Suppose the investor sells only if the price of his stock falls for 5 consecutive days, or rises for 5 consecutive days. Now what is a trial, and what is the set of all outcomes?

2.2A Consider the following statements. Does each reflect an *objective assessment, frequentist,* or *subjective assessment* view of probability? Explain. (a) "It snowed on 10 days last January. So the probability of snow in January is 10/31." (b) "These 200 crates look identical. But precisely 10 of them contain contraband. So if you open one at random, the probability of discovering a bogus shipment is 1/20." (c) "Probability theory is easy. My chance of getting an 'A' in this course is better than 90%."

2.3A Alex has 20 blue socks and 10 brown ones in his drawer. (a) If he picks two socks at random, what is the probability of selecting a matching pair (both blue or both brown)? (b) If he picks 3 at random, what is the chance of having a matching pair?

2.4A An automotive firm is about to introduce a new sports coupe. The new model will succeed only if the economy continues to expand *and* consumers rediscover a penchant for fast cars. The marketing department expects that the chance of a strong economy is .7, and that for rediscovering speedy cars is .8. If these two developments are independent, what is the probability of success?

2.5A A bank has two offices in its check-processing division. Each week the first office handles twice as many checks as the second. The probability of committing an error when a check goes through the first office is 1 out of 400,000. The corresponding probability at the second office is 1 out of 800,000. If a check is known to be defective, what is the chance that it went through the first office?

2.6A Consider the uniform density function given by

$$f(x) = \begin{cases} \dfrac{1}{b-a} & a \le x \le b \\ 0 & \text{otherwise} \end{cases}$$

(a) Draw the *density* function on a chart. (b) Draw the *distribution* function on a chart.

2.6B A scientist in California requires tranquil conditions to conduct his experiments. Too often, however, tremors arise that disturb his work and force him to start anew. Suppose his experiment requires 30 days to run, and that the time between tremors is a random variable from an exponential density with parameter $\lambda = 1/90$ days. What is the probability that he can finish an experiment in peace?

2.7A   The revenues at Tony's Pizza Parlor depend on the weather. On a sunny day the revenues are \$500; on a cloudy day, \$600; on a rainy day, \$200. If the chance of sun, cloud, and rain is .6, .3, and .1, respectively, what is the expected daily revenue?

2.7B   Let X be a random variable from a binomial distribution with parameters n and p. The variable Y is similar to X, except that its mean is twice that of X. What is the mean of their sum $X + Y$?

2.8A   Let X and Y be random variables with the joint density $f(x, y) = a^2 \exp[-a(x + y)]$ for positive values of the arguments, and $f(x, y) = 0$ otherwise. Suppose X has exponential density with parameter a. (a) Are X and Y independent? (b) What is the simultaneous probability of having $X > 2/a$ and $Y \leq 1/a$?

2.9A   Let X be a random variable from a uniform density on the interval $[a, b]$. Show that the variance of X is $(b - a)^2/12$.

2.9B   Let X and Y be independent random variables taking values 1, 3, and 5 with equal probability. Calculate the variance of $X + Y$.

2.9C   Let X be a variable having density $f(X)$. Consider the deviation from the mean

$$\delta(X) = X - E(X)$$

Why is δ meaningless as a general measure of dispersion for X? (*Hint*: Apply the expectation operator.)

2.10A   Suppose $p = 1/20$ is the probability that a particular apple from an orchard is bad. Let X denote the number of bad apples in a basket of $n = 10$. (a) What is the probability that $X = 3$ apples are bad? (b) What is the probability that at least 1 apple is bad?

2.10B   Let X be the number of meteors observed each night. Suppose X has a Poisson density with parameter $\lambda = 3$. (a) What is the probability that $X = 7$ meteorites are observed on a given night? (b) What about the probability of no sightings at all?

2.10C   Let X be the height of European women. Suppose X is distributed normally with mean $\mu = 163$ cm and standard deviation $\sigma = 2$ cm. What fraction of the population has heights exceeding 167 cm?

2.10D   Verify the formulas for the mean and variance of the binomial distribution by differentiating the binomial expansion formula.

a. Begin with the relation

$$1 = [(1 - p) + p]^n = \sum_{k=0}^{n} \binom{n}{k} p^k (1 - p)^{n-k}$$

and differentiate the left-hand and right-hand sides with respect to p. Then work on the result to obtain an expression involving

$$\sum_{k=0}^{n} k \binom{n}{k} p^k (1 - p)^{n-k}$$

which is the mean. Show that $E(X) = np$.

b. By using the same approach as above, show that $Var(X) = np(1 - p)$.

2.10E   Let X be a random variable from a normal density:

$$f(X) = \frac{1}{\sigma \sqrt{2\pi}} \exp\left(\frac{-(x - \mu)^2}{2\sigma^2}\right)$$

Show that the mean and variance of X are μ and σ^2, respectively.

FURTHER READING

Bayes, Thomas. 1763. An essay toward solving a problem in the doctrine of chance. *Philosophical Transactions of the Royal Society.*

Ross, Sheldon M. 1985. *Introduction to Probability Models.* 3rd ed. New York: Academic Press.

Savage, Leon J. 1954 *The Foundations of Statistics.* New York: Wiley.

3

Theory of Probability

3.1 OUTCOMES AND EVENTS

Random activities constitute the backbone of the theory of probability. An *experiment* or *trial* is an activity that can lead to one of several results. Each potential result is called an *outcome, observation, sample,* or *datum.* The set of all outcomes is known as the *outcome space* and is denoted by the symbol Ω. The outcome space is also called the *sample space* or *data space.*

A subset of the outcome space is called an *event* and is denoted E. We say that the event occurs if the outcome of the experiment is a member of the event E.

As an illustration, consider the experiment involving the roll of a die. The outcome space is $\Omega = \{1, 2, 3, 4, 5, 6\}$. An event might be defined as the subset $E = \{2, 4, 6\}$; an informal characterization of this event is the set of even numbers resulting from the roll of a die.

The outcome space may be continuous as well as discrete. An experiment could involve the wavelength of photons impinging on a detector, for which the outcome space would be the positive real axis. Or it could relate to the location of the next distress signal from a ship, in which case the outcome would be a multidimensional quantity that could assume negative values, depending on the coordinate system used.

To each event E of an outcome space Ω, we assign a number or measure called the *probability.* We define the probability $P(E)$ of an event E by assuming that it satisfies the following conditions or axioms:

1. *Fractionality.* P lies between 0 and 1, inclusively:

$$0 \leq P(E) \leq 1$$

2. *Unity.* The outcome space has unit probability:

$$P(\Omega) = 1$$

3. *Additivity.* The probability of the union of disjoint events is the sum of their individual probabilities:

$$P\left(\bigcup_{i=1}^{n} E_i\right) = \sum_{i=1}^{n} P(E_i)$$

since the E_i are disjoint and have no elements in common. This equality holds even when the number of events is infinite; namely, when $n \to \infty$.

These axioms lead to a number of ready consequences. One of these results holds that the *probability of the null event $\varnothing$ is zero.* We first note that Ω and $\varnothing$ are disjoint events whose union is Ω. As a result,

$$1 = P(\Omega) = P(\Omega \cup \varnothing) = P(\Omega) + P(\varnothing) = 1 + P(\varnothing)$$

The first and last equalities follow from Axiom 2 (Unity); the second from the fact that $\Omega = \Omega \cup \varnothing$; and the third equality from Axiom 3 (Additivity). We can now conclude that $P(\varnothing) = 0$.

More generally, we can show that the *probability of a complement is the complement of the probability.* To be precise, let E^c denote the complement of an event E, namely, the outcomes in Ω which are not in E. In symbolic notation, $E^c \equiv \Omega - E$. We claim the following result:

$$P(E^c) = 1 - P(E)$$

To show this, we note that E and E^c are disjoint events whose union is the entire outcome space Ω. Consequently,

$$1 = P(\Omega) = P(E \cup E^c) = P(E) + P(E^c)$$

The first equality relies on Axiom 1 (Unity), and the third on Axiom 3 (Additivity). The desired result follows immediately.

A third property of the probability measure is that the *probability of a subset does not exceed that of a superset.* In algebraic notation,

$$P(E) \leq P(F) \quad \text{whenever } E \subset F \tag{3-1}$$

To show this, we partition F into the two subsets E and D. In other words,

$D \equiv F - E$. Because D and E are disjoint,

$$P(F) = P(E \cup D) = P(E) + P(D)$$

where the second equality relies on the Additivity Axiom of probability. We can write the result in terms of $P(E)$ as

$$P(E) = P(F) - P(D) \leq P(F)$$

The inequality above depends on the Fractionality Axiom. Since $P(D) \geq 0$ for any event D, we have just verified Eq. (3–1).

Note that the result holds whether or not E is a proper subset of F. In particular, if $E = F$, then D is the null set $\varnothing$ for which $P(\varnothing) = 0$. In this case, Eq. (3–1) holds with strict equality.

We can also show that the *probability of the union* of any two events is given by the sum of their probabilities reduced by the probability of their intersection. This result holds whether or not the events are disjoint. More specifically, let E_1 and E_2 be two events. Then,

$$P(E_1 \cup E_2) = P(E_1) + P(E_2) - P(E_1 \cap E_2) \qquad (3–2)$$

To verify this result, we first partition $E_1 \cup E_2$ into three subsets as depicted in Figure 3–1. We let B denote the intersection of the primary events: $B = E_1 \cap E_2$. Moreover, A is the complement of E_1 with respect to E_2: $A \equiv E_1 - E_2$. The converse set is C, consisting of elements of E_2 not in E_1: $C \equiv E_2 - E_1$.

The sets A, B, and C are disjoint events which collectively constitute $E_1 \cup E_2$. We can, therefore, write

$$P(E_1 \cup E_2) = P(A \cup B \cup C) = P(A) + P(B) + P(C)$$

$$= P(A) + P(B) + P(C) + [P(B) - P(B)]$$

$$= P(A \cup B) + P(C \cup B) - P(B) = P(E_1) + P(E_2) - P(E_1 \cap E_2)$$

For the third equation, we have simply added and subtracted $P(B)$. The fourth equality relies on the fact that A, B, and C are disjoint events. The last equation follows from the definition of A, B, and C as they relate to E_1 and E_2. We have thus demonstrated the verity of Eq. (3–2).

Note that Eq. (3–2) is valid regardless of the relationships between E_1 and E_2. If, say, E_2 is a subset of E_1, then $E_1 \cap E_2 = E_2$. In that case, Eq. (3–2) boils down to $P(E_1 \cup E_2) = P(E_1)$. On the other hand, if E_1 and E_2 are disjoint, then $P(E_1 \cap E_2) = P(\varnothing) = 0$. The result is $P(E_1 \cup E_2) = P(E_1) + P(E_2)$, which corresponds to the Axiom of Additivity.

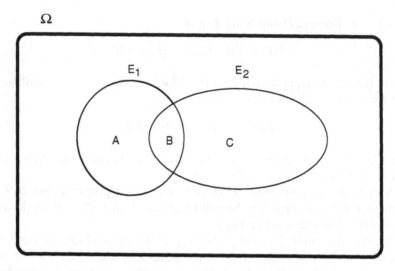

FIGURE 3–1. Example of two events E_1 and E_2 within the outcome space Ω. B is the intersection of E_1 and E_2; A is E_1 without E_2; C is E_2 without E_1.

Another consequence of the axioms of probability states that the probability of a union of arbitrary events does not exceed their associated sum. In algebraic notation,

$$P\left(\bigcup_{i=1}^{n} E_i\right) \leq \sum_{i=1}^{n} P(E_i)$$

The proof is left as an exercise.

3.2 CONTINUITY OF PROBABILITY

A useful characteristic of the probability function lies in its continuity. The probability $P(E)$ of an event increases smoothly with the size of its argument E.

Our first task in this direction is to define the notation of a limiting event. Let $E_1, E_2, \ldots$, be a sequence of events. We say that the sequence is increasing if each event is a superset of the previous one:

$$E_{n+1} \supset E_n$$

An example of an increasing sequence is $E_n =$ "Birth of the nth generation after Adam and Eve". Another illustration is "Greatest speed traveled by a human."

Conversely, the sequence is *decreasing* if each is a subset of its predecessor:

$$E_{n+1} \subset E_n$$

Examples of a decreasing sequence are "Record time for running a marathon" or "The years until a cure for cancer is found." Based on the nature of the sequence, we may specify new events in the following way. If $E_1, E_2, \ldots$ is an increasing sequence of events, we define the upper limit of the sequence as

$$E^* = \lim_{n \to \infty} E_n = \bigcup_{i=1}^{\infty} E_i$$

On the other hand, if the sequence is decreasing, we define the lower limit as

$$E_* = \lim_{n \to \infty} E_n = \bigcap_{i=1}^{\infty} E_i$$

Using this notation, we can show that the probability of a limiting event is the limit of the associated probability. We state this more precisely as a proposition.

Proposition (Limiting Event). Let $E_1, E_2, \ldots$ be a sequence of events that may be either increasing or decreasing. The probability of the limit is the limit of the probability:

$$P\left(\lim_{n \to \infty} E_n\right) = \lim_{n \to \infty} P(E_n)$$

Proof. Assume for the moment that $E_1, E_2, \ldots$ is an increasing sequence. We define a sequence of disjoint events F_n based on the E_n. The first of these events is identical to E_1; that is, $F_1 \equiv E_1$. The second event is that part of E_2, which is not in E_1:

$$F_2 = E_2 - E_1$$

This is the pattern for all the F_n except the first:

$$F_n \equiv E_n - E_{n-1}, \quad n > 2$$

Since the primary events E_n are increasing, the incremental events F_n are disjoint. Further, the F_n collectively cover the E_n:

$$\bigcup_{i=1}^{n} F_n = \bigcup_{i=1}^{n} E_n$$

for all $n \geq 1$ and as $n \to \infty$.

The probability of the upper limit E^* is given as

$$P(E^*) = P\left(\bigcup_{i=1}^{\infty} E_i\right) = P\left(\bigcup_{i=1}^{\infty} F_i\right) = \sum_{i=1}^{\infty} P(F_i) = \lim_{n \to \infty} \sum_{i=1}^{n} P(F_i) \qquad (3\text{--}3)$$

The third equality relies on the Additivity Axiom of the probability function. The last equality depends on the fact that an infinite sum is simply shorthand for an n-fold sum as $n \to \infty$.

Consider the last sum in Eq. (3–3). It can be written as

$$\sum_{i=1}^{n} P(F_i) = P\left(\bigcup_{i=1}^{n} F_i\right) = P\left(\bigcup_{i=1}^{n} E_i\right) = P(E_n)$$

The final equality again springs from the Additivity Axiom. The second relies on the equivalence of the unions of the events F_n and E_n, whereas the last conversion depends on the fact that E_n is an increasing sequence. Inserting these results into the limit operator in Eq. (3–3), we obtain the desired result for an increasing sequence:

$$P\left(\lim_{n \to \infty} E_n\right) = \lim_{n \to \infty} P(E_n)$$

For the second part of the proof, we assume that $E_1, E_2, \ldots$ is a decreasing sequence. We define an increasing sequence $G_1, G_2, \ldots$ as the complementary events; that is, $G_n \equiv E_n^c$ for $n \geq 1$.

Since the G_n are an increasing sequence, we can define a sequence of disjoint events based on them:

$$H_1 \equiv G_1$$
$$H_n \equiv G_n - G_{n-1} \quad \text{for } n \geq 2$$

Since E_n and G_n are complementary, we have $E_n \cup G_n = \Omega$ for every n. Whatever is the limit of E_n as $n \to \infty$, G_n is the complement. Hence, we have

$$P\left(\lim_{n \to \infty} E_n\right) + P\left(\lim_{n \to \infty} G_n\right) = 1 \qquad (3\text{--}4)$$

Consider the second term in this equation. Since G_n can be written as $\bigcup_1^n H_i$, we can justify the final equality by

$$P\left(\lim_{n\to\infty} G_n\right) = P\left(\lim_{n\to\infty} \bigcup_{i=1}^n H_n\right) = P\left(\bigcup_{i=1}^\infty H_n\right)$$

$$= \sum_{i=1}^\infty P(H_i) = \lim_{n\to\infty} \sum_{i=1}^n P(H_i)$$

$$= \lim_{n\to\infty} P\left(\bigcup_{i=1}^n H_i\right)$$

The third and fifth equalities spring from the Additivity Axiom of probability.

Now the union of $H_1, \ldots, H_n$ is equal to G_n, which in turn equals E_n^c. Hence, the last expression in the above equation becomes

$$\lim_{n\to\infty} P\left(\bigcup_1^n H_i\right) = \lim_{n\to\infty} P(E_n^c) = 1 - \lim_{n\to\infty} P(E_n)$$

The second equivalence depends on the fact that $P(E^c) = 1 - P(E)$. By inserting the last expression into Eq. (3–4), we obtain

$$P\left(\lim_{n\to\infty} E_n\right) = \lim_{n\to\infty} P(E_n)$$

as desired. This proves the Limiting Event Proposition.   ∎

As an indication of how this proposition can be applied, consider a gambler who goes to the casino with his entire savings. He plans to remain there indefinitely, or until all his funds are depleted. What is the probability that he will eventually become bankrupt?

Let X_n denote the size of his funds at the nth trial or game. Once he is bankrupt, he will remain so. Consequently $X_n = 0$ implies $X_{n+1} = 0$ for all n. Let E_n denote the condition $\{X_n = 0\}$, namely, bankruptcy at trial n. Then E_1, $E_2, \ldots$, is an increasing sequence.

The probability of eventual ruin is given by

$$P\left(\lim_{n\to\infty} E_n\right) = \lim_{n\to\infty} P(E_n)$$

In other words, the probability of eventual ruin is equal to the limiting probability of the event $X_n = 0$ as $n \to \infty$. The precise value of this limiting probability will, of course, depend on the values associated with each event $X_n = 0$.

For instance, suppose that the probability of ruin at step n is $P(E_n) = 1 - 1/2^n$. Then the probability of eventual ruin, namely, $P(E_\infty)$, is given by $\lim_{n \to \infty} (1 - 1/2^n) = 1$.

Infinite Recurrence

The Limiting Event Proposition is instrumental for deriving a result relating the probabilities of events to their long-term behavior. In particular, if the total probability of a sequence of events is finite, then there is zero probability of observing the events infinitely often.

Lemma (Borel–Cantelli). Consider a sequence of events $E_1, E_2, \ldots$. If the sum of their probabilities is finite,

$$\sum_{i=1}^{\infty} P(E_i) < \infty \tag{3-5}$$

then the probability that an infinite number of the E_i will occur is nil:

$$P(\text{Infinitude of } E_i \text{ occurring}) = 0$$

Proof. Let D_n be the event that tracks the partial union of all the E_i starting with E_n. In other words,

$$D_n = \bigcup_{i=n}^{\infty} E_i \tag{3-6}$$

Then, event D_n occurs if and only if at least one of $E_n, E_{n+1}, \ldots$ occurs. We note that the set $\{D_n\}$ is a decreasing sequence.

We define another event C which is the infinite intersection of the D_n:

$$C = \bigcap_{n=1}^{\infty} D_n \tag{3-7}$$

The event C occurs if and only if all the D_n occur.

Our first task is to show that the occurrence of C is equivalent to the event in which an infinite number of the E_i occur. To show this in the forward direction, assume that C occurs. From Eq. (3–7), we infer that event D_n occurs for all values of n. From Eq. (3–6), we know that for each D_n, at least one of the E_i occurs. Hence, the E_i occur infinitely often.

For the backward demonstration, assume that the E_i occur infinitely often. From Eq. (3–6), we know that D_n occurs for each value of n. Consequently, we know from Eq. (3–7) that the event C occurs. Hence, we have proved that the occurrence ol

$$\bigcap_{n=1}^{\infty} \bigcap_{i=n}^{\infty} E_i$$

is equivalent to the infinite occurrence of the E_i.

The probability of this event is given by

$$P(C) = P\left(\sum_{n=1}^{\infty} D_n\right)$$

$$= P\left(\lim_{n \to \infty} D_n\right)$$

$$= \lim_{n \to \infty} P(D_n) \qquad (3-8)$$

The second equality relies on the fact that D_n is a decreasing sequence; as a result,

$$\bigcap_{n=1}^{m} D_n = D_m$$

Taking the limit as $m \to \infty$ yields the required identity. The last transformation in Eq. (3–8) depends on the Limiting Event Proposition.

We continue from the latest result by substituting for D_n in terms of the E_i:

$$P(C) = \lim_{n \to \infty} P\left(\bigcup_{i=n}^{\infty} E_i\right) \leq \lim_{n \to \infty} \sum_{i=n}^{\infty} P(E_i) = 0$$

The inequality follows from the fact that the probability of a union cannot exceed the sum of the individual probabilities. The last equality follows from Eq. (3–5): since the probability of all events is finite, there must exist some upper bound $n = N$ beyond which the partial sum $\sum_{i=N, \infty} P(E_i)$ must vanish. We have, therefore, shown that the probability of observing an infinite number of the E_i is 0. ∎

The use of the Borel–Cantelli Lemma is illustrated in the following scenario. Let $X_1, X_2, \ldots$ be a sequence of random variables. The variable X_n takes value 1 with probability p^n for some p between 0 and 1; and the value -1 with probability $1 - p^n$.

Suppose E_n is the event $\{X_n = 1\}$. Then the total probability of E_n is

$$\sum_{n=1}^{\infty} P(E_n) = \sum_{n=1}^{\infty} p^n = \sum_{n=0}^{\infty} p^n - 1 = \frac{1}{1-p} - 1 = \frac{p}{1-p} < \infty$$

We invoke the Borel–Cantelli Lemma to conclude that the probability of X_n equaling 1 an infinite number of times is 0. For large values of n, X_n must equal -1 with probability 1:

$$\lim_{n \to \infty} X_n = -1$$

The Borel–Cortelli Lemma states the conditions under which the probability of infinite recurrence is 0. The converse proposition stipulates how the probability of infinite recurrence can equal 1. For this proposition, however, we must draw on the assumption of independence among the events.

Proposition (Infinite Recurrence). Let $E_1, E_2, \ldots$ be a sequence of independent events whose total probability is infinite:

$$\sum_{i=1}^{\infty} P(E_i) = \infty$$

Then the probability that an infinite number of the E_i occur is 1:

$$P\{\text{Infinitude of } E_i \text{ occurring}\} = 1$$

Proof. The probability of infinite recurrence is given by

$$P\{\text{Infinite occurrence of } E_i\} = P\left\{ \lim_{n \to \infty} \bigcup_{i=n}^{\infty} E_i \right\} = \lim_{n \to \infty} P\left(\bigcup_{i=n}^{\infty} E_i \right)$$

$$= \lim_{n \to \infty} \left[1 - P\left(\bigcap_{i=n}^{\infty} E_i^c \right) \right] \qquad (3\text{–}9)$$

The second equality relies on the Limiting Event Proposition. The third depends on the fact that the complement of a union is the intersection of the complements:

$$\left(\bigcup_{i=n}^{\infty} E_i \right)^c = \bigcap_{i=n}^{\infty} E_i^c$$

By the assumption of independence, the probability of an intersection is the product of the probabilities. We, therefore, obtain

$$P\left(\bigcap_{i=n}^{\infty} E_i^c \right) = \prod_{i=n}^{\infty} P(E_i^c) = \prod_{i=n}^{\infty} P[1 - P(E_i)] \le \prod_{i=n}^{\infty} \exp[-P(E_i)]$$

$$= \exp\left(-\sum_{i=n}^{\infty} P(E_i) \right) = 0$$

The inequality follows from the fact that $1 - x \leq e^{-x}$. The last equality depends on the initial assumption

$$\sum_{i=1}^{\infty} P(E_i) = \infty$$

Each term is finite because the probability $P(E_i)$ cannot exceed 1. If the total sum is infinite, there is no upper index $n = N$ for which the partial sum is finite. As a result,

$$\sum_{i=n}^{\infty} P(E_i)$$

is infinite for all values of n.

Injecting the result into the last part of Eq. (3–9) yields the desired result. ∎

To illustrate the use of the Infinite Recurrence Proposition, consider the sequence $X_1, X_2, \ldots$ of random variables. Each variable X_n can assume the value 1 with probability $1/n$ and value -1 with probability $(n - 1)/n$. Let E_n be the event $\{X_n = 1\}$. The total probability does not converge:

$$\sum_{i=n}^{\infty} P\{E_n\} = \sum_{i=n}^{\infty} \frac{1}{n} = \infty$$

From the Infinite Recurrence Proposition, E_n will occur with infinite frequency.

On the other hand, the converse is also true. We know that

$$\sum_{n=1}^{\infty} P\{E_n^c\} = \sum_{i=n}^{\infty} \frac{n-1}{n} = \infty$$

Therefore, the complementary event will also occur infinitely often. In short, X_n will bounce back and forth between -1 and 1 indefinitely with probability 1. It will converge to no limiting value as n goes to infinity.

3.3   RANDOM VARIABLES

A random variable is a function that associates a real number with each outcome in the outcome space. In other words, a *random variable* X maps the outcome space Ω into the set $\mathscr{R}$ of reals.

The probability that X assumes a value in a given subset of $\mathscr{R}$ depends on probability of the corresponding outcome within Ω. Suppose that E is a subset of the outcome space Ω, and R the subset of $\mathscr{R}$ resulting from the mapping X; that is, $R = \{x \mid x = X(e) \ \& \ e \in E\}$. Then the probability of X being in R matches the probability of E:

$$P\{X \in R\} = P(E)$$

The *distribution function* F of the random variable X is defined by the cumulative probability of X lying at or below x for any real value x. In symbolic notation,

$$F(x) = P\{X \le x\} = P\{X \in (-\infty, x)\}$$

The complement of F, denoted $\overline{F}$, is the converse probability of X lying above any threshold x:

$$\overline{F}(x) \equiv P\{X > x\} = 1 - F(x)$$

The continuity of a random variable depends on its numerosity or cardinality. The random variable X is called *discrete* if its space of values is countable or countably infinite. The distribution function for a discrete variable is given by

$$F(x) = \sum_{y \le x} P\{X = y\}$$

The values $p_y \equiv P\{X = y\}$ are collectively known as the *probability mass function* for random variable X.

On the other hand, a random variable X is said to be continuous if there is a function whose integral yields the appropriate probabilities. More precisely, the function f must have the following property:

$$P\{X \in R\} = \int_R f(x)\, dx$$

for any set R. The function f is called the *probability density function* of X, or simply the *density*.

For a continuous random variable, the distribution function can be expressed as an integral:

$$F(x) = \int_{-\infty}^{x} f(X)\, dX$$

It is clear that the density is simply the derivative of the distribution function:

$$f(x) = \frac{d}{dx}F(x)$$

The *joint distribution function* F_{XY} of two random variables X and Y is defined by their cumulative probabilities. In algebraic form,

$$F_{XY}(x, y) \equiv P\{X \leq x, Y \leq y\}$$

Let F_X and F_Y denote the individual distribution functions, namely,

$$F_X(x) = P\{X \leq x\}$$
$$F_Y(y) = P\{Y \leq y\}$$

These individual distributions can be derived from the joint distribution F_{XY} by drawing on the fact that the probability measure is continuous.

To this end, let $y_1, y_2, \ldots$ be an increasing sequence of values growing to infinity. Then the sequence of events $\{X \leq x, Y \leq y_n\}$ are increasing sets. We can write

$$\lim_{n \to \infty} \{X \leq x, Y \leq y_n\} = \bigcup_{n=1}^{\infty} \{X \leq x, Y \leq y_n\}$$

We have assumed that $y_n \to \infty$ as $n \to \infty$. The last expression in the above equation becomes $\{X \leq x, Y \leq \infty\}$. Since Y can take any value, this event is the same as the event $\{X \leq x\}$. From the continuity property of the probability function,

$$\lim_{n \to \infty} P\{X \leq x, Y \leq y_n\} = P\{X \leq x\}$$

We conclude that

$$F_X(x) = \lim_{y \to \infty} F(x, y) = F_{XY}(x, \infty)$$

In a similar way, we can show that $F_Y(y) = F_{XY}(\infty, y)$.

We say that two random variables X and Y are *independent* if their joint distribution is the product of their individual distributions:

$$F_{XY}(x, y) = F_X(x)F_Y(y)$$

for all values of x and y.

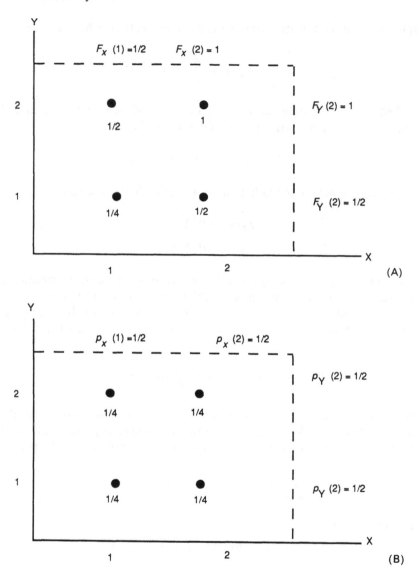

FIGURE 3–2. (A) Illustration of independent variables. The joint density is $F(x, y) = xy/4$, as indicated next to the dark circles. Their individual distributions F_x and F_y are shown to the top and right, respectively. Since $F(x, y) = F_X(x)F_Y(y)$, the variables are independent. (B) Probability mass function corresponding to (A). For the distribution $F(x, y) = xy/4$, the mass function is $p(x, y) = 1/4$ for all x and y.

To illustrate, consider two random variables X and Y. Each of these variables can take the value 1 or 2, and their joint distribution is $F(x, y) = xy/4$.

The values of F for each combination of the variables is shown in Figure 3–2A. For instance, $F(1, 2) = 1(2)/4 = 1/2$. The individual distributions for X and Y are shown respectively to the top and right of the chart. For instance,

$$F_X(1) = \max_y F(1, y) = \max \{ \tfrac{1}{4}, \tfrac{1}{2} \} = \tfrac{1}{2}$$

We can verify that $F(x, y) = F_X(x)F_Y(y)$ for all values of x and y. Therefore, the variables X and Y are independent.

The probability mass function corresponding to Figure 3–2A is shown in Figure 3–2B. The joint density is $p(x, y) = 1/4$ for all x and y. Since the individual mass functions are $p_X(x) = p_Y(y) = 1/2$ for each value of x and y, the joint mass function is factorable:

$$p(x, y) = p_X(x)p_Y(y)$$

This phenomenon applies more generally. In other words, if two discrete random variables possess a joint mass function that is factorable, then they are independent. In a similar way, when two continuous variables exhibit a density function that is factorable, then they are independent. We state this claim as a theorem.

Theorem (Factorability). Two random variables X and Y are independent if and only if their joint mass or density functions can be factored into their individual functions. More precisely, suppose that X and Y are discrete variables with joint mass function p and individual mass functions p_X and p_Y. If the mass functions relate as

$$p(x, y) = p_X(x)p_Y(y)$$

then X and Y are independent.

Similarly, let X and Y be two continuous random variables with joint density function f and individual densities f_X and f_Y. If the densities are governed by the relation

$$f(x, y) = f_X(x)f_Y(y)$$

then X and Y are independent.

Proof. Consider the case where the variables are continuous. We first show the result in the forward direction. The joint distribution of X and Y is given by

$$F(x, y) = \int_{-\infty}^{x} \int_{-\infty}^{y} f(X, Y) \, dx \, dy \qquad (3\text{–}10)$$

By the assumption of independence, $F = F_x F_y$. Therefore,

$$F(x, y) = F_X(x) F_Y(y) = \int_{-\infty}^{x} f_X(x) \, dx \int_{-\infty}^{y} f_Y(y) \, dy$$

$$= \int_{-\infty}^{x} \int_{-\infty}^{y} f_X(x) f_Y(y) \, dx \, dy \qquad (3\text{–}11)$$

Taking the derivatives of the last expressions in Eqs. (3–3) and (3–4) leads to the result $f(x, y) = f(x, y) = f_X(x) f_Y(y)$.

Now consider the backward direction. The joint distribution function can be written as

$$F(x, y) = \int_{-\infty}^{x} \int_{-\infty}^{y} f(x, y) \, dy \, dx = \int_{-\infty}^{x} \int_{-\infty}^{y} f_X(x) f_Y(y) \, dy \, dx$$

$$= \int_{-\infty}^{x} f_X(x) \, dx \int_{-\infty}^{y} f_Y(y) \, dy = F(x) F(y)$$

Hence, X and Y are independent.

The proof for the discrete case is analogous and left as an exercise. ∎

Many of these concepts generalize to multiple variables. For instance, the joint distribution of random variables $X_1, \ldots, X_n$ is given by

$$F(x_1, \ldots, x_n) \equiv P\{X_1 \le x_1, \ldots, X_n \le x_n\}$$

The individual distribution for each random variable X_i can be obtained by driving all the other arguments to infinity in the joint distribution:

$$F_{X_i}(X_i) = F(\infty, \ldots, x_i, \ldots, \infty).$$

As with the two-variable case, we say that the random variables are independent if their joint density can be factored into individual densities:

$$F(x_1, \ldots, x_n) = F_{X_1}(X_i) \cdot F_{X_2}(X_2) \cdots F_{X_n}(X_n)$$

3.4   EXPECTATION

The expectation of a random variable locates its center of mass. The center of weight is determined by the values of the variable, as weighted by its mass function or density function. For a random variable X, its *expectation* or *mean* is

$$E(X) \equiv \int_{-\infty}^{\infty} x \, dF(x) \qquad (3-12)$$

This notation dF refers to the derivative of the distribution function F of X. If X is continuous, the Eq. (3–12) simply becomes

$$E(X) \equiv \int_{-\infty}^{\infty} xf(x) \, dx$$

where f is the density function of X. On the other hand, if X is discrete, Eq. (3–12) becomes

$$E(X) \equiv \sum_{x=-\infty}^{\infty} x p_X(x)$$

where p_X is the mass function of X.

The expectation of any function of X is defined in a similar way. Since X is a random variable, any function $h(X)$ is itself a random variable. The expectation of $h(X)$ is given by

$$E[h(X)] \equiv \int_{-\infty}^{\infty} h(x) \, dF(x)$$

The expectation is a linear operator. In other words, if a and b are constants, then

$$E(aX + b) = aE(X) + b$$

for any random variable X. This can be shown as

$$E(aX + b) = \int (ax + b) \, dF(x)$$

$$= a \int x \, dF(x) + b \int dF(x) = aE(X) + b$$

The limits on each integral are the usual infinities, from $-\infty$ to $+\infty$. The second equality springs from the linearity of the integral operator. The third equality relies on the definition of $E(X)$ and the fact that $\int dF(x) = F(\infty) = 1$.

When two variables are independent, the expectation of the product is the product of the expectations: $E(XY) = E(X)E(Y)$. We show this for the continuous case as follows.

Let f be the joint density for random variables X and Y, whose individual densities are f_X and f_Y, respectively. The expectation of their product is

$$E(XY) = \int\int xy\, f(x, y)\, dx\, dy = \int\int xy\, f_X(x) f_Y(y)\, dx\, dy$$

$$= \int x\, f_X(x)\, dx \int y\, f_Y(y)\, dy = E(X)E(Y)$$

Each integral is taken over the entire real line, from minus to plus infinity. The second equation depends on the independence of X and Y, for which $f = f_X f_Y$. The proof for the discrete case is analogous.

3.5   VARIANCE AND COVARIANCE

A measure of the volatility of X is found in the expected value of its squared deviation from the center of mass. This is called the *variance* of X:

$$\mathrm{Var}(X) \equiv E\{[X - E(X)]^2\}$$

Since $E(X)$ is the center of mass for variable X, the transformed quantity $X' \equiv X - E(X)$ is called the *centralized* form of X. In other words, the centralized version of a random variable is its deviation from the mean.

In addition, the mean value of a power of a random variable is called its *moment*. More specifically, $E(X^k)$ is known as the kth *moment* of X.

With these terminologies, we can define the variance of a random variable as its *second central moment*. In algebraic notation, $\mathrm{Var}(X) = E[(X')^2]$ where X' is the centralized form of variable X.

We can obtain a handy equivalence for the variance by expanding the right-hand term:

$$\mathrm{Var}(X) = E[X - E(X)]^2 = E[X^2 - 2XE(X) + E^2(X)]$$

$$= E(X^2) - 2E(X)E(X) + E^2(X) = E(X^2) - E^2(X)$$

The third equation relies on the linearity of the expectation operator and the fact that $E(X)$ is a constant for any variable X.

The variance is a nonlinear operator. In particular, it squares multiplicative constants and eliminates additive constants:

$$\text{Var}(aX + b) = a^2\,\text{Var}(X)$$

where X is any random variable and a and b are constants. This can be shown as follows:

$$\text{Var}(aX + b) = E[(aX + b) - E(aX + b)]^2$$
$$= E\{a^2[X - E(X)]^2\} = a^2 E[X - E(X)]^2$$
$$= a^2\,\text{Var}(X)$$

The second equality relies on the fact that $E(aX + b) = aE(X) + b$.

The variance is a measure of variability around the center of mass. It, therefore, ignores the additive constant b, which simply shifts the location of X to the right by an amount b. But this translation has no effect on the dispersion of X.

In contrast, the multiplicative constant a is squared by the variance operator. This springs from the fact that the variance is a measure of the *squared* deviation of X from its center of mass.

The standard deviation σ, on the other hand, is the square root of the variance. Therefore, σ is also a measure of dispersion. The standard deviation happens to be a linear operator for a:

$$\sigma(aX) = [\text{Var}(aX)]^{1/2} = [a^2\,\text{Var}(X)]^{1/2}$$
$$= a[\text{Var}(X)]^{1/2} = a\sigma(X)$$

Intuitively, this makes sense because a is a scaling factor for X in the transformation $X' = aX$. If X represents the height of individuals in meters, then $X' = 100X$ is the height of individuals in centimeters. The standard deviation for X' is 100 times that for X and has the same units. This contrasts with the variance for X', which is 10,000 times that for X and has the units of centimeters squared.

Two random variables may influence each other's behavior. Suppose X and Y are two random variables having a joint distribution. Their *covariance* is given by the expected value of their cross-product after adjusting for their respective means:

$$Cov(X, Y) \equiv E\{[X - E(X)][Y - E(Y)]\}$$

In other words, the covariance of X and Y is the mean cross-product of their centralized values. By expanding the quantities within curly braces and simplifying, we can obtain another formula for the covariance:

$$Cov(X, Y) = E(XY) - E(X)E(Y)$$

Note that the covariance vanishes when the variables are independent. If X and Y are independent, then $E(XY) = E(X)E(Y)$, from which $Cov(X, Y) = 0$.

However, the converse is false. If the covariance of X and Y is 0, we cannot conclude that X and Y are independent. The numbers for the distributions of X and Y might just happen to yield a zero covariance even when the variables are, in fact, dependent.

The variance of two variables is the sum of their variances plus twice the covariance:

$$Var(X + Y) = Var(X) + Var(Y) + 2\,Cov(X, Y)$$

This can be shown as follows. To simplify the algebra slightly, let μ_X and μ_Y denote the means for X and Y, respectively. Then,

$$\begin{aligned}
Var(X + Y) &= E(X + Y - \mu_X - \mu_Y)^2 \\
&= E[(X - \mu_X) + (Y - \mu_Y)]^2 \\
&= E[(X - \mu_X)]^2 + E[(Y - \mu_Y)^2] + 2E[(X - \mu_X)(Y - \mu_Y)] \\
&= Var(X) + Var(Y) + 2\,Cov(X, Y)
\end{aligned}$$

If X and Y are independent, then $Cov(X, Y) = 0$. We infer that

$$Var(X + Y) = Var(X) + Var(Y)$$

for two independent random variables.

Example (Matching Problem)

Each person brings a gift to a Christmas party and places it under the tree. The gifts are later selected at random. What is the expected number of people who end up with their own gifts? What is the variance of this number?

Suppose the party consists of n people. Let X_i be the indicator variable for the ith person selecting his own gift: X_i equals 1 if the match occurs and is 0 otherwise. The number of people who select their own gift is

$$X = \sum_{i=1}^{n} X_i$$

Since there are n gifts, the probability that the ith person chooses his own is

$$E(X_i) = \frac{1}{n}$$

The expected number of matches among all partygoers is

$$E(X) = E\left(\sum_{i=1}^{n} X_i\right) = \sum_{i=1}^{n} E(X_i) = n\left(\frac{1}{n}\right) = 1$$

The variance of each indicator X_i is

$$\mathrm{Var}(X_i) = E(X_i^2) - E^2(X_i)$$

$$= [0^2 \cdot P\{X_i = 0\} + 1^2 \cdot P\{X_i = 1\}] - \left(\frac{1}{n}\right)^2$$

$$= \left[0 + 1\left(\frac{1}{n}\right)\right] - \left(\frac{1}{n}\right)^2 = \frac{n-1}{n^2}$$

To compute the covariance of each X_i, we will need the mean of the cross-products:

$$E(X_i X_j) = 0 \cdot P(X_i = 0, X_j = 0) + 0 \cdot P(X_i = 1, X_j = 0)$$

$$+ 0 \cdot P(X_i = 0, X_j = 1) + 1^2 \cdot P(X_i = 1, X_j = 1)$$

$$= P(X_i = 1) \cdot P(X_j = 1 | X_i = 1)$$

$$= \frac{1}{n}\left(\frac{1}{n-1}\right)$$

The last equality depends on the fact that if the ith person has already selected his own gift, then the jth person has equal chance of doing likewise from the remaining $n - 1$ gifts. The covariance can now be computed:

$$\text{Cov}(X_i, X_j) = E(X_i X_j) - E(X_i)E(X_j) = \frac{1}{n}\left(\frac{1}{n-1}\right) - \left(\frac{1}{n}\right)^2$$

$$= \frac{1}{n^2(n-1)}$$

Finally, the variance of X is

$$\text{Var}(X) = \text{Var}(\Sigma_i X_i) = \Sigma_i \text{Var}(X_i) + 2\sum_{i=1}^{n}\sum_{j=1}^{i-1}\text{Cov}(X_i, X_j)$$

$$= n\left(\frac{n-1}{n^2}\right) + 2\left(\frac{(n-1)n}{2}\right)\left(\frac{1}{n^2(n-1)}\right) = 1$$

In short, the number of matches has mean and variance both equal to 1.   ◆

3.6   TRANSFORMS

The language of mathematics provides a formal vehicle for reasoning about a problem. For instance, a verbal problem described in an algebra text might be translated into an algebraic equation and then solved. In a similar way, characteristics of random variables can often be deduced more easily by way of a transformation rather than directly from the distribution functions.

Moment Generating Function

One class of properties often of interest is the expected value of a random variable raised to some power. The mean of X to the nth power—namely, $E(X^n)$—is called the nth *moment* of X. For example, $E(X^2)$ is called the second moment of X.

A random variable is said to be *centralized* when it is adjusted around its mean. More specially, the transformed variable $X' = X - E(X)$ is called the centralized form of the underlying variable X. The expectation of X' to the nth power is called the nth central moment of X:

$$E(X')^n = E[X - E(X)]^n$$

As an illustration, the variance is the second central moment of X: $\text{Var}(X) = E[X - E(X)]^2$.

The central moments of a random variable may also be expressed in terms of the basic moments. We have seen, for instance, that the variance is equal to the second basic moment reduced by the square of the first basic moment:

$$\text{Var}(X) = E(X^2) - E^2(X)$$

The utility of moments for summarizing the properties of a random variable underscores the desirability of simple ways to calculate the moments. One such method is the *exponential transform* or *moment generating function*, defined in the following way. Let X be a random variable with distribution F. The moment generating function of X is given by

$$M(t) \equiv E(e^{tX}) = \int e^{tX} \, dF(X)$$

When $M(t)$ is differentiated n times and the parameter t set to zero, the result is the nth moment of X. The nth derivative of $M(t)$ is

$$\frac{d^n}{dt^n} M(t) = \frac{d^n}{dt^n} \int X^n e^{tX} \, dF(X) = E(X^n e^{tX})$$

Setting t to 0 yields the nth moment:

$$\left. \frac{d^n}{dt^n} M(t) \right|_{t=0} = E(X^n)$$

As an illustration, consider the random variable X from an exponential density with parameter λ. The moment generating function for X is

$$M(t) = E(e^{tX}) = \int_0^\infty (e^{tx})(\lambda e^{-\lambda x}) \, dx$$

$$= \lambda \int_0^\infty e^{-(\lambda - t)x} \, dx = \frac{\lambda}{\lambda - t}$$

The first moment of X is given by differentiating $M(t)$ and evaluating at $t = 0$:

$$E(X) = \left. \frac{d}{dt} M(t) \right|_{t=0} = \left. \frac{\lambda}{(\lambda - t)^2} \right|_{t=0} = \frac{1}{\lambda}$$

Depending on the density function, a moment generating function may or may not exist. If one exists, it is unique and, therefore, uniquely defines the underlying distribution. In a formal sense, a moment generating function characterizes a random variable as much as the distribution function itself. The distribution function has the advantage, of course, of providing more intuitive insight into the behavior of the random variable.

Shown in Table 3–1 are some discrete distribution functions and their properties, including the moment generating functions. A similar presentation for some continuous distributions is given in Table 3–2.

Moment generating functions can also be useful for determining the properties of multiple variables. For instance, suppose X and Y are independent random variables. Then the moment generating function of the sum is the product of the individual moment generating functions:

$$M_{X+Y}(t) = E(e^{t(X+Y)}) = E(e^{tX}e^{tY}) = E(e^{tX})E(e^{tY})$$
$$= M_X(t)M_Y(t)$$

The third equality springs from the assumption of independence. The use of this result is illustrated below.

Example

Suppose X_1 and X_2 are independent random variables from a normal density. Let μ_i and σ_i^2 be the respective mean and density for X_i. We know that the moment generating function for the sum $X_1 + X_2$ is the product of the functions:

$$M_{X_1+X_2}(t) = M_{X_1}(t)M_{X_2}(t) = \exp\left(\mu_1 t + \frac{\sigma_1^2 t^2}{2}\right)\exp\left(\mu_2 t + \frac{\sigma_2^2 t^2}{2}\right)$$
$$= \exp\left((\mu_1 + \mu_2)t + \frac{(\sigma_1^2 + \sigma_2^2)t^2}{2}\right)$$

The first equality depends on the assumption of independence. The second equality can be derived from the definition of $M(t)$, or simply obtained from Table 3–2.

Since the moment generating function uniquely characterizes the underlying distribution, we see from the last expression above that the sum $X_1 + X_2$ is distributed normally. Further, the sum has mean $\mu = \mu_1 + \mu_2$ and variance $\sigma^2 = \sigma_1^2 + \sigma_2^2$. ◆

Other Transforms

The moment generating function may not always exist. On theoretical grounds, it is more appropriate to define a transformation that is guaranteed to exist. This is the *characteristic function* of X defined as

$$C(X) \equiv E(e^{itX})$$

where t is any real number and $i = \sqrt{-1}$ is the imaginary unit. It is possible to show that $C(X)$ exists. In addition, the characteristic function uniquely defines a distribution function in the same way that the moment generating function does.

Certain random variables assume no negative values, as exemplified by the binomial distribution. For these variables, it is often more convenient to replace the characteristic function with another type of transformation called the *Laplace transform*. The Laplace transform $\tilde{F}$ of a distribution function F

TABLE 3–1. Some discrete probability distributions and their properties.

Name	Parameter(s)	Mass function, $p(x)$	Moment Generating Function, $M(t)$	Mean	Variance
Binomial	$n, p, 0 \le p \le 1$	$\binom{n}{x} p^x (1-p)^{n-x}$ $x = 0, 1, \ldots, n$	$[pe^t + (1-p)]^n$	np	$np(1-p)$
Poisson	$\lambda > 0$	$e^{-\lambda} \dfrac{\lambda^x}{x!}$ $x = 0, 1, 2, \ldots$	$\exp\{\lambda(e^t - 1)\}$	λ	λ
Geometric	$0 \le p \le 1$	$p(1-p)^{x-1}$ $x = 1, 2, \ldots$	$\dfrac{pe^t}{1 - (1-p)e^t}$	$\dfrac{1}{p}$	$\dfrac{1-p}{p^2}$
Negative binomial	r, p	$\binom{x-1}{r-1} p^r (1-p)^{x-r}$ $x = r, r+1, \ldots$	$\left(\dfrac{pe^t}{1 - (1-p)e^t}\right)^t$	$\dfrac{r}{p}$	$\dfrac{r(1-p)}{p^2}$

TABLE 3–2. Some continuous probability distributions and their properties.

Name	Parameters	Outcome Space, Ω	Density Function, $f(x)$	Moment Generating Function, $M(t)$	Mean	Variance
Uniform	a, b	$a < x < b$	$\dfrac{1}{b-a}$	$\dfrac{e^{tb} - e^{ta}}{t(b-a)}$	$\dfrac{a+b}{2}$	$\dfrac{(b-a)^2}{12}$
Exponential	$\lambda > 0$	$x \ge 0$	$\lambda e^{-\lambda x}$	$\dfrac{\lambda}{\lambda - t}$	$\dfrac{1}{\lambda}$	$\dfrac{1}{\lambda^2}$
Gamma	$n, \lambda; \lambda > 0$	$x \ge 0$	$\dfrac{\lambda e^{-\lambda x}(\lambda x)^{n-1}}{(n-1)!}$	$\left(\dfrac{\lambda}{\lambda - t}\right)^n$	$\dfrac{n}{\lambda}$	$\dfrac{n}{\lambda^2}$
Normal	μ, σ^2	$-\infty < x < \infty$	$\dfrac{1}{\sigma\sqrt{2\pi}} \exp\left\{\dfrac{(x-\mu)^2}{2\sigma^2}\right\}$	$\exp\left\{\mu t + \dfrac{\sigma^2 t^2}{2}\right\}$	μ	σ^2
Beta	$a, b; a > 0,$ $b > 0$	$0 < x < 1$	$cx^{a-1}(1-x)^{b-1}$ $c = \dfrac{\Gamma(a+b)}{\Gamma(a)\Gamma(b)}$		$\dfrac{a}{a+b}$	$\dfrac{ab}{(a+b)^2(a+b+1)}$

is defined as

$$\tilde{F}(s) \equiv \int_0^\infty e^{-sx} \, dF(x)$$

where $s = a + bi$ is a complex parameter. The real component a must be non-negative; b can assume any real number. The Laplace transform uniquely characterizes a distribution function, much like the characteristic and moment generating functions.

3.7 CONDITIONAL FUNCTIONS

Given two random variables, we are often interested in the behavior of one of them under the constraint that the other is fixed at a particular value. This is the idea behind the conditional distribution of a random variable.

The conditional distribution can be used to determine the properties of a constrained variable. Perhaps the most basic of these properties is the expectation of the conditional distribution.

Conditional Distribution

Suppose X and Y are discrete random variables. The *conditional mass function* of X for a given value of $Y = y$ is

$$P\{X = x | Y = y\} = P\{X = x, Y = y\}/P\{Y = y\}$$

Clearly, this quantity is defined only for active values of Y—namely, those for which $P\{Y = y\}$ is positive.

The conditional mass function serves as the basis for the *conditional distribution* which takes the following form:

$$F(x|y) = P\{X \le x | Y = y\}$$

for discrete variables X and Y.

The conditional functions for continuous variables are specified in similar ways. Let f be the joint density function for random variables X and Y having respective densities f_X and f_Y. The *conditional density function* of X, given the value $Y = y$, is

$$f(x|y) = \frac{f(x, y)}{f_Y(y)} \tag{3-13}$$

Once again, this quantity is defined only for active values of y: namely, those values for which the density function $f_Y(y)$ is positive.

The *conditional distribution function* is given by

$$F(x|y) = P\{X \le x | Y = y\} = \int_{-\infty}^x f(x|y) \, dx$$

Conditional functions are instrumental in characterizing a number of properties of interest. One such property is the conditional expectation.

Conditional Expectation

As in the unconstrained case, we are interested in the properties of conditioned variables. One of these properties is the mean of the conditional distribution. The *conditional expectation* of variable X, given $Y = y$, is

$$E(X|Y = y) = \int x \, dF(x|y) \tag{3-14}$$

If X is discrete, this quantity is tantamount to

$$E(X|Y = y) = \sum_x xP\{X = x|Y = y\}$$

If X is continuous, on the other hand, the conditional expectation for X when Y is fixed at value y is

$$E(X|Y = y) = \int_{-\infty}^{\infty} xf(x|y) \, dx$$

The conditional expectation is a constant once Y is specified. More specifically, $E[X|Y = y]$ is the mean value of X given the value y. Since the variable Y can take different values, the quantity $E(X|Y = y)$ can also assume different values.

How often does $E(X|Y = y)$ take one value rather than another? That depends on how often Y takes the value y rather than some other value y'. In short, the conditional mean of random variable X is fixed when Y is fixed, but varies as Y changes.

The conditional distribution of X is, therefore, a random variable when the conditioning parameter Y is viewed as a variable. We let $E(X|Y)$ denote the random variable whose distribution is defined by the distribution of Y. In other words, $E(X|Y)$ is the random variable whose value at $Y = y$ is the number given by $E(X|Y = y)$.

A useful property of the conditional expectation is that its mean matches the overall mean of the conditioned variable:

$$E[E(X|Y)] = E(X)$$

To show this result, we first note that the variability in $E(X|Y)$ is due to the variability of Y. Identifying the mean of $E(X|Y)$ is, therefore, tantamount to taking its expectation with respect to the distribution of Y:

$$E[E(X|Y)] = \int E(X|Y = y) \, dF_Y(y)$$

Let us consider the case where Y is a continuous variable. Then the preceding equation becomes

$$E[E(X|Y)] = \int E(X|Y = y)f_Y(y)\,dy \qquad (3\text{–}15)$$

The conditional expectation within the integral can be written as

$$E(X|Y = y) = \int x\,dF_Y(x|y) = \int xf(x|y)\,dx$$

$$= \int x\left(\frac{f(x, y)}{f_Y(y)}\right)dx$$

The first equality relies on Eq. (3–14), and the third on Eq. (3–13).

We now substitute the result so far for the conditional expectation on the right-hand side of Eq. (3–15) to obtain

$$E[E(X|Y)] = \int\left(\int x\frac{f(x, y)}{f_Y(y)}\,dx\right)f_Y(y)\,dy$$

$$= \int x\left[\int f(x, y)\,dy\right]dx = \int xf_X(x)\,dx = E(X)$$

which is the desired result. The proof for the discrete case is similar and is left as an exercise.

We reinforce these concepts through a numerical exercise.

Example

Let X and Y be discrete random variables, each taking values 0 or 1. Their joint mass function $p(x, y)$ takes the form $p(0, 0) = p(1, 1) = 1/8$, whereas $p(1, 0) = p(0, 1) = 3/8$.

The outcome space is depicted in Figure 3–3. The boxes at the bottom of the figure contain the unconditional probabilities for X. For instance,

$$p_X(1) = p(1, 0) + p(1, 1) = 1/8 + 3/8 = 1/2$$

The boxes to the left of the chart show the unconditional masses p_Y for variable Y.

The boxes to the right and above the chart contain the conditional expectations of X and Y, respectively. For instance,

$$E(X|Y = 0) = \sum_x xP\{X = x|Y = 0\} = \sum_x x\frac{P\{X = x, Y = 0\}}{p\{Y = 0\}}$$

$$= 0\cdot\frac{1/8}{1/2} + 1\cdot\frac{3/8}{1/2} = \frac{3}{4}$$

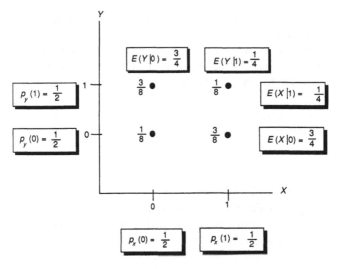

FIGURE 3-3. Illustration of conditional expectation.

The expectation of the conditional expectation is

$$E[E(X|Y)] = E(X|Y = 0)P(Y = 0) + E(X|Y = 1)P(Y = 1)$$
$$= \tfrac{3}{4}(\tfrac{1}{2}) + \tfrac{1}{4}(\tfrac{1}{2}) = \tfrac{1}{2}$$

The overall mean of X may be computed directly:

$$E(X) = 0 \cdot p_X(0) + 1 \cdot p_X(1) = 0 + 1(\tfrac{1}{2}) = \tfrac{1}{2}$$

We see that the relation $E[E(X|Y)] = E(X)$ is upheld.   ◆

The idea of conditional expectation is used in the next example, which derives a formula for the sum of a random collection of random variables. The result is relevant to many practical contexts. In designing an elevator, for instance, the total load is determined by the number of passengers as well as their individual weights. The weight of an unspecified person is a random variable, as is the number of people who decide to board the elevator.

A similar scenario arises in a marketing context. The total sales for a store or firm will depend on the number of customers as well as their individual purchases, both of which are random variables. One could easily concoct similar examples in other contexts.

Example (Sum of a Random Number of Random Variables)

Suppose $X_1, X_2, \ldots$ is a sequence of independent random variables from a common distribution. Let N be a random variable taking values among the whole numbers 0, 1, 2, ... Further, N is independent of each of the X_i

variables. We are interested in the behavior of the random sum given by
$S = X_1 + \cdots + X_N$.

We first define a new quantity $R \equiv e^{tS}$. Since S is a random variable, so is
R. In addition, because the behavior of S can be conditioned on particular
values of N, so can R. The conditional expectation of R, given $N = n$, is

$$E(R|N = n) = E(e^{tS}|N = n) = E[e^{t(X_1 + \cdots + X_n)}]$$

$$= \prod_{i=1}^{n} E(e^{tX_i}) = [M_X(t)]^n \qquad (3\text{--}16)$$

The second equality follows from setting N to the value n in the definition of
S. The third equality depends on the independence of the X_i variables. The
last transformation follows from the definition of the moment generating
function $M_X(t)$ for each variable X_i.

Equation (3–16) holds for each value of N: 0, 1, 2, etc. Therefore, we may
write

$$E(R|N) = [M_X(t)]^N \qquad (3\text{--}17)$$

With this equivalence, the expectation becomes

$$E(R) = E[E(R|N)] = E[M_X(t)^N] \qquad (3\text{--}18)$$

The first equality relies on the by-now familiar result that the expectation of
a conditional expectation is simply the mean of the conditioned variable. The
second conversion springs from Eq. (3–17).

Another formula for the mean of R can be obtained as

$$E(R) = E(e^{tS}) = M_S(t) \qquad (3\text{--}19)$$

By linking Eqs. (3–18) and (3–19), we obtain the moment generating function
for S in terms of N and $M_X(t)$:

$$M_S(t) = E[M_X(t)^N]$$

We can now take derivatives to our heart's delight and determine the
moments of S. The first and second derivatives of the function, denoted $M'_S(t)$
and $M''_S(t)$, respectively, are

$$M'_S(t) = E\{N[M_X(t)]^{N-1} M'_X(t)\}$$

$$M''_S(t) = E\{N(N-1)[M_X(t)]^{N-2}[M'_X(t)]^2 + N[M_X(t)]^{N-1} M''_X(t)\}$$

By setting t to zero, we obtain the first two moments of S:

$$E(S) = M'_S(0) = E[N(1)^{N-1}E(X)] = E(N)E(X)$$

$$E(S^2) = M''_S(0) = E[N(N-1)(1)^{N-2}E^2(X) + N(1)^{N-1}E(X^2)]$$

$$= E\{N[E(X^2) - E^2(X)] + N^2E^2(X)\}$$

$$= E(N)\text{Var}(X) + E(N^2)E^2(X)$$

Finally, the variance of S is

$$\text{Var}(S) = E(S^2) - E^2(S)$$

$$= E[N]\text{Var}(X) + E^2(X)\text{Var}(N)$$

The results reduce nicely to familiar formulas for some simple cases. For instance, suppose N is fixed at value n. Then the mean of S is $nE(X)$, whereas the variance is $n\,\text{Var}(X)$.

On the other hand, if X is fixed at value x, then the mean of S is $xE(N)$, whereas the variance is $x^2\,\text{Var}(N)$. The last expression reflects the squaring property of the variance operator for scaling factors. For instance, if the mean of X jumps from 1 to 5, then the standard deviation of X would increase proportionately by a factor of 5 and the variance would rise by a factor of 25.   ◆

Conditional Events

So far, we have focused on conditional mass and density functions. The conditioning approach can also be applied to compound events which involve sets containing two or more events. This can be achieved as follows.

Let X be the indicator function of some event E. In other words, $X = 1$ if the event occurs and $X = 0$ otherwise. Since E is a set of outcomes from a random experiment, X is a random variable.

The mean of X is given by the probability that event E occurs:

$$E(X) = 0 \cdot P(\text{not } E) + 1 \cdot P(E) = P(E)$$

Given another random variable Y, the conditional mean of X given $Y = y$ is

$$E(X|Y = y) = 0 \cdot P(\text{not } E|Y = y) + 1 \cdot P(E|Y = y)$$

$$= P(E|Y = y)$$

The probability of event E can be obtained by weighting it over all possible values of Y:

$$P(E) = \int P(E|Y = y) \, dF_Y(y) \qquad (3\text{--}20)$$

As an illustration, consider the distribution for the sum of two variables. Let X and Y be independent random variables with respective distributions F and G. The distribution of the sum $S \equiv X + Y$ is called the *convolution* of F and G, and is symbolized as $F \circ G$.

The convolution can be expressed as follows:

$$(F \circ G)(s) \equiv P\{X + Y \le s\} = \int_{-\infty}^{\infty} P\{X + Y \le s | Y = y\} \, dG(y)$$

$$= \int_{-\infty}^{\infty} P\{X + y \le s\} \, dG(y)$$

$$= \int_{-\infty}^{\infty} P\{X \le s - y\} \, dG(y) = \int_{-\infty}^{\infty} F(s - y) \, dG(y)$$

The second equality depends on Eq. (3–20). For the third equality, the variable Y is replaced by its value y.

The convolution of n random variables from the same distribution F is called the n-fold convolution of F with itself and denoted by F_n. For instance, the two-fold convolution is $F_2 = F \circ F$. More generally, the n-fold convolution is the convolution of F with the $(n - 1)$th convolution: $F_n = F \circ F_{n-1}$.

3.8 LIMIT THEOREMS

Random variables exhibit some remarkable properties when they congregate in large numbers. The results relating to the properties of variables as they form aggregates of unbounded size are called limit theorems.

One result relates to the asymptotic tendency of the average of random variables to converge toward their mean.

Theorem (Strong Law of Large Numbers). Let $X_1, X_2, \ldots$ be a sequence of independent random variables from a common distribution with mean μ. In the limit as the number of variables expands, the average value of the variables converges to their common mean with unit probability:

$$P\left\{ \lim_{n \to \infty} \left(\frac{X_1 + X_2 + \cdots + X_n}{n} \right) = \mu \right\} = 1$$

A stronger result claims that the average value of *any* collection of random variables converges to a particular density function, as long as the variance of each variable is finite.

The normal distribution is a pervasive function that tends to swallow all others. It is the limiting distribution for the aggregation of indefinitely large clusters of random variables, regardless of their respective densities.

Theorem (Central Limit). Suppose X_1, X_2, ... are independent, identically distributed random variables with mean μ and variance σ^2. Let $S_n \equiv X_1 + \cdots + X_n$ denote their sum. Then the standardized form of S_n tends toward the normal density function:

$$\lim_{n \to \infty} P\left\{ \frac{S_n - n\mu}{\sigma\sqrt{n}} \le z \right\} = \int_{-\infty}^{z} \frac{1}{\sqrt{2\pi}} e^{-x^2/2} \, dx$$

Note that the mean of S_n is given by $n\mu$, whereas the variance is $n\sigma^2$. Consequently, the quantity $(S_n - n\mu)/\sigma\sqrt{n}$ is the "standardized" form of S_n.

3.9   EVENTS WITHOUT MEMORY

Problems become relatively simple when they can be compartmentalized and each morsel treated in isolation. Along the temporal dimension, such chunking is tantamount to events without memory. A memoryless event is one whose behavior is unaffected by past events. In other words, the behavior of a system depends on its current status and is independent of its history— the path or sequence of states which has led to the present condition.

We say that a random variable X is memoryless if it possesses the property

$$P\{X > s + t \mid X > s\} = P\{X > t\} \tag{3-21}$$

for all $s, t \ge 0$. To illustrate, suppose that X denotes the life span of a system. Then the probability that it survives beyond $s + t$ hours, given that it already survived up to s hours, is equal to the original probability of surviving beyond t hours. The future prospects for this system are unaffected by its age.

The memoryless assumption is inapplicable to biological systems for any sizable values of s and t. On the other hand, the assumption is often appropriate for inanimate systems, such as the life span of a light bulb or time to decay for an atomic particle.

When the underlying process is a sequence of discrete trials, a memoryless phenomenon is called a *binomial process*. The corresponding phenomenon for the continuous case is called a *Poisson process*.

As an illustration, the time required for an event of interest to occur in the continuous case follows an exponential function. The cumulative distribution for the exponential density is given by

$$F(x) = \int_{-\infty}^{x} f(X)\,dX = \begin{cases} 1 - e^{-\lambda x} & x \geq 0 \\ 0 & x < 0 \end{cases}$$

Let $\bar{F}$ denote the complementary distribution indicating the probability that random variable X exceeds a threshold value x. It is defined by

$$\bar{F}(X) \equiv P\{X > x\} = 1 - F(x) = e^{-\lambda x}$$

for any value of $x \geq 0$. We can verify Eq. (3–21) in this way:

$$P\{X > s + t \mid X > s\} = \frac{P\{X > s + t, X > s\}}{P\{X > s\}}$$

$$= \frac{P\{X > s + t\}}{P\{X > s\}} = \frac{e^{-\lambda(s+t)}}{e^{-\lambda s}}$$

$$= e^{-\lambda t} = P\{X > t\}$$

The exponential distribution, therefore, satisfies the memoryless property.

In the next two sections, we will examine the nature of memoryless processes for both discrete and continuous cases. We will also derive formulas specifying the chance of occurrence of events over extended stretches of time.

3.10 BINOMIAL PROCESS

A binomial random variable is one that assumes two outcomes. The canonical example is that of tossing a coin, for which the variable X might be assigned the value 0 for a tail and the value 1 for a head.

Since a binomial random variable has two outcomes, one may be viewed as a "success" and the other a "failure." An obvious situation where this interpretation applies may be found in the win or loss of a soccer game, or the acceptance or rejection of a new product in the marketplace.

Because binomial trials are so prevalent in practical situations, we are interested in the properties of a sequence or concatenation of binomial trials. Such a sequence is called a *binomial process*.

An important characteristic of a binomial process is the number of trials before one or more successes occur. In the sports arena, a football coach may be interested in the number of games before a win is likely to occur. In the business environment, a division manager may desire to know the number of new products that should be launched to assure a reasonable chance of attaining three successes in the marketplace.

We examine these topics in greater detail below.

Interarrival Time for Binomial Process

We are often interested in the number of times an experiment is repeated before a particular event occurs. An example lies in the number of tails that turn up in the toss of a coin before a head appears.

A sequence of random trials is called a *process*. An event of interest in a trial is also known as a *success* or an *arrival*. The *interarrival time* of an event is the number of times that an experiment is repeated before that event occurs.

For the coin example, an arrival might be the occurrence of a head. The interarrival time would then be the number of tosses before a head appears for the first time.

The number of trials required to observe the first occurrence of an arrival is called the first-order interarrival time. Let R_1 denote the first-order interarrival time for a binomial process. Then R_1 can take any positive integer value.

Suppose the chance of observing a head is p and that for a tail is $1 - p$. Then the probability that the first-order interarrival time equals 1 is simply the chance of obtaining a head on a toss:

$$p_{R_1}(1) = p$$

On the other hand, the first-order interarrival time equals 2 if the first trial is a tail and the second is a head. Since the trials are independent, the probability for this sequence of outcomes equals the product of the individual outcomes:

$$p_{R_1}(2) = P\{\text{Tail on first toss \& Head on second toss}\}$$
$$= P\{\text{Tail}\}P\{\text{Head}\} = (1 - p)p$$

More generally, the event $R_1 = k$ occurs if the first $k - 1$ outcomes are tails and the last is a head. The probability of this turn of events is

$$p_{R_1}(k) = P\{\text{Tails on } k - 1 \text{ trials}\}P\{\text{Head}\}$$
$$= (1 - p)^{k-1}p$$

This result is illustrated in Figure 3–4.

The mass function p_{R_1} for the first-order interarrival time decreases in

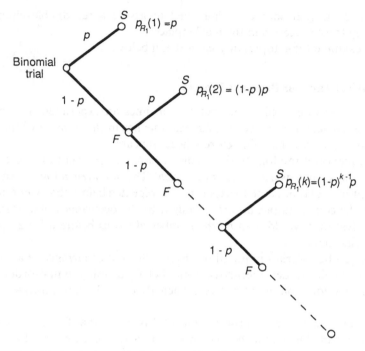

FIGURE 3–4. First-order interarrival time for a binomial process with parameter p. A success or an arrival can occur with probability p in each trial. Interarrival time R_1 equals k if each the first $k-1$ outcomes is a failure (F) and the last is a success (S).

geometric progression with increasing k. Consequently, the mass function is known as a *geometric* function.

Suppose we are interested in the time that transpires for multiple arrivals. The number of trials up to, and including, the jth arrival is called the jth-order *interarrival time* and denoted as R_j.

Consider the event in which the jth success for a binomial process occurs on the kth trial. For this state of affairs, there must have been exactly $j-1$ successes by the $(k-1)$th trial and a success on the kth trial. The corresponding probability is

$$p_{R_j}(k) = P\{j-1 \text{ successes in first } k-1 \text{ trials}\}$$

$$\cdot P\{\text{successes on } k\text{th trial, given } j-1 \text{ successes in } k-1 \text{ trials}\}$$

$$(3-22)$$

Here $p_{R_j}(k)$ denotes the probability that the jth arrival occurs on the kth trial.

The probability of $j-1$ successes in $k-1$ trials is given by the binomial formula

$$P\{j - 1 \text{ successes in } k - 1 \text{ trials}\} = \binom{k-1}{j-1} p^{j-1}(1 - p)^{k-j}$$

Each trial is independent of the others. Consequently, the conditional probability of a success on the kth trial, given $j - 1$ successes in $k - 1$ trials, is simply the unconditional probability of a success on any trial—namely, p. Inserting these results into Eq. (3–22) yields

$$P_{R_j}(k) = \binom{k-1}{j-1} p^{j}(1 - p)^{k-j}; \quad j = 1, 2, 3, \ldots; k = j, j + 1, j + 2, \ldots$$

which defines the probability mass function for the jth-order interarrival time for a binomial process. This formula is also known as the *Pascal* mass function. When $j = 1$, the preceding formula reduces to the mass function for the first-order interarrival time R_1.

Example

Ty Koon, a 23-year-old graduate from a notorious business school, is a budding entrepreneur. His plan for success is to invest in a promising business project each year. Suppose the chance of success on each project is 1/2. Ty believes that a successful project will make him a millionaire within 5 years after its inception, with probability 1/3. (a) What is the chance that his first success occurs on the third project? (b) Given that he has had exactly three successes in his first five projects, what is the chance of encountering the sixth success on the ninth project? (c) What is the probability of his first becoming a millionaire at age 30?

The probability of success is $p = 1/2$. The chance of the first success arising from the third project is given by failure in the first two and success in the third. The corresponding probability is $(1 - p)(1 - p)p = (1 - .5)(1 - .5)(.5) = 1/8$. This result also can be obtained directly from the mass function for the first-order interarrival time for $k = 3$:

$$p_{R_1}(3) = (1 - p)^{3-1} p = (1 - \tfrac{1}{2})^2(\tfrac{1}{2}) = \tfrac{1}{8}$$

Given that Ty has had three successes in five projects, we desire the probability of the sixth success on the ninth project. For this to occur, he must obtain the third success on the fourth upcoming project: since binomial trials are independent, the past does not influence the future. The desired probability is simply the third-order interarrival time for $k = 4$:

$$p_{R_3}(4) = \binom{4-1}{3-1}\left(\frac{1}{2}\right)^3\left(1 - \frac{1}{2}\right)^{4-3} = \frac{3}{16}$$

The chance of a project making Ty a millionaire in 5 years, given that the project is successful, is 1/3. Consequently, the probability that any project will make him a millionaire in half a decade is

$$q \equiv P\{\text{Millionaire}\} = P\{\text{Success}\}P\{\text{Millionaire}|\text{Success}\}$$
$$= (\tfrac{1}{2})(\tfrac{1}{3}) = \tfrac{1}{6}$$

Since a successful project takes 5 years to fully blossom, only the projects initiated by age 25 will count toward his becoming a millionaire by the age of 30. Since he is now 23, his third project will be initiated at age 25. The first order interarrival time with $k = 3$ is

$$p_{R_1}(3) = \left(1 - \frac{1}{6}\right)^2\left(\frac{1}{6}\right) = \frac{25}{216}$$

This is the chance of first becoming a millionaire at age 30.   ◆

One property of the Pascal mass function is that the total probability for all interarrival times equals p for each value of the index k. More specifically, we have

$$\sum_{j=1}^{\infty} p_{R_j}(k) = p \quad \text{for } k = 1, 2, \ldots$$

This is the probability that a success occurs at the kth trial. A success, if it occurs on the kth trial, must be the first, or second, or third, etc. From the independence of binomial trials, the probability of a success at any step is fixed and equals p.

3.11   POISSON PROCESS

The binomial process was defined as a sequence of independent binomial trials. These trials were considered to occur at discrete intervals $k = 1, 2, 3, \ldots$, and so on.

When the discrete intervals become increasingly small in parallel with the probability of success, the binomial process can be viewed as a phenomenon occurring along a continuous parameter. It is often convenient to think of this

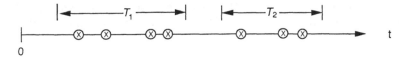

FIGURE 3–5. A Poisson process consists of events occurring along a continuous parameter t. Each encircled cross represents a success or arrival. Events relating to disjoint intervals T_1 and T_2 are independent.

parameter as time. The limit of a binomial process as the interval between trials decreases is called a *Poisson process.*

In other words, suppose we regard a binomial process as a sequence of trials, each occurring during a time interval Δt. As the interval Δt decreases toward 0, the binomial trials can be viewed as experiments along a continuous temporal parameter. An event of interest in a Poisson process is called a *success* or an *arrival.*

Figure 3–5 illustrates a Poisson process. Each success occurring over the continuous parameter t is indicated by an encircled cross. The Poisson process is a limiting case of a binomial process for which trials are independent. Hence, events occurring over disjoint time intervals T_1 and T_2 are independent. As an illustration, the event "The number of arrivals occurring in time period T_1" is independent of the event "More than 7 arrivals in the second half of time period T_2."

Suppose the probability of success were fixed at λ' for each time period, regardless of the size of the interval Δt. In that case, the number of successes over any finite time interval τ would be infinite as $\Delta t \to 0$. For instance, suppose the chance of success were $\lambda' = .5$ for each trial; then as Δt becomes infinitesimally small, we would expect an infinite number of successes during 1 hour, or 1 second, or any other finite time period τ. This type of model has little use in practical applications because we tend to view such infinite phenomena as deterministic rather than probabilistic processes.

A more useful model requires the probability p of success to be proportional to the time interval Δt. In other words, we define a parameter λ which defines the rate of successes, such as $\lambda = 2$ per second. Then the probability of success over any interval Δt is given by $p = \lambda \Delta t$.

Properties of Poisson Process
Before deriving the probability function for the Poisson process, we explore a number of properties it should exhibit. To this end, let $p(k|t)$ denote the probability that exactly k arrivals occur during any time interval t.

Although the quantity $p(k|t)$ incorporates t as a continuous parameter, the argument k is a discrete variable. During any fixed time interval t, the number

k of arrivals is 0, 1, 17, or some other whole number. Consequently $p(k|t)$ is a mass function for any fixed value of the parameter t. Its sum over all values of k is unity:

$$\sum_{k=0}^{\infty} p(k|t) = 1$$

On the other hand, $p(k|t)$ is *not* a probability density function for parameter t. For instance, we cannot attempt to integrate $p(3|t)$ over all values of t and claim that the probability of 3 arrivals is unity; there is a positive chance of observing fewer or more than 3 arrivals over the entire time line. Another way to discern that $p(k|t)$ is not a density function for t relies on the fact that $p(k|s)$ and $p(k|t)$ are not mutually exclusive events. For example, the quantities $p(k|9)$ and $p(k|9.1)$ do not indicate disjoint outcomes; in fact, these two quantities denote almost the same phenomenon.

Some other properties of the Poisson process are the following. Since the probability of success over any time interval Δt is given by $p = \lambda \Delta t$, there is no chance of observing any arrivals in zero time—that is, $p = \lambda \cdot 0 = 0$. Conversely, having zero arrivals in zero time is a certainty. Thus, $p(0|0) = 1$ and $p(k|0) = 0$ for each positive integer k.

What about the possibility of having no arrivals as a function of time? We have seen that the chance of having no arrivals is unity when $t = 0$. As t increases, however, the chance of observing no arrivals should decrease. Hence, the quantity $p(0|t)$ should begin at unity for $t = 0$ and fall as t increases.

For any positive integer k, the probability of observing that number of arrivals should start at 0 for $t = 0$, then increase with t. On the other hand, $p(k|t)$ should eventually decrease as t becomes extremely large. In other words, the chance of observing exactly k arrivals is small if the time interval is too short or too long.

We summarize the properties of the Poisson process in a definition.

- Let X be a random variable taking values $k = 0, 1, 2, \ldots$. Suppose that any events occurring on disjoint intervals are independent. Furthermore, assume that the mass function for X over any small interval Δt is given by

$$p(k|\Delta t) = \begin{cases} \lambda \Delta t + \vartheta(\Delta t^2) & k = 1 \\ 1 - \lambda \Delta t + \vartheta(\Delta t^2) & k = 0 \\ \vartheta(\Delta t^2) & k = 2, 3, \ldots \end{cases}$$

Under these conditions, the sequence of trials is called a *Poisson process*.

The notation $\vartheta(\Delta t^2)$ indicates a term which is second order in Δt; namely,

$c\Delta t^2$, where c is some constant. The definition stipulates that the chance of observing one arrival within time interval Δt is approximately $\lambda\Delta t$; that for zero arrivals is roughly the complement, $1 - \lambda\Delta t$; and that for any other number of arrivals is very small.

Poisson Mass Function

The defining properties of the Poisson process can be used to establish a series of relations, which in turn can be solved to obtain an analytic formula.

To this end, consider the quantity $p(k|t + \Delta t)$, which specifies the probability of having precisely k arrivals during the interval $[0, t + \Delta t]$. There are two possible pathways for the process to follow during the interval from time t to time $t + \Delta t$:

- k arrivals have occurred by time t, and no arrivals occur during the next interval Δt. Since the two time intervals are disjoint, the probability for this pair of events is given by the product of the respective probabilities: $p(k|t)p(0|\Delta t)$.
- $k - 1$ arrivals have occurred by time t, and exactly 1 arrival occurs during the next interval Δt. Again the two intervals are disjoint, and the corresponding probability is given by the product $p(k - 1|t)p(1|\Delta t)$.

These two pathways are depicted in Figure 3–6.

Since the two chains of events are disjoint, the chance of observing k arrivals by time $t + \Delta t$ is given by the sum of the probabilities for the two other pathways. The result is

$$p(k|t + \Delta t) = p(k|t)p(0|\Delta t) + p(k - 1|t)p(1|\Delta t)$$
$$= p(k|t)[1 - \lambda\Delta t + \vartheta(\Delta t^2)] + p(k - 1|t)[\lambda\Delta t + \vartheta(\Delta t^2)]$$

The second equality relies on the characterization of the Poisson process from the last definition. By rearranging some terms and dividing by Δt, we obtain

$$\frac{p(k|t + \Delta t) - p(k|t)}{\Delta t} + \lambda p(k|t) = \lambda p(k - 1|t) + [p(k|t) + p(k - 1|t)]\frac{\vartheta(\Delta t^2)}{\Delta t}$$

By taking the limit as $\Delta t \to 0$, the first term on the left becomes a derivative. Since the quantity $\vartheta(\Delta t^2)$ can be written as $c\Delta t^2$ for some constant c, we note that $\vartheta(\Delta t^2)/\Delta t = c\Delta t$; this quantity vanishes as $\Delta t \to 0$.

We, thus, obtain the differential equation

$$\frac{d}{dt}p(k|t) + \lambda p(k|t) = \lambda p(k - 1|t) \tag{3-23}$$

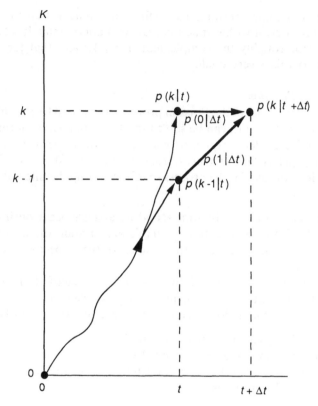

FIGURE 3–6. Derivation of the Poisson mass function. Suppose k arrivals occur by time $t + \Delta t$. This state can be reached from two mutually exclusive states: k arrivals can have occurred by time t, followed by no arrivals within Δt; or $k - 1$ arrivals can have occurred by time t, followed by 1 arrival during Δt.

This equation may be solved iteratively for $k = 0$, $k = 1$, and so on, in conjunction with the initial conditions $p(0|0) = 1$ and $p(k|0) = 0$ for any other integer k. For instance, consider the case where $k = 0$. Equation (3–23) becomes

$$\frac{d}{dt}p(0|t) + \lambda p(0|t) = 0 \tag{3–24}$$

The relationship relies on the fact that $p(k - 1|t) = 0$ for $k = 0$. The solution to Eq. (3–24) is

$$p(0|t) = ap(0|0)e^{-\lambda t}$$

where a represents some constant of integration. From the initial condition $p(0|0) = 1$, we know that $a = 1$, and the solution becomes $p(0|t) = e^{-\lambda t}$. More

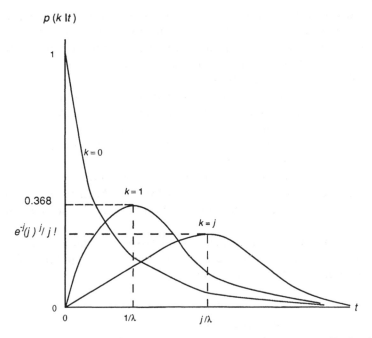

FIGURE 3–7. Probability as a function of parameter t for a Poisson process. For $k = 0$, the function is $e^{-\lambda t}$ for $t \geq 0$, and 0 for negative values of t. For other values $k = j$, the maximum value of $e^{-j}(j)^j/j!$ occurs at $t = j/\lambda$.

generally, the solution to Eq. (3–23) is

$$p(k|t) = \frac{e^{-\lambda t}(\lambda t)^k}{k!} \tag{3–25}$$

where $t \geq 0$ and $k = 0, 1, 2, \ldots$. This solution may be verified by direct substitution into Eq. (3–23).

The probability function specified in Eq. (3–25) is shown in Figure 3–7 as a function of the parameter t. When $k = 0$, the function is merely the exponential $e^{-\lambda t}$ for non-negative values of t, and equals 0 for negative values of t. For any positive integer $k = j$, the maximum occurs at $t = j/\lambda$. At this point the peak probability is given by $e^{-j}(j)^j/j!$.

The function in Eq. (3–25) is more commonly expressed by substituting μ for λt. The result is

$$p_X(k) = \frac{e^{-\mu}\mu^k}{k!} \quad k = 0, 1, 2, \ldots$$

This is known as the *Poisson function* for the discrete random variable X.

The Poisson function was derived in the context of the parameter t, which was regarded as a meausre of time. The function applies, however, to many other situations. Since the Poisson process is a limiting case of the binomial process, it arises in many applications where the events of interest occur with finite probability over some continuous parameter. An example may be found in the number of meteorites that impinge upon small, disjoint regions of the lunar surface. Because the locations of impact may be viewed as independent variables, the numbers of meteorite craters in various regions should take the form of a Poisson random variable. We assume that the regions are relatively small and close to each other, but do not overlap.

It can be shown that the mean and variance of the Poisson distribution are both μ. In other words, suppose X is a random variable from a Poisson mass function with parameter λ. Then $E(X) = \text{Var}(X) = \mu$. The proof is left as an exercise.

The recognition of μ as the mean of a Poisson distribution permits us to interpret the parameter λ in the corresponding Poisson process. We recall from the definition of the Poisson process:

$$p(k|\Delta t) = \begin{cases} 1 - \lambda \Delta t & k = 0 \\ \lambda \Delta t & k = 1 \\ 0 & k = 2, 3, \ldots \end{cases}$$

where second-order effects in Δt are ignored for the sake of simplicity. Since we know that $E(X) = \mu = \lambda t$, we obtain $\lambda = \mu/t$. We see that λ is the expected number of arrivals per unit time for a Poisson process. The parameter λ is known as the *average arrival rate* for the process.

Interarrival Times for Poisson Process

The Poisson mass function focuses on the rate of arrivals during a particular time period. Often we are interested in the converse quantity, namely, the time required for a given number of arrivals to occur.

Let R_k denote the time required for the kth success in a Poisson process. Then R_k is a continuous random variable known as the kth-order *interarrival time* or *waiting time*.

We derive the density functions $f_{R_k}(r)$ for the interarrival time of order k by using an argument which should look familiar by now. Consider a Poisson process having an arrival rate of λ. What is the chance that the kth arrival occurs within a time interval $(r, r + \Delta r]$? For this event to arise, $j - 1$ arrivals must have occurred in the preceding period from 0 to r. The situation is depicted in Figure 3–8.

FIGURE 3–8. Interarrival times for Poisson process. For the kth arrival to occur in period $[r, r + \Delta r]$, there must have been $k - 1$ arrivals during $(0, r]$ and 1 arrival within $(r, r + \Delta r]$.

The probability of observing $k - 1$ arrivals during the interval $(0, r]$ is given by Eq. (3–25) with parameter r.

$$p(k - 1 | r) = \frac{e^{-\lambda r}(\lambda r)^{k-1}}{(k - 1)!}$$

Further, the probability of observing precisely one arrival in the period $(r, r + \Delta r]$ is given by $p(1 | \Delta r) = \lambda \Delta r$ from the definition of the Poisson process.

Since the two intervals $(0, r]$ and $(r, r + \Delta r]$ are disjoint, their related events are independent. The overall probability is the product of the individual probabilities:

$$P\{r < R_k \leq r + \Delta r\} = p(k - 1 | r) p(1 | \Delta r)$$

The left-hand side of the equation is approximately equal to $f_{R_k}(r) \Delta r$ for small values of Δr. We can, therefore, write the preceding equation as

$$f_{R_k}(r) \Delta r \approx \left(\frac{e^{-\lambda r}(\lambda r)^{k-1}}{(k - 1)!} \right)(\lambda \Delta r)$$

where the two factors on the right-hand side have been replaced by their explicit expressions.

In the limit as $\Delta r \to 0$, the above equation becomes exact. After canceling the common factor Δr, we obtain the density function for the kth-order interarrival time for a Poisson process with parameter λ:

$$f_{R_k}(r) = \frac{e^{-\lambda r} \lambda^k r^{k-1}}{(k - 1)!} \tag{3–26}$$

where $k = 1, 2, \ldots$ and r is non-negative. This function is also known as the *Erlang* family of probability density functions.

The first-order interarrival time corresponding to $k = 1$ yields the familiar exponential density function

$$f_{R_1}(r) = \lambda e^{-\lambda r}$$

for $r \geq 0$. We have seen in Chapter 2 that the mean and variance of the exponential density function are $E(R_1) = 1/\lambda$ and $\text{Var}(R_1) = 1/\lambda^2$.

The behavior of R_1 is independent of its history. If we are told that 37 seconds have passed since the last arrival, the density for the next arrival is still given by $\lambda e^{-\lambda r}$ for non-negative values of r. This memoryless property of the exponential function can be traced to the definition of the Poisson process, in which the rate λ of an arrival is constant, independent of previous events.

The same characterization applies to the kth-order waiting time R_k. Even if we are told that some time r' has elapsed since the last arrival, the density function for R_k is still given by Eq. (3–26).

The family of interarrival times for a Poisson process is depicted in Figure 3–9. As mentioned previously, first-order waiting time, R_1, is defined by the exponential function $\lambda e^{-\lambda r}$. The value of the function at this point is $e^{-(k-1)}(k-1)^{k-1}\lambda/(k-1)!$.

Higher-order interarrival times are simply the sum of independent first-order arrival times. Hence, the mean and variance for R_k are directly propor-

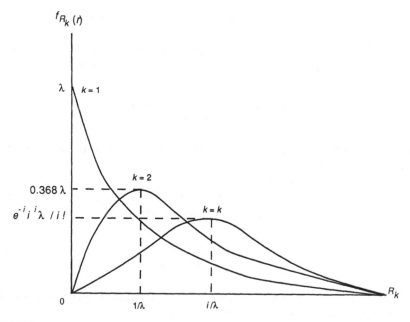

FIGURE 3–9. Interarrival times of a Poisson process. The density for R_1 is the exponential function $\lambda e^{-\lambda r}$. For second-order and higher waiting times R_k, the peak occurs at i/λ where $i \equiv k - 1$.

tional to those for the first-order waiting time:

$$E(R_k) = kE(R_1) = \frac{k}{\lambda}$$

$$\text{Var}(R_k) = k\,\text{Var}(R_1) = \frac{k}{\lambda^2}$$

The Poisson process arises in many practical situations. The most straight-forward way to determine whether a physical process should be modeled as a Poisson process is to check whether the first-order interarrival times take the form of the exponential density $\lambda e^{-\lambda r}$ for some parameter λ.

In fact, this represents an alternative—and common—way of defining a Poisson process. More specifically, a *Poisson process* is one in which first-order interarrival times are *independent, identically distributed* (i.i.d.) exponential random variables.

Example

A biologist is engaged in a study of desert animals. For this study she must cross a rugged desert in a 4-wheeled vehicle. The trip will take 10 days, including stops to make observations. Suppose that flat tires occur according to a Poisson process with parameter 0.1/day. (a) If she carries one spare tire, what is the chance of completing the trip without mishap? (b) How many spare tires should she carry to ensure that the chance of finishing the study as scheduled is better than 90%?

The arrival rate of flat tires is $\lambda = 0.1$/day. The biologist can finish the 10-day trip if she encounters 0 or 1 flat tire. This is tantamount to saying that the second-order interarrival time for flats must exceed $t = 10$ days. The probability is given by

$$P\{R_2 > 10\} = \int_{10}^{\infty} \frac{e^{-\lambda r}\lambda^2 r}{1!}\,dr$$

$$= (\lambda t)e^{-\lambda t} + e^{-\lambda t} = (.1)10e^{-.1(10)} + e^{-.1(10)}$$

$$= .7358$$

Hence, the probability of completing the trip without mishap is about 74%.

We can also obtain this result by considering the corresponding Poisson random variable. Let X be the number of failures during $t = 10$ days. The trip will succeed if there is 0 or 1 arrival during the period. The associated probability is

$$P\{\text{Success for 1 spare}\} = p(0|t) + p(1|t) = \frac{e^{-\lambda t}(\lambda t)^0}{0!} + \frac{e^{-\lambda t}(\lambda t)}{1!}$$

$$= e^{-.1(10)} + e^{-.1(10)}(.1)10 = .7358$$

which matches the earlier result.

Suppose the biologist carries two spare tires. Then her trip will succeed if there are 0, 1, or 2 flats. Letting X denote the Poisson random variable for the number of flats, we obtain

$$P\{\text{Success for 2 spares}\} = e^{-.1(10)} + e^{-.1(10)}(.1)10 + \frac{e^{-.1(10)}[(.1)10]^2}{2!}$$

$$= .9197$$

We conclude that two spare tires will suffice: the probability of success will then exceed 90%. ◆

Sum of Poisson Random Variables

The independent nature of successes in a Poisson process is unaffected by merging two or more similar processes. As a result, the sum of Poisson random variables is itself a Poisson random variable.

In particular, suppose that X and Y are Poisson random variables with parameters λ_X and λ_Y, respectively. If the two processes are merged, we obtain another Poisson process whose rate of success equals the sum of λ_X and λ_Y. The probability of any arrival during time period Δt is

$$p(k|\Delta t) = \begin{cases} (\lambda_X + \lambda_Y)\Delta t + \vartheta(\Delta t^2) & k = 1 \\ 1 - (\lambda_X + \lambda_Y)\Delta t + \vartheta(\Delta t^2) & k = 0 \\ \vartheta(\Delta t^2) & k \geq 2 \end{cases}$$

for sufficiently small values of Δt. As a result, if X and Y are Poisson random variables, then the sum $W = X + Y$ is another Poisson process with parameter $\lambda = \lambda_X + \lambda_Y$.

Example

A plane with four engines can remain in flight if at least two of its engines are working. The reliability of each engine is independent of the others'. For each engine, the time to failure is distributed exponentially with parameter $\lambda = 1/600$ hours. What are the mean and variance of the time until the plane is forced to land due to engine problems?

Let T_{ij} denote the random interval between the ith and jth engine failures. For instance, T_{01} is the interval from $T = 0$ to the first malfunction. The plane can fly while two or more engines work—in other words, until the third failure. We are, therefore, interested in the behavior of the random variable

$$X = T_{01} + T_{12} + T_{23}$$

During the period T_{01}, four engines are in operation. Since the engines operate independently of each other, the overall failure rate is given by the sum of four uncoupled Poisson processes. This failure rate is another Poisson process with parameter 4λ, for which the mean time to failure is $E(T_{01}) = 1/4\lambda$. During the period T_{12}, three engines are in operation. The mean time to failure during this period is $E(T_{12}) = 1/3\lambda$. In a similar way, the mean time between the second and third failures is $E(T_{23}) = 1/2\lambda$. The average time until the plane is forced out of the air is 650 hours:

$$E(X) = E(T_{01}) + E(T_{12}) + E(T_{23}) = \frac{1}{4\lambda} + \frac{1}{3\lambda} + \frac{1}{2\lambda} = \frac{13}{12}(600) = 650$$

To obtain the variance of X, we first note that the T_{ij} are disjoint. For Poisson processes, events over disjoint intervals are independent. The variance of the time until the third failure is the sum of the variances for the T_{ij}:

$$\text{Var}(X) = \text{Var}(T_{01}) + \text{Var}(T_{12}) + \text{Var}(T_{13})$$

$$= \left(\frac{1}{4\lambda}\right)^2 + \left(\frac{1}{3\lambda}\right)^2 + \left(\frac{1}{2\lambda}\right)^2 = \frac{61}{144}(600)^2 = 152,500$$

The standard deviation of flying time is about 391 hours.   ◆

3.12 STOCHASTIC PROCESSES

A *stochastic* or *random process* is a sequence of random variables that depends on one or more parameters. The parameter can assume discrete or continuous values. A discrete parameter is illustrated by the sequence of steps relating to a computer algorithm, whereas a continuous one relates to the time of day in characterizing the humidity in a factory.

A stochastic process is denoted by the notation $X(t, \omega)$, where t is a parameter such as time or distance, and ω is an outcome. The parameter ω is a member of the space Ω of all possible outcomes.

For a fixed value of t, say $t = t_0$, the quantity $X(t_0, \omega)$ is a random variable. When ω is fixed, say at value ω_0, the quantity $X(t, \omega_0)$ is a function of t and is called a *realization* of the process. For instance, $X(t, \omega_0)$ may define the number of customers in a queue, or perhaps the condition of a robot, as a function of time t.

Two stochastic processes which often arise in applications are called Markov and stationary. These are described below.

Markov Processes and Chains

A *Markov process* is one which depends only on the present or immediate past. The future is determined only by the present, not by the distant past.

More specifically, let $X(t_0, \omega)$, $X(t_1, \omega)$, ..., $X(t_n, \omega)$ be a sequence of random variables over a time $t_0 < t_1 < \cdots < t_n$. Even if the distribution of each X up to time t_n is known, the distribution of the variable $X(t_{n+1}, \omega)$ at the next step depends only on the current value of X at t_n. To illustrate, let $X(t_n, \omega)$ be the number of customers in a restaurant at time t_n. Suppose $A(t_n, \omega)$ is the random variable denoting the number of customers arriving at the restaurant during time period t_n, and $L(t_n, \omega)$ the number of customers leaving. Then the number of customers at the start of the next period is given by

$$X(t_n + 1, \omega) = X(t_n, \omega) + A(t_n, \omega) - L(t_n, \omega)$$

The numbers of customers arriving and leaving, namely, A and L, are random variables that fluctuate independently of each other and from one time period to another.

A *Markov chain* is a Markov process for which the parameter t takes a discrete set of values. An *ergodic process* is a Markov process where the time parameter extends to infinity. Ergodic theorems deal with the properties of ergodic processes.

Stationary Processes

A *stationary process* is a stochastic process in which the sources of variation are independent of time. This occurs, for instance, in characterizing the maximum pressure in an engine from one cycle to the next. Another example lies in the diameter of holes drilled in a series of workpieces passing through a stamping station.

Suppose that a stationary process is defined by the quantity $X(t, \omega)$, where the parameter t is regarded as time. Then for each value of t, the quantity X is a random variable characterized by a fixed distribution function.

One class of results states that the correlation between two random variables depends only on the difference in their temporal parameter. In particular, let $X(t, \omega)$ be a stochastic process with constant and finite values for the expectation μ and variance σ^2. Suppose that the time point s precedes t. Then the correlation between $X(s, \omega)$ and $X(t, \omega)$ is a function only of the difference in times:

$$\rho[X(s, \omega), X(t, \omega)] \equiv E\{[X(s, \omega) - \mu][X(t, \omega) - \mu]\}/\sigma^2$$

$$= R(t - s)$$

for some function R. Stationary processes arise in many fields of engineering as well as the natural and social sciences.

PROBLEMS

The problems below are numbered according to the sections to which they correspond. For instance, Problem 3.1A is the first ("A") exercise pertaining to Section 3.1.

3.1A   An experiment involves the roll of 3 dice. (a) What is the outcome space Ω? (b) An event E is described verbally as "Sum of numbers equals 4." List the elements of E.

3.1B   Prove that the probability of a union of events does not exceed the sum of the individual probabilities:

$$P\left(\bigcup_{i=1}^{n} E_i\right) \leq \sum_{i=1}^{n} P(E_i)$$

Also, does this result hold as $n \to \infty$? (*Hint:* Use induction by first showing the result to be valid for $n = 2$. Next assume that the result holds for n, then show that it works for $n + 1$.)

3.1C   Let E, F, and G be any three events from a given sample space. Show that the following formula holds:

$$P(E \cup F \cup G) = P(E) + P(F) + P(G) - P(E \cap F) - P(E \cap G)$$

$$- P(F \cap G) + P(E \cap F \cap G)$$

3.3A   Let X and Y be discrete variables taking values in $\{0, 1, 2\}$. Suppose their cumulative distribution is $F(x, y) = (x + y)/4$. (a) Chart the values for $F(x, y)$, F_X, and F_Y as exemplified by Figure 3–2A. (b) Do likewise for $p(x, y)$, p_X, and p_Y as in Figure 3–2B. (c) Are X and Y independent? Explain.

3.3B   Let X and Y be continuous random variables with joint density

$$f(x, y) = \lambda^2 e^{-\lambda(x+y)} \quad x, y \geq 0$$

and $f(x, y) = 0$ for other values of x and y. (a) What are $F(x, y)$, F_X, and F_Y? (b) What are f_X and f_Y? (c) Are the variables independent? Explain.

3.3C   Prove the Factorability Theorem for the discrete case. That is, suppose p_X and p_Y are mass functions for X and Y, respectively, and that their joint mass is factorable: $p(x, y) = p_X p_Y$. Then prove that the variables are independent.

3.4A   Suppose X and Y are *discrete* random variables. Prove that $E(XY) = E(X)E(Y)$.

3.5A   Let $X_1, X_2, \ldots, X_n$ be independent random variables. Assume that the variance of the sum of two random variables is given by the sum of their variances. In other words, assume that if $Y = X_i + X_j$ for any pair $i \neq j$, then

$$\text{Var}(Y) = \text{Var}(X_i) + \text{Var}(X_j)$$

Show that the result generalizes to all n variables; that is,

$$\text{Var}(X_1 + \cdots + X_n) = \text{Var}(X_1) + \cdots + \text{Var}(X_n)$$

(*Hint*: Use induction by (a) showing that the formula holds for the base case of $n = 2$ variables and (b) proving that the formula holds for case $n + 1$ whenever it holds for case n.)

3.5B   If X and Y are independent, then $\text{Cov}(X, Y) = 0$. However, the converse need not be true. Show that this is the case for the mass function $p(-1, 1) = p(0, 0) = p(1, 1) = 1/3$, where X and Y are discrete.

3.5C   Let $\{X_i\}$ be a set of random variables. Show that

$$\text{Var}\left(\sum_{i=1}^{n} X_i\right) = \sum_{i=1}^{n} \text{Var}(X_i) + 2 \sum_{i<j}\sum \text{Cov}(X_i X_j)$$

3.6A   Consider Table 3–1. (a) Derive the moment generating function $M(t)$ for the binomial distribution. (b) Use $M(t)$ to determine the mean and variance.

3.6B   Consider Table 3–2. (a) Derive the moment generating function for the normal distribution. (b) Use $M(t)$ to deduce the mean and variance.

3.6C   Let X_1 and X_2 be two random variables from a Poisson distribution with respective parameters λ_i. Let $Y = X_1 + X_2$. Use the moment generating function to determine the distribution of Y.

3.7A   Let X and Y be discrete variables with mass function $p(x, y)$. Assume $p(0,0) = p(1,1) = 1/10$ and $p(1,0) = p(0,1) = 4/10$. Calculate: $E(X|Y)$ for each value of Y, $E(Y|X)$ for each value of X; as well as $E(X)$ and $E(Y)$. Then show that $E[E(X|Y)] = E(X)$ and that $E[E(Y|X)] = E(Y)$.

3.7B   Let X and Y be discrete. Prove that $E[E(X|Y)] = E(X)$.

3.7C   A marketing firm has modeled the number of orders each month according to a Poisson distribution with parameter 500. Moreover, the amount of each order is distributed exponentially with parameter $\lambda = 1/\$9,000$. What are the mean and standard deviation of the total revenues on a monthly basis?

3.8A   Let $X_1, X_2, \ldots$ be independent, identically distributed random variables with mean $\mu = 5$ and variance $\sigma^2 = 1$. Consider the statistic $\bar{X}_n \equiv (X_1 + \cdots + X_n)/n$. As $n \to \infty$, what is the probability that $\bar{X}_n$ exceeds 7?

3.10A   Eve Ning is a party beast. The probability that Eve meets a boy she likes—and monopolizes—at a party is 3/4. The chance that a boy she has targeted will ask her for a date is 2/3. Based on years of experience, Eve knows that she will accept half of the invitations. (a) What is the chance that Eve meets the first boy she likes at the third party? (b) Given that she has met 10 likeable boys in 10 parties, what is the chance of meeting the twelfth at the fourteenth party? (c) What is the chance that the second victim—person, that is—she dates will be snared at the fifth party?

3.11A   Consider the Poisson mass function $p_X(k) = e^{-\lambda}\lambda^k/k!$ for $k = 0$, 1, 2, $\ldots$. (a) Derive the moment generating function $M_X(t)$ for the Poisson variable X. (b) Use $M_X(t)$ to determine the mean and variance of X.

3.11B   A sailor is shipwrecked and stranded on a strange island. The time until he falls prey to a hungry beast is distributed exponentially with parameter $\lambda = 1/13$ days. The time until he is noticed by a passing ship is distributed exponentially with parameter $\mu = 1/39$ days. What is the chance that he survives this adventure?

3.11C   An immunologist is investigating the effect of infections on newborn rabbits. Suppose that the rate of viral infections is a Poisson process with parameter $\lambda = 0.2$ days. The health of a rabbit is unaffected by a single virus. On the other hand, any rabbit will die if it suffers 2 infections within 10 days after birth. If a surviving rabbit falls prey to 2 or more viruses between the 10th and 20th days, it will become severely ill; 3 or more viruses will paralyze it for a while. (a) What is the probability that a newborn rabbit will live? (b) What is the chance that a newborn rabbit will become severely ill? (c) What is the chance that a healthy 10-day-old rabbit will become severely ill? (d) A rabbit can suffer from bacterial infections, independently of viral strikes, in a Poisson process with parameter $\lambda' = 0.3$ days. Suppose that a rabbit will die of 2 or more infections—whether viral or bacterial—if they are contracted within the first 10 days. What is the chance that a newborn rabbit survives?

FURTHER READING

Ash, Robert. 1972. *Real Analysis and Probability*. New York: Academic Press.

Breiman, L. 1968. *Probability*. Reading, MA: Addison-Wesley.

Doob, J.L. 1953. *Stochastic Processes*. New York: Wiley.

Feller, W. 1957. *An Introduction to Probability Theory and Its Applications*. Vol. 1. New York: Wiley.

Hoel, Paul, Sidney Port, and Charles Stone. 1971. *Introduction to Probability Theory*. Boston: Houghton Mifflin.

Ross, Sheldon. 1985. *Introduction to Probability Models*. 3rd ed. New York: Academic Press.

II

Statistics

4

Concepts of Statistics

The field of statistics deals with the analysis of data and the generation of plausible inferences. The discipline may be broadly classified into two areas: exploration and inference. Exploration refers to the informal evaluation of a set of data that are regarded as characteristics of some population. In contrast, inference pertains to more formal procedures for deducing results from complete or partial data sets.

The first section presents a brief discussion of the key issues and objectives in exploratory data analysis. The rest of the chapter, as well as the book as a whole, focuses on the topics of statistical inference and decision making, which represent the foundations of statistics.

4.1 EXPLORATORY ANALYSIS

The task of *exploratory analysis* is to examine data and condense them into meaningful summaries. To illustrate, consider the set of data shown in Figure 4–1, which might represent grades on a quiz. For this data set, the low value is 3 and the high is 10; the range is their difference, or 7. The *median* of 6 is the value that splits the sample: as many data points lie above as below the median. The *mode* of 7 is the most frequently occurring value. The *mean* is about 6.18.

An important task in exploratory analysis is the determination of the extent to which aggregation does make sense. Figure 4–2 depicts a distribution of observations that shows multiple peaks. The data may relate to the life spans of tigers, spectral emissions from astronomical objects, or some other phenomenon. It is very likely that the peaks at x and y indicate two different populations, such as tigers living in distinct habitats. The mode at z

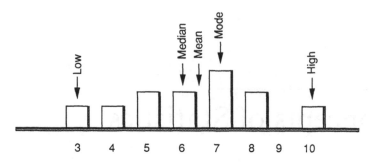

FIGURE 4–1. Histogram and some summary statistics for the data set
{3, 4, 5, 5, 6, 6, 7, 7, 7, 8, 10}. The median is 6; the mode 7; the mean approximately 6.18. The
range is the difference between high and low values: $10 - 3 = 7$.

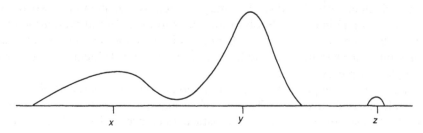

FIGURE 4–2. Example of a multimodal distribution: the local peaks occur at x, y, and z. The
data at z are outliers: special cases or even errors.

indicates the existence of *outliers* or extreme values; it may represent yet a
third category of objects, or may even be due to errors in instrumentation or
observation. Whichever the case, the data suggest further exploration rather
than blind summarization such as the calculation of a single mode or mean
for the entire population.

Exploratory analysis is often a prelude to statistical inference. *Inference*
refers to the generation of reasonable conclusions from data using probabilistic arguments.

4.2   OVERVIEW OF
STATISTICAL INFERENCE

Most of the work in statistics relates to models and methodologies for drawing inferences. The task of inferencing may, in turn, be classified into three
main subjects: estimation, localization, and testing. We now turn to these
subjects.

Point Estimation

Estimation refers to the determination of a single point or value. For instance, the mean income of all Europeans might be estimated from the payroll data of 500 companies. Sometimes the estimation involves a multidimensional task: if the mean as well as the standard deviation of European incomes were desired, then the estimate would involve two values.

Interval Localization

Sometimes we are concerned more about the range of a parameter than its precise value. *Localization* refers to the determination of an interval likely to contain the parameter of concern. For instance, a vacationer might wish to know that the range of temperatures at her resort is 22 to 33 degrees Celsius. In a similar way, an experimenter might need to know that his range finder is accurate to 5 nanometers, or that the recovery rate of guinea pigs in response to a new drug is 72 to 81%.

In our uncertain universe, localization is a relative rather than absolute activity. For instance, the temperature at the beachside resort is 22 to 33 degrees Celsius most of the time–or 98% of the time, to be precise, but the temperature moves lower or higher the remaining 2% of the time. In a similar way, the recovery rate of guinea pigs is believed to be 72 to 81%; but this assessment could be incorrect with probability .05.

The lack of certitude implies that a probability assessment should accompany a localization task. The resulting interval is, therefore, known as a *confidence interval* to which a degree of belief is attached. For example, the recovery rate of guinea pigs is 72 to 81% with a confidence level of 95%. Alternatively, we also say that the 95% confidence interval for the recovery rate is 72 to 81%.

A confidence interval may be regarded intuitively as a long-term batting average. Suppose that we determine that the 95% confidence interval for the recovery rate is 72 to 81%; then if we make such pronouncements 100 times, we would expect to be correct on 95 occasions.

Hypothesis Testing

The third major activity in statistical inference is testing. *Hypothesis testing* involves the determination of the truth or falsity of a proposition. Examples of hypotheses are "It will not rain tomorrow" or "This drug has no effect on curing cancer."

The default proposition is called the *null hypothesis* and denoted as H_0. The converse is the *alternate hypothesis* and symbolized as H_1. For instance, the hypotheses for the drug problem are

H_0:   Drug has no effect.
H_1:   Drug has some effect.

Suppose that, without any treatment, 50% of the patients get better, whereas 50% fare poorly. Then the hypotheses may be stated in quantitative terms as follows:

H_0:   The recovery rate is 50%.
H_1:   The recovery rate is *not* 50%.

The *critical region* is the set of values for which we would accept the alternate hypothesis. In the above example, as stated, the critical region is any value between 0 and 100, inclusive, except 50.

This is an example of a *two-sided test* in which the critical region falls both above and below the null region. In other words, any value above 50 is bad news for the null hypothesis, as well as any value below. A two-sided test is appropriate in many circumstances, such as testing whether a spacecraft will intercept a planet on cue: the craft should be neither too early nor too late if it is to intercept its target.

A *one-sided test*, on the other hand, is an experiment in which the critical region lies solely on one side of the null value. In the pharmaceutical example, we might not be interested in the possibility that the drug may *worsen* the condition of subjects. If we care only about the balmy possibilities of the drug, a one-sided test can be stated as follows:

H_0:   The recovery rate is 50%.
H_1:   The recovery rate exceeds 50%.

It is clear that hypothesis testing will often involve estimation and/or localization. In the pharmaceutical scenario, how is one to judge the efficacy of the drug? One has to conduct experiments, collect data, and estimate the effect.

The result of such activities may be the following: the estimate of the effect on subjects is 60%, with a 95% confidence interval of 55 to 65%. Since the interval lies entirely within the critical region of (50%, 100%], we accept the alternate hypothesis H_1 rather than the null hypothesis H_0.

The efficacy of a test is usually specified in terms of a significance level. A *level of significance* α defines the probability of error in accepting a hypothesis—or, equivalently, rejecting its converse. For instance, if the chance of committing an error is 10%, then the significance level is $\alpha = 0.10$.

Returning to the drug example, suppose our level of confidence that the true recovery rate lies in the critical region is precisely 95%. Then the level of significance is 0.05. In other words, we accept the alternate hypothesis (and reject the null) at significance level $\alpha = 0.05$. This implies that we would expect to be incorrect in about 5 out of 100 scenarios that have the identical level of significance.

4.3   POPULATIONS AND SAMPLES

The population for a statistical inquiry is defined by a collection of objects. To be more precise, a *population* is a set of observations or trials concerning some aspect of each object. For instance, if the set consists of all Americans over the age of 65, then these individuals exhibit a number of attributes that may be associated with random variables. Such attributes may relate to the height, weight, income, or other aspects of each individual.

If the random variable is denoted as X, then the distribution function $F(X)$ defines the pattern of the random variable within the population of Americans. A statistical study usually involves some finite subset of the population. The actual set of observations is called the *sample*. If the sample consists of values $X_1, X_2, \ldots, X_n$, then n is called the *sample size*.

Sample Mean

As with any random variable, a sample can be summarized into composite characteristics. Two of the most common attributes are the mean and variance.

Suppose that a sample consists of values $X_1, X_2, \ldots, X_n$. Then the sample mean, denoted $\overline{X}$, is given by

$$\overline{X} \equiv \frac{1}{n} \sum_{i=1}^{n} X_i$$

Sometimes the data may be arranged into classes or categories. Suppose that couples are surveyed for the number of children they have. If the data set is $\{0, 1, 1, 1, 2\}$, then there are 3 distinct classes, namely, $\{0, 1, 2\}$. The frequency distribution is as follows:

Class	Frequency
0	1
1	3
2	1

When the frequency distribution is presented in the form of a chart, it is called a *histogram*. From the data, we can calculate the average number of children by summing the product of each class and its frequency, then dividing by the sample size:

$$\bar{x} = \frac{0(1) + 1(3) + 2(1)}{5} = 1$$

More generally, let $y_1, y_2, \ldots, y_k$ be the classes or distinct values of the data set $x_1, x_2, \ldots, x_n$. If n_i is the number of entries having value y_i, then the sample mean is given by

$$\bar{x} = \frac{1}{n} \sum_{i=1}^{k} n_i y_i$$

Let f_i denote the fraction of samples in the ith class; that is, $f_i \equiv n_i/n$. Then the sample mean can be written as the average value of the classes y_i, weighted by the proportions f_i:

$$\bar{x} = \sum_{i=1}^{k} f_i y_i$$

The fraction f_i—or equivalently, the frequency n_i—in a frequency distribution indicates the relative likelihood that an observation takes value y_i. In other words, the fraction f_i corresponds to the probability value p_i in a mass function for random variable Y.

In some situations, the sample *is* the entire population. This is exemplified by the set of grades in a particular course. Suppose that the grades are x_1, $x_2, \ldots, x_n$, among which there are district values $y_1, y_2, \ldots, y_k$. If n_i is the frequency of scores in category y_i, then the ratio n_i/n gives the relative frequency of the score y_i. If X is the random variable denoting the grade of a student selected at random, then the probability mass function is defined by n_i/n. In other words, $p_i = n_i/n$ is the probability that X takes the value y_i.

Sample Variance

Once again, let $x_1, x_2, \ldots, x_n$, denote the sample data. Then the *sample variance* is denoted as s^2 and defined as the average squared derivation from the average:

$$s^2 \equiv \frac{1}{n} \sum_{i=1}^{n} (x_i - \bar{x})^2$$

$$= \frac{1}{n} \left(\sum_{i=1}^{n} x_i^2 - n\bar{x}^2 \right)$$

The square root of the variance, indicated by the symbol s, is called the *standard deviation* of the sample.

Suppose the data are arranged as a frequency distribution with distinct values $y_1, y_2, \ldots, y_k$. As before, let n_i correspond to the numerosity of y_i. Then the sample variance is obtained as

$$s^2 = \frac{1}{n} \sum_{i=1}^{k} n_i(x_i - \bar{x})^2$$

$$= \frac{1}{n} \left(\sum_{i=1}^{k} n_i x_i^2 - n\bar{x}^2 \right)$$

Once again, n_i defines the relative likelihood of obtaining value y_i.

4.4 POINT ESTIMATION

Samples are often obtained in order to draw inferences about the underlying population. An example is found in a marketing survey of individuals in a shopping mall as a way to assess the purchasing habits of consumers in the society at large.

The statistical study may involve the determination of the entire distribution for the random variable of interest. Many applications, however, require only the estimation of one or a handful of parameters rather than the entire distribution. Examples of such estimates are the mean and standard deviation of the random variable. The procedure for estimating a parameter should possess a number of properties, such as unbiasedness, consistency, and efficiency.

Let $\hat{\theta}$ be the estimate for a parameter θ from a space of possible values Θ. For instance, if θ is the parameter for the exponential density function, then Θ is the positive real line. We say that the estimate $\hat{\theta}$ is *unbiased* if the expected value of $\hat{\theta}$ is the actual value θ of the parameter. In other words, the estimate $\hat{\theta}$ is unbiased if $E(\hat{\theta}) = \theta$.

To illustrate, let $\{X_i\}$ be a set of observations for a random variable having a density function with mean μ. The sample mean is given by $\bar{X} = \sum_i X_i/n$. The expectation of this quantity is

$$E(\bar{X}) = E\left(\frac{1}{n} \sum_{i=1}^{n} X_i \right) = \frac{\sum_{i=1}^{n} E(X_i)}{n} = \frac{n\mu}{n} = \mu$$

The second equality springs from the fact that the expectation is a linear operator; consequently any constant such as $1/n$ can be extracted from the expectation. The third equality relies on the fact that the expectation of a sum is the sum of the expectations. From the result $E(\overline{X}) = \mu$, we see that the sample mean $\overline{x}$ is an unbiased estimator for the population mean μ.

Consistency refers to the notion that an estimate should approach its target with arbitrary accuracy as the sample size increases without bound. More specifically, let $\hat{\theta}(n)$ be the estimator based on a sample size of n, for an unknown parameter θ. Then $\hat{\theta}$ is *consistent* if its probability of approaching θ goes to 1 with increasing n:

$$P\{|\hat{\theta} - \theta| < \varepsilon\} \to 1 \quad \text{as } n \to \infty$$

for any arbitrary $\varepsilon > 0$.

To illustrate, the arithmetic mean $\overline{X}$ is a consistent estimator for the population mean μ. We can show this as follows. The σ^2 denotes the variance of each observation X_i. Then the variance of the sample mean is

$$\mathrm{Var}(\overline{X}) = \mathrm{Var}\left(\frac{1}{n}\sum_{i=1}^{n} X_i\right) = \frac{1}{n^2}\mathrm{Var}\left(\sum_{i=1}^{n} X_i\right)$$

$$= \frac{1}{n^2}\sum_{i=1}^{n} \mathrm{Var}(X_i) = \frac{n\sigma^2}{n^2} = \frac{\sigma^2}{n}$$

The second equality depends on the squaring property of multiplicative constants within the variance operator. The third equality follows from the independence of the observations X_i. Since the variance of $\overline{X}$ is σ^2/n, the dispersion of the sample average falls to 0 as n increases indefinitely. Consequently, $\overline{X}$ is a consistent estimator of the sample mean μ.

Efficiency refers to the notion that an estimator should pursue its target with precision rather than fluctuate wildly. More precisely, an estimator $\hat{\theta}$ of an unknown parameter θ is *efficient* if its variance is minimal. In other words,

$$\mathrm{Var}(\hat{\theta}) = \min_{i} \mathrm{Var}(\widehat{\theta_i})$$

among all estimators $\widehat{\theta_i}$.

In comparing the arithmetic mean $\overline{x}$ with the median $\tilde{x}$, the variance of $\overline{x}$ can be shown to be smaller than that of $\tilde{x}$. The mean is, therefore, a more efficient estimator than the median.

The most popular approach used in estimation is to seek the parameter whose value maximizes the chance of generating the sample data. Let $x_1, \ldots, x_n$ be a set of observations of a random variable X having density f with parameter θ. The *likelihood function* $\mathscr{L}(x_1, \ldots, x_n|\theta)$ is the joint density function of the sample given the parameter q—namely, $f(x_1, \ldots, x_n|\theta)$. The *maximum likelihood estimator* $\hat{\theta}$ of an unknown parameter θ from a parameter space Θ is the value which maximizes the likelihood function.

If the data are incarnations of independent observations, then their joint density is given by the product of their individual densities. The likelihood function is then given by

$$\mathscr{L}(x_1, \ldots, x_n | \theta) = f(x_1 | \theta) f(x_2 | \theta) \cdots f(x_n | \theta)$$

Example

Let $x_1, x_2, \ldots, x_n$ be a set of independent observations of a random variable X from a normal density. The mean μ and variance σ^2 are unknown; determine the maximum likelihood estimators for those parameters.

The unknown parameter is a vector $\theta = (\mu, \sigma^2)$ having two components. The likelihood function can be written as

$$\mathscr{L}(X_1, \ldots, X_n | \mu, \sigma^2) = \prod_{i=1}^{n} \frac{1}{\sigma \sqrt{2\pi}} \exp\left(-\frac{(x_i - \mu)^2}{2\sigma^2} \right)$$

$$= (2\pi\sigma^2)^{-n/2} \exp\left(-\frac{1}{2\sigma^2} \sum_i (x_i - \mu)^2 \right)$$

Our task is to find the values for parameters μ and σ^2 which maximize $\mathscr{L}$. Since the logarithm of $\mathscr{L}$ is a monotonic function of $\mathscr{L}$, maximizing $\ln \mathscr{L}$ is equivalent to maximizing $\mathscr{L}$ directly. Differenting $\ln \mathscr{L}$ with respect to each parameter and setting the results to zero, we obtain

$$\frac{\partial \ln \mathscr{L}}{\partial \mu} = \frac{1}{2\sigma^2} 2 \sum_i (x_i - \mu) = 0$$

$$\frac{\partial \ln \mathscr{L}}{\partial \sigma^2} = \frac{-n}{2} \frac{2\pi}{2\pi\sigma^2} + \frac{\sum_i (x_i - \mu)^2}{2\sigma^2} = 0$$

Solving the two equations in parallel leads to $\hat{\mu} = \sum_i x_i / n$ and $\widehat{\sigma^2} = \sum_i (x_i - \hat{\mu})^2 / 2n$. These are the maximum likelihood estimators for the mean and variance, respectively.   ◆

4.5   INTERVAL LOCALIZATION

In point estimation, the goal is to obtain a single number to serve as the proxy for the unknown parameter. However, the probability that the single value will exactly match the unknown parameter might be small, or even zero when the parameter space Θ is an interval on the real line. For instance, if a data set indicates that the mean time to failure for an engine is 96.7 hours, how trustworthy is this assessment? Could this be a statistical aberration, when the actual mean is 83.2, or perhaps 96.701 hours?

A way to circumvent this problem is to define an interval as the estimate. Instead of saying that the mean time to failure is 96.7 hours, we might say that the actual rate should lie in the interval 93.7 to 99.7 hours.

More generally, *localization* refers to the determination of an interval $(\hat{\theta} - \delta, \hat{\theta} + \delta)$ which is presumed to contain the unknown parameter θ. The task of localization is also called *interval estimation*.

But the story does not end here. The goal of localization is to identify some *proper* subset of the potential parameter space Θ. The value of an estimate is based on a set of observations. Since the observations are instantiations of a random variable, there is still some chance that the actual value θ falls outside the estimated interval.

We, therefore, associate a *confidence level* σ with the interval. If $\sigma = 0.90$, for instance, then we would expect the actual value of the parameter to fall within the interval in about 90 out of 100 situations where the same confidence level is given.

To illustrate, consider the random variable X from a normal density with parameters μ and σ^2. The variance is assumed to be known, but the mean μ is not. Our task is to construct an interval for μ from the sample data for $x_1, \ldots, x_n$.

Suppose we use the sample mean $\bar{x}$ as the estimator of μ. Since each x_i is an independent observation of a random variable X_i from a normal distribution, the function

$$\overline{X} \equiv \frac{1}{n} \sum_{i=1}^{n} X_i$$

is also distributed normally. We recall from the previous section that the mean of the random variable $\overline{X}$ is μ, whereas the variance is σ^2/n.

Since $\overline{X}$ is distributed normally with mean μ and variance σ^2/n, we can construct an interval centered at the estimated mean of $\bar{x}$. This is achieved by selecting the value of the standard normal density z such that the probability of falling within z standard deviations of μ is γ:

$$P\left(|\bar{x} - \mu| < z \frac{\sigma}{\sqrt{n}} \right) = \gamma$$

The term $z\sigma/\sqrt{n}$ follows from the fact the standard deviation for $\overline{X}$ is $\sigma/\sqrt{n}$. The event "$\bar{x}$ and μ lie within $z\sigma/\sqrt{n}$ units of each other" is tantamount to the event "μ lies in the interval $\bar{x} - z\sigma/\sqrt{n}$ and $\bar{x} + z\sigma/\sqrt{n}$." We can, therefore, write the previous equation as

$$P\left(\bar{x} - z \frac{\sigma}{\sqrt{n}} < \mu < \bar{x} + z \frac{\sigma}{\sqrt{n}} \right) = \gamma$$

The strip $(\bar{x} - z\sigma/\sqrt{n}, \bar{x} + z\sigma/\sqrt{n})$ is an interval of confidence level γ for the parameter μ.

Example

A random variable is known to have a normal density with a variance of 16. The mean, however, is unknown. If a set of 25 observations yields a sample mean of 46, determine a 95% confidence interval for the population mean.

The confidence interval for the population mean lies within z standard deviations of the sample mean, where z is the value of the standard normal variable covering 95% of the area around the mean. This is the same value z for which the cumulative function $\Phi(z)$ consumes 95% of the total, plus half of the complement, namely, $5/2 = 2.5\%$. In algebraic form, we have

$$\Phi(z) = \gamma + \frac{1 - \gamma}{2} = .95 + \frac{1 - .95}{2} = .975$$

From the tables of the normal distribution, we see that the appropriate value of z is 1.96. In other words, the chance of lying above $z = 1.96$ is 2.5%, as is the chance of lying below -1.96. Consequently, the chance of falling within 1.96 standard deviations of the mean is 95%.

Since we are told that the variance is $\sigma^2 = 16$ and the sample size is $n = 25$, we obtain

$$\gamma = .95 = P\left\{\bar{x} - z\frac{\sigma}{\sqrt{n}} < \mu < \bar{x} + z\frac{\sigma}{\sqrt{n}}\right\}$$

$$= P\left\{46 - 1.96\frac{4}{\sqrt{25}} < \mu < 46 + 1.96\frac{4}{\sqrt{25}}\right\}$$

Performing the arithmetic yields (44.432, 47.568) as the 95% confidence interval for μ. In 95 cases out of 100, we would expect the actual value of μ to lie within this strip. ◆

4.6 HYPOTHESIS TESTING

Estimation refers to the determination of a single value for an unknown parameter. This type of issue is reflected in the question "What is the incidence of brain tumors in the population at large?" Estimation is the limiting case of localization, relating to the identification of a confidence interval. A problem of localization is implied by the query "What are upper and lower limits for the incidence of brain tumors?"

Testing, on the other hand, refers to the determination of a categorical result. An example of a test is the question "Does caffeine affect the incidence of brain tumor?" The appropriate response to this question is a simple "Yes" or "No." Another type of test is illustrated by the query "Which of treatments A, B, and C is best for alleviating brain tumors?" The answer is found among the alternatives A, B, C, or some subset thereof.

The essence of testing lies in a comparative analysis: evaluating a treated group against a control group, comparing several groups against each other, or scrutinizing a sample against alternatives which may have occurred under similar conditions. The last situation—analyzing a sample against potential alternatives—is the basic scenario in the theory of tests. Here, the experimenter seeks to determine whether the deviations reflected in a set of data are caused by some significant underlying factor, or are due merely to randomness.

To illustrate, consider a particular birth defect which occurs in 2 out of 1000 babies. A new drug to prevent this defect is administered to 5000 pregnant women. Among babies from treated mothers, the rate of birth defects is only 1.4 per 1000. Can the result be attributed to the drug, or is it due simply to chance?

The default proposition behind a test is called the *null hypothesis* and denoted H_0. Its complement is the *alternate hypothesis*, denoted H_1. In the drug example, the null hypothesis would be $H_0 =$ "Drug has no effect" and the alternative $H_1 =$ "Drug has some effect." The default hypothesis is accepted or rejected based on the values of the data sample. The *critical region* $\mathscr{K}$ is the set of values for which the null hypothesis is rejected. In other words, if $y = y(x_1, x_2, \ldots, x_n)$ is the test statistic based on the observations x_i, then H_0 is rejected for $y \in \mathscr{K}$. Conversely, the alternate hypothesis is accepted if $y \in K$.

Since the observations x_i are manifestations of a random variable, extreme values *could* be due to chance rather than the result of significant causes. Conversely, an innocuous sample could arise even when a differentiating factor underlies the observations.

In view of this uncertainty, the result of a test is associated with a probability of error. *False rejection* occurs when the null hypothesis is rejected despite the fact that it is, in reality, valid. This kind of mistake is also called a Type I error as indicated in Table 4–1. The probability of false rejection is called the *level of significance* α.

In contrast, *false acceptance* occurs when the null hypothesis is accepted despite its being incorrect. This mistake is also called a Type II error.

Guarding against false rejection is a conservative strategy: only overwhelming evidence will move the decision maker to discard his default assumption. An analogy is found in the legal system of the United States, in

TABLE 4–1. Consequences of hypothesis testing.
False rejection is called Type I error; and false acceptance,
Type II.

	Decision	
Validity of H_0	Accept	Reject
True	Correct	Type I error
False	Type II error	Correct

which the accused is presumed to be innocent. The null hypothesis of inno-cence is rejected only when the evidence against it is so strong that no reasonable doubt remains.

In contrast, guarding against false acceptance is an aggressive strategy. This approach represents the situation where a decision maker can little afford to be caught unprepared, taking comfort in the old ways when changes are afoot.

The data samples used in hypothesis testing are instantiations of a random variable. The decision procedure based on these variables is also a random variable characterized by its own distribution function. The most common functions for test variables are the normal, t-, F-, and χ^2-distributions.

Normal Distribution

When the normal distribution is used in hypothesis testing, the critical region is given by one or both tails on either side of the mean. Let μ be the mean and σ^2 be the variance for a random variable under the null hypothesis. Then the test statistic Y can be converted to standard normal form by the transforma-tion $Z = (Y - \mu)/\sigma$.

Consider a test to determine whether the mean m of a sample has the same mean μ as that under the default assumption. The test takes the following form:

H_0: $m = \mu$
H_1: $m \neq \mu$

Then the critical region $\mathcal{K}$ involves both the left and right tails of the normal density. If the level of significance is α, the right-hand threshold is given by the value $z = z_c$ such that the area to its right is $\alpha/2$. Similariy, the left-hand threshold is given by $z = -z_c$, for which the area to the left is also $\alpha/2$. The critical region is the union of these two intervals: $\mathcal{K} = (-\infty, -z_c) \cup (z_c, \infty)$.

For instance, if the significance level is 0.05, then $z_c = 1.96$ is the right-hand threshold. The critical region is $\mathscr{K} = (-\infty, -1.96) \cup (1.96, \infty)$. The null hypothesis is rejected if the test statistic falls in this region. However, this could happen with probability .05 even if the null hypothesis were true; so the chance of a Type I error is 5%.

A test may also involve a single tail of the normal density. Consider a test to determine whether the mean m of a sample is greater than the mean μ under the default hypothesis. The test is of the following form:

H_0: $m = \mu$
H_1: $m > \mu$

The critical region $\mathscr{K}$ covers only the right tail of the normal density. In particular, $\mathscr{K} = (z_c, \infty)$, where z_c is the value of the standard normal variable for which the area to the right is α. In other words, z_c is obtained from the relation $\Phi(z_c) = 1 - \alpha$.

Example

A high school principal claims that students in her school perform better than the national average on collegiate entrance exams. As evidence, she cites records showing that the average score for the last 400 graduates is 510. Suppose that the score on an entrance exam is distributed normally with a mean of 500 and standard deviation of 100. Assuming that the standard deviation among her students is the same as that for the population at large, can the principal's claim be justified?

Let m be the mean among the students. This is a one-sided test for which the hypotheses are

H_0: $m = 500$
H_1: $m > 500$

Under the default hypothesis, the score X on the exam is a random variable from a normal density with $\mu = 500$ and $\sigma = 100$. The test procedure is the sample mean given by $\overline{X} = \sum_i X_i/n$. We have seen earlier that $\overline{X}$ is also distributed normally with mean μ and standard deviation $\sigma/\sqrt{n}$.

For a significance level of 0.05, the threshold for the critical region is z_c such that 5% of the area lies to its right, namely, $z_c = 1.65$. In terms of the exam scores, the corresponding value is

$$\overline{x}_c = \mu + z_c \frac{\sigma}{\sqrt{n}} = 500 + 1.65 \frac{100}{\sqrt{400}} = 508.25$$

Since the observed value of 510 exceeds the threshold, we can reject the null hypothesis at the 0.05 level of significance. In other words, we conclude that the students, in fact, perform better than the national average. Since $\alpha = 0.05$, however, this conclusion has a 5% chance of being incorrect.   ◆

Other Distributions

The procedures used in hypothesis testing follow many different density functions. Among these, the most common functions are the t-, F-, and χ^2-distributions.

t-Distribution

Use of the normal distribution for testing requires knowledge of the population mean μ and standard deviation σ. If these parameters must be estimated from the sample statistics $\bar{x}$ and s, the normal density function is no longer applicable. For small sample sizes, the estimate s could differ greatly from that of the population. Consequently, the expected error increases as well.

The appropriate function in this case is the *student's t-distribution*. This function depends on the sample size n as well as the significance level α.

The density function for the t-distribution looks somewhat like the normal function. Due to the large error in the standard deviation s, however, the function tends to be flatter at the center and thicker at the tails. As n increases, however, the curve becomes sharper at the center and, in fact, converges to the normal density in the limit as $n \to \infty$. Suppose that $x_1, \ldots, x_n$ constitute a sample from which m parameters must be estimated. The t-function actually depends on the *degrees of freedom* in the sample, given by $v = n - m$. The value v defines the number of dimensions along which the data can vary. For instance, if the mean is unknown and must be calculated by $\bar{x} = \sum_i x_i/n$, then this relation constrains the dimensionality of the sample values $x_1, \ldots, x_n$. If there are 20 data values, the degree of freedom is

$$v = n - m = 20 - 1 = 19$$

In other words, only 19 of the data values can take arbitrary values: the twentieth is constrained by the relation

$$x_{20} = n\bar{x} - \sum_{i=1}^{19} x_i$$

As indicated above, the t-statistic approaches the normal density for increasing degrees of freedom. The approximation to the normal density is exact in the limit as $v \to \infty$.

To illustrate the use of the t-distribution, suppose that a sample of size n is taken from a normal density. The test involves a comparison of the sample average against the mean μ of a normal density. If the sample has a mean of $\bar{x}$ and variance s^2, the quantity

$$t = \frac{\bar{x} - \mu}{s/\sqrt{n}}$$

follows a t-distribution with degree of freedom $v = n - 1$.

The critical region at significance level α is defined by the value t_c beyond which the area under the t-density is α. In other words, the critical region is $\mathcal{K} = (-\infty, -t_c) \cup (t_c, \infty)$. If the calculated value of t falls in $\mathcal{K}$, the null hypothesis is rejected. We would conclude that the sample comes from a normal density whose mean is unequal to μ.

F-Distribution

Another distribution comes into play when the variances from two samples are to be compared. Consider two samples of size n_1 and n_2 from a normal density. If s_1^2 and s_2^2 are the respective sample variances, their ratio given by

$$F = \frac{s_1^2}{s_2^2}$$

follows the so-called F-distribution. This distribution is characterized by two degrees of freedom corresponding to the respective samples; these are $v_1 = n_1 - 1$ and $v_2 = n_2 - 1$.

Since F is the ratio of two positive quantities, it is also positive. The structure of the density function for the F-distribution looks approximately like the curve shown in Figure 4–3. The precise shape will depend, of course, on the parameters v_1 and v_2.

The F-statistic may be subjected to a one- or two-tailed test. For a two-tailed test, the critical region of significance level α is identified by the values F_a and F_b beyond which the total area under the curve is α. In other words, the critical region at level a is given by $\mathcal{K} = (0, F_a) \cup (F_b, \infty)$.

The procedure for a one-tailed test involves finding a threshold F_c beyond which α of the total area lies. The null hypothesis—that the sample variances s_1^2 and s_2^2 are equal—is rejected if the computed value of F falls within $\mathcal{K}$.

χ^2-Distribution

The sum of squares of variables from a normal density function follows a pattern called the χ^2-distribution. More specifically, let $X_1, \ldots, X_n$ be random variables from a normal density with mean μ and variance σ^2. Then

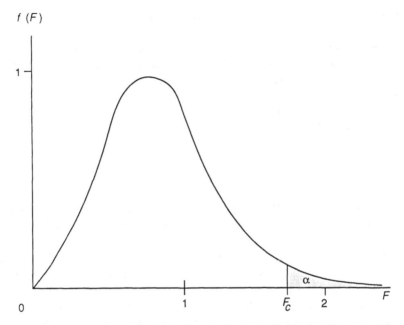

FIGURE 4–3. General form of the density function for the F-distribution. The value F_c beyond which α of the area lies, is the threshold for the F-test at significance level α. The density function for the χ^2-distribution is roughly similar.

the quantity

$$\chi^2 = \frac{1}{\sigma^2} \sum_{i=1}^{n} (X_i - \mu)^2$$

has a χ^2-distribution. This function is parameterized by the degree of freedom $v = n - m$, where n is the number of observations and m the number of parameters estimated from the sample.

Since χ^2 is defined as the sum and ratio of non-negative quantities, it is itself non-negative. The χ^2-density looks somewhat like that for the F-statistic shown in Figure 4–3. The precise shape of χ^2 depends on its parameter, the degrees of freedom v.

For a one-tailed test at significance level α, the critical region is defined by the value χ_c^2 beyond which the area under the curve is α. In other words, the critical region is $\mathcal{K} = (\chi_c^2, \infty)$. The procedure for a two-tailed test is analogous. If the computed value of χ^2 based on the sample falls within $\mathcal{K}$, the null hypothesis is rejected; otherwise, it is accepted.

Although the χ^2-distribution was first derived for the sum of squares of normal variables, it is useful for testing the goodness of fit for arbitrary distributions.

More precisely, let $x_1, x_2, \ldots, x_n$ be a set of independent observations of a random variable X having an arbitrary but known distribution F_X. Suppose $y_1, \ldots, y_k$ are the set of distinct values or classes into which the data x_i fall. Further, let n_i be the number of actual entries in class y_i, based on the observed values x_i; and e_i the expected number of entries in class y_i based on the distribution F_X. The quantity

$$\chi^2 = \sum_{i=1}^{k} \frac{(n_i - e_i)^2}{e_i}$$

has a χ^2-distribution with parameter v. The degree of freedom is given by $v = k - m - 1$, where k is the number of classes and m the number of estimated parameters.

If the calculated value of χ^2 exceeds the artificial threshold χ_c^2, then the null hypothesis is rejected: the sample is assumed to result from a distribution differing from that of F_X. Otherwise, the null hypothesis is accepted. To use this test with assurance, the sample must be large enough and the classes broad enough, so that the expected value e_i in each class is 5 or more.

PROBLEMS

The problems below are numbered according to the sections to which they correspond. For instance, Problem 4.4A is the first ("A") exercise pertaining to Section 4.4.

4.3A   Once upon a time in Calgary, there lived a cave girl named Kalory. Kalory once estimated the effective temperature in her dining room by counting the number of logs used each day. The experiment, using logs of a standard size, resulted in the following data: 5.3, 4.2, 6.7, 8, 9.2, 4, 7.8, 6, 5.7, 7.1. (a) What are the mean and variance of the sample? (b) Draw a histogram, after rounding numbers to the nearest integer. Based on the histogram, what are the sample mean and variance?

4.3B   The sample variance s^2 is often defined as the quantity on the left-hand side of the equation

$$\frac{1}{n-1} \sum_{i=1}^{n} (x_i - \bar{x})^2 = \frac{1}{n-1} \left(\sum_{i=1}^{n} x_i^2 - n\bar{x}^2 \right)$$

Show that it reduces to the quantity on the right-hand side of the formula.

4.4A   Let $x_1, \ldots, x_n$ be a set of independent observations from an exponential density with parameter λ. What is the maximum likelihood estimator for λ?

4.5A   A set of 100 observations are made from a normal density function with a standard deviation of $\sigma = 20$. If the sample average is $\bar{x} = 60$, construct a 99% confidence interval for the mean μ.

4.6A   Consider the example of the high school principal in Section 4.4. What is the critical region for significance level $\alpha = 0.01$? Can the null hypothesis be rejected at the 1% significance level?

4.6B   Consider the data in Problem 4.3A, and assume they originate from a normal density function. (a) Construct a 90% confidence interval for the mean. (b) Do likewise for a 99% confidence interval.

FURTHER READING

Hoel, Paul G., Sidney C. Port, and Charles J. Stone. 1971. *Introduction to Statistical Theory*. New York: Houghton Mifflin.

Savage, L.J. 1954. *The Foundations of Statistics*. New York: Wiley.

Savage, L.F., et al. 1962. *The Foundation of Statistical Inference*. London: Methuen.

Tukey, J.W. 1971. *Exploratory Data Analysis*. Preliminary edition. Reading, MA: Addison-Wesley.

5

Theory of Statistics: Introduction

Statistics involves decision making of decisions based on empirical data. From the statistical perspective, the data are considered to be outcomes of a random variable X. The variable X is regarded as a member of a family of distributions: the specific member of the family is defined by the value of a parameter θ. As an illustration, the variable X may be viewed as belonging to a distribution with parameter $\theta = 3.2$ from an exponential family.

Both the variable X and the parameter θ may be regarded as multidimensional. When n observations are available, the data can be viewed as the outcome of a random variable with dimension n. More specifically, the list of observations $(x_1, x_2, \ldots, x_n)$ is considered to be a value of the n-dimensional vector $X \equiv (X_1, X_2, \ldots, X_n)$.

The multidimensional nature of the parameter θ is apparent when the family of distributions is characterized by more than one quantity. For instance, a member of the normal distribution is defined by its mean μ as well as its standard deviation σ^2. The parameter is, therefore, a two-dimensional vector: $\theta = (\mu, \sigma^2)$.

The basic problem in statistics is that of *inference*: determining which member, or members, of a family of distributions is most likely to account for the data. In the context of quality assurance, for instance, the task of statistical inference might be to assess the true proportion of defective items in a shipment based on a small sample. In this case, the random variable is $X_i = 1$ if the ith item is defective, and $X_i = 0$ otherwise. The parameter is $\theta \equiv p$, the true proportion of defectives. The goal is to deduce the value of p based on the data $(X_1, \ldots, X_n) = (x_1, \ldots, x_n)$.

The problem in quality assurance is one of *estimation*: determining the most likely value of an unknown parameter. A reasonable estimate for p is the average number of defectives in the sample, given by $\hat{p} = \sum_i x_i / n$.

The result of a problem of estimation is a specific number. Perhaps 2 out of 100 evaluated items are defective, leading to the estimate of $\hat{p} = 0.02$.

We have seen in Chapter 3 that two other problems are fundamental to statistics, namely, localization and testing. *Localization* deals with the determination of an interval for an unknown parameter. In the quality assurance example, we might infer that the interval for p is $(0.01, 0.04)$, with a confidence level of 5%.

The third major task of statistical inference is *testing*: determining whether a proposition is true or false. In the context of quality assurance, we might begin with the null hypothesis as the proposition $H_0 = $ "The proportion of defectives is less than 0.02." The alternate hypothesis would then be $H_1 = $ "The proportion of defectives matches or exceeds 0.02." The statistical inference procedure might result, perhaps, in the decision to reject the null hypothesis at a significance level of 0.05.

This chapter presents a more formal introduction to the basic techniques of statistical inference that were discussed informally in Chapter 4. We begin with some criteria for evaluating alternative procedures for decision making. This is followed by a discussion of estimation and localization in Chapter 6, then testing in Chapter 7.

5.1   NATURE OF STATISTICAL PROBLEMS

In this section, we examine the notion of a decision function for mapping observations into actions. This topic is followed by the idea of a prior distribution that can be used to characterize the nature of the unknown parameter in problems of statistical inference. The final topic of the section is a discussion of the relationships between implicit and explicit factors in statistical decision making.

Decision Functions

In a statistical context, a *decider* is an agent who makes a decision based on empirical data. Each potential decision is also called an *action*. The procedure which maps the data space into the action space is called a *decision function* or a *decision procedure*. We encapsulate these ideas into a list of definitions.

- A *decider D* is an agent who makes a decision based on a set of observations $X_1, \ldots, X_n$. Each random variable X_i takes a value x_i from the sample space Ω.
- The *action space* **A** is the set of decisions available to the decider.
- A *decision function* or *procedure* δ is a mapping from the observations $X_1, \ldots, X_n$ into the action space **A**. In other words, the procedure represents a function $\delta(X_1, \ldots, X_n) = a_j$ for each combination of $X_1, \ldots, X_n$ and some a_j in **A**.

The nature of the action space and the decision function depends on the problem at hand. For problems of estimation, the action space **A** is equal to the parameter space Ω.

As an illustration, consider the problem of estimating the proportion θ of individuals who favor the legalization of all drugs. The parameter space is the interval between 0 and 1; that is, $\Theta = [0, 1]$. The action space is equal to the parameter space: $\mathbf{A} = [0, 1]$. Let the variable X_i take the value 1 when the ith respondent is in favor of legalization, and 0 otherwise. The outcome space for each X_i is binary: $\Omega_i = \{0, 1\}$. The outcome space for a set of n observations is given by the set of all combinations of values in Ω_i:

$$\Omega = \prod_{i=1}^{n} \Omega_i = \{0, 1\} \times \{0, 1\} \times \cdots \times \{0, 1\}$$

$$= \{0, 1\}^n$$

The outcome space Ω is therefore the n-fold Cartesian product of $\{0, 1\}$.[1]

Suppose that the decision function for the polling example is given by the sample average:

$$\delta(X_1, \ldots, X_n) = \frac{\sum_{i=1}^{n} X_i}{n}$$

Then the action $a = \delta(x_1, \ldots, x_n)$ takes some value in the closed interval $[0, 1]$.

For problems of localization, the goal is to construct some interval that is likely to contain the unknown parameter θ. In this situation, each action a is some subset of the parameter space Θ. For the polling example, a potential action is $a = (.2, .4)$.

When the inference problem involves hypothesis testing, the action space is binary. In this context, the two potential decisions are acceptance or rejection of the null hypothesis, for which the action space is

$$\mathbf{A} = \{\text{Accept, Reject}\}$$

Other types of decision problems lead to different action spaces. One example is the *multiple-decision* problem in which one among three or more distinct candidates must be selected. This scenario arises with great frequency in many practical contexts. A student may have to select one out of five

---

[1] If **A** and **B** are sets, their *Cartesian product*, denoted by $\mathbf{A} \times \mathbf{B}$, is given by $\{(x, y) : x \in \mathbf{A} \ \& \ y \in \mathbf{B}\}$. For instance, if $\mathbf{A} = \{a, b\}$ and $\mathbf{B} = \{c, d\}$, then $\mathbf{A} \times \mathbf{B} = \{(a, c), (a, d), (b, c), (b, d)\}$. A Cartesian product formed by n instances of a set **A** is called the n-fold Cartesian *product* of **A** and denoted $\mathbf{A}^n$. As an illustration, $\mathbf{A}^2 = \mathbf{A} \times \mathbf{A} = \{(a, a), (a, b), (b, a), (b, b)\}$.

possibilities on a multiple-choice test; a physician must decide whether to increase, decrease, or maintain the level of medication prescribed to a patient; a mobile robot must determine whether an object in its visual field is a vehicle, an animal, a person, or a permanent fixture; or an executive must determine which combination of products to introduce into the marketplace.

Because of their prevalence, multiple-decision problems deserve to be addressed at greater length. We will return to this topic in Chapter 10.

Prior Distributions

As with any mathematical model, statistical models deal with our perception of reality rather than the reality itself. Consequently, there is no such thing as a *correct* model; there are only good ones and bad ones.

The appropriate question is not "Is this model correct?" No model will perfectly represent the underlying reality. Even in an elementary model of a rock falling under the influence of gravity, we must ignore some details to avoid dealing with a model more cumbersome than its referent system. For the case of a falling rock, we might ignore factors such as the color of the object, its temperature distribution, and the change in mass due to collisions with dust particles in the air. The best model depends on the uses to which it will be put.

The same philosophy applies to statistical models. The goodness of a model lies in the mind of the beholder and the purpose of the exercise.

One aspect of statistical modeling that involves more personal judgment than most relates to prior distributions for the parameter under consideration. In many situations, the parameter θ is viewed as an unknown, but fixed, quantity. This happens, for instance, in considering the fraction of citizens who favor a reduction in government expenditures for national defense. The result of a poll is viewed as an estimate of the actual proportion p of the population that supports the policy of military cutbacks. The value of p is regarded as fixed, whether at the value 0.38, or 0.7209, or some other value.

In other situations, however, the value of a parameter can itself be viewed as a random variable. When the parameter θ is viewed as a random variable, its distribution is called the *prior distribution* and denoted as $\pi(\theta)$. The distribution $\pi(\theta)$ is called *prior* since its pattern is specified before the experiment is conducted to estimate the value of θ.

For the case of quality assurance, suppose that a shipment of goods might have originated from one of three production lines. Based on previous experience, the first two production lines are known to have a defective rate of 0.01, and the third line a corresponding rate of 0.03. If a shipment is likely to originate from any of the three lines, then the prior distribution for θ is given by $\pi(0.01) = 2/3$ and $\pi(0.03) = 1/3$. This information may be used in conjunction with the sample data to determine the most likely value of the parameter p for a particular shipment.

The use of a prior distribution might be justified even in the polling example. One might argue, for instance, that the proportion p of citizens in favor of military cutbacks exceeds 1/2 in times of peace, and drops below 1/2 in tension or conflict. Or one might refer to historical precedents and past polls to obtain a set of values for the prior distribution π. In this context, the argument for a prior distribution is more tenuous than in the quality assurance example; but it might be justified nevertheless.

In many arenas of the social and life sciences, the prior distribution π may be unknown or even inappropriate. When prior information is available, however, it can be used to advantage together with observational data to deduce the value of the parameter θ.

Implicit and Explicit Factors

The framework for statistical decision making involves both implicit and explicit factors. The relationships among these quantities are depicted in Figure 5–1.

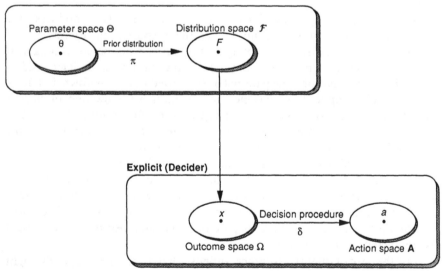

FIGURE 5–1. Implicit and explicit factors in decision making. A parameter value θ is selected from the parameter space, based on a prior distribution π. This parameter value determines the sample distribution function F from the space of distribution functions. The sample distribution determines the nature of the outcome x from the outcome space. The decision function δ maps the outcome x into an action a from the action space.

The *implicit* factors are quantities that the decider cannot observe directly, and over which he has little or no influence. These quantities may be regarded as lying within the jurisdiction of Mother Nature. In particular, we may think of nature selecting a value θ of a particular parameter from a space Θ of possibilities, having a prior distribution function $\pi(\theta)$. As we have seen earlier, in many applications the value of θ is best regarded as fixed: this corresponds to a degenerate probability distribution in which the chance of a particular value is 1, and vanishes for all other values. Some examples of degenerate prior distributions are the electric charge on a proton or the average height of Americans in 1990. In other contexts, the decision maker may be able to influence the parameter even though he cannot observe it directly. The results of a poll on a political issue, or an evaluation of investors' confidence in the stock market, can affect the very parameters they are designed to measure once the evaluations are made public. In other cases, the act of observation can itself influence the results. This happens in the physical arena when an attempt to determine the position of an electron is likely to affect its location; or in the psychological arena where the knowledge of impending evaluation can undermine the creativity of an individual; or in a social context when the existence of observers spurs athletes to enhanced performance.

The selection of a value for the parameter θ determines the distribution F from a family $\mathscr{F}$ of distributions. For instance, $\mathscr{F}$ may represent the exponential family, whereas a parameter value of $\theta = 7.3$ determines a particular member of that family. The distribution F, in turn, affects the explicit quantities in the decision-making situation.

The *explicit* factors are those that the decider can observe directly. These quantities are the outcomes X and the actions $\mathbf{A}$. The distribution F results in the selection of a particular datum x from the outcome space Ω. The decision function δ then maps the datum into some action a from the action space $\mathbf{A}$. This action will, hopefully, fulfill whatever might be the goal of the decision maker.

5.2 EVALUATION CRITERIA
FOR PROCEDURES

An action taken by a decider is defined by the decision procedure. This procedure, however, is based on observations whose inherently probabilistic nature may lead the decider astray. In this section we examine criteria for evaluating alternative procedures based on their ability to yield good or bad actions.

Loss Function

A decision function does not always yield the optimal action. For certain problems, in fact, the probability of attaining a perfect result is zero. This happens, as an illustration, in estimating a parameter θ from an exponential density. Here θ can take on any non-negative real number. Since θ is a continuous parameter, the chance of a *perfect* match between the estimate and the parameter is nil.

When a decider can assign numerical values to the severity of mistakes due to the decision function, these numbers define a *loss function*. In other words, a loss function $L(\theta, a)$ tracks the penalty for selecting action a when the true value of the parameter is θ.

The action a depends on the data sample $x_1, \ldots, x_n$ through the decision function $\delta(x_1, \ldots, x_n)$. This relationship can be indicated explicitly by writing the loss function as $L(\theta, \delta(x_1, \ldots, x_n))$.

The structure of the loss function L will, of course, depend on the decider and the application. If the decider considers the penalty to be proportional to the discrepancy between the action and the parameter, then an appropriate function is the *absolute deviation*:

$$L(\theta, a) = |a - \theta|$$

To illustrate, suppose that a decider wished to determine the mean θ for a normal density based on the sample average. If the data values are $x_1, \ldots, x_n$, then the decision function is

$$\delta(x_1, \ldots, x_n) = \frac{\sum_{i=1}^{n} X_i}{n}$$

The corresponding loss function is

$$L(\theta, \delta(x_1, \ldots, x_n)) = \left| \frac{\sum_{i=1}^{n} X_i}{n} - \theta \right|$$

In many instances, the penalty increases more than proportionately with the magnitude of the error. A plausible function in this case is the squared error loss given by

$$L(\theta, a) = (a - \theta)^2$$

This loss function is commonly used in estimation problems due to its convenient mathematical properties, as illustrated later in this section.

For certain applications, the appropriate loss function takes a simple binary form. This occurs in hypothesis testing, where the loss is 0 for a correct decision and a positive number c for an incorrect one. Since there is only one nonzero number, and the objective is to minimize the loss, the precise value of c is immaterial. For convenience, the value $c = 1$ is chosen. The loss function for hypothesis testing then takes the following binary form:

$$L(\theta, a) = \begin{cases} 0 & \text{for a correct decision} \\ 1 & \text{for an incorrect decision} \end{cases}$$

The binary loss function can also be used for multiple-decision problems. In this case, the loss is considered to be 1 regardless of which incorrect alternative is chosen. If a shuttle in an automated factory delivers a workpiece to the wrong assembly station, then the loss is regarded to be the same no matter which erroneous station is chosen.

A more general form of the loss function is

$$L(\theta, a) = c(\theta)|a - \theta|^b$$

Here $c(\theta)$ is some positive function of θ, and b is a positive number. This general format encompasses the loss functions we have discussed previously. For instance, taking $c(\theta) \equiv 1$ and $b \equiv 2$ yields the squared error loss function.

Since a critical objective of a decider is to minimize the penalty, the value of a loss must be driven toward zero regardless of the precise structure of the loss function. As a result, alternative choices for the loss function often lead to no significant changes in the nature of the decision function. This is especially true when the sample size is large.

Further, our primary concern in this book focuses on the foundations of decision theory. For these reasons, we will employ simple loss functions appropriate to the subject under discussion. For the most part, the same general arguments would apply if more intricate loss functions were to be used.

Risk Function

The loss function specifies the penalty associated with a particular action and parameter value. The action, on the other hand, is defined by the decision function, which, in turn, depends on a set of observations. Since the observed data represent the outcome of a random variable, so do the corresponding action and the resulting penalty. The penalty takes a number of values depending on the nature of the decision function as well as on the character of the underlying sample variable.

More specifically, the loss $L(\theta, a)$ is a specific number for particular values of the parameter θ and the action a. But the action a depends on the decision function $\delta(X)$, which in turn relies on the random variable X. Consequently, the loss function $L(\theta, \delta(X))$ is itself a random variable, as illustrated in Figure 5-2.

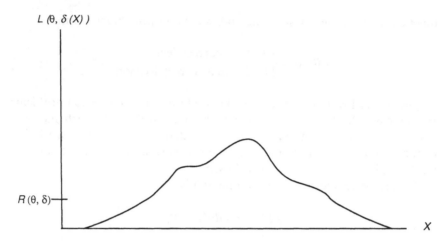

FIGURE 5–2. The loss $L(\theta, \delta(X))$ as a function of the sample variable X. For a given value of the parameter θ, since X is a random variable, so is L. Here X is assumed to be a continuous variable; the precise curve for L depends on the decision procedure δ as well as the algebraic structure of the loss function L. The mean loss is called the risk R for a given value of θ and decision procedure δ.

As with any random variable, it is often convenient to characterize the loss function L in terms of summarizing quantities. The most popular summary procedure is the average penalty, known as the risk. We reiterate this notion in the form of a definition.

- Let $L(\theta, \delta(X))$ be a loss function for a given value of the parameter θ and a decision procedure δ that maps a random variable X into a set of actions. The *risk* function $R(\theta, \delta)$ is the *mean loss* for a procedure δ when the value of the parameter is θ:

$$R(\theta, \delta) \equiv E_\theta L(\theta, \delta(X))$$

Here the expectation is taken over the distribution of X.

The subscript θ on E_θ indicates that the expectation operator is to be applied while regarding θ as fixed. The random variable X may be a scalar or a vector quantity. The variable is a vector, for instance, when n observations of a random variable are taken. The vector variable is then $X = (X_1, X_2, \ldots, X_n)$, where X_i refers to the ith observation of variable X.

We can be more explicit about the procedure for calculating the risk. Suppose X is a continuous random variable having density $f(x|\theta)$, where the symbol θ indicates that the parameter of the density is regarded as fixed. The risk for a procedure $\delta(X)$ is given by

$$R(\theta, \delta) = \int L(\theta, \delta(x)) f(x|\theta) \, dx \qquad (5\text{--}1)$$

where $L(\theta, \delta(x))$ is the loss due to selecting action $\delta(x) = a$ when the parameter has value θ.

As usual, the random variable X may be viewed as a vector rather than a scaler quantity. This situation arises for multiple observations of the variable X having density $f(x|\theta)$. The random variable for the entire sample is given by $X = (X_1, \ldots, X_n)$. Since each observation is independent of the others, the density for the whole equals the product of the individual densities:

$$f(x_1, \ldots, x_n|\theta) = \prod_{i=1}^{n} f(x_i|\theta)$$

Equation (5–1) can then be written

$$R(\theta, \delta) = \int_{x_1} \cdots \int_{x_n} L(\theta, \delta(x_1, \ldots, x_n)) \prod_{i=1}^{n} f(x_i|\theta) \, dx_1 \ldots dx_n$$

Similar arguments apply when X is a discrete random variable with mass function $p(x|\theta)$. The risk is given by

$$R(\theta, \delta) = \sum_i L(\theta, \delta) p(x_i|\theta)$$

where the summation is taken over all values of the variable, $X = x_i$. When the random variable X is a vector quantity, then $X = (X_1, \ldots, X_n)$ as usual and the summation must be taken over all the variables, from X_1 through X_n.

The risk function can be used to evaluate the performance of its associated decision function. This idea is illustrated in the next example.

Example

The manager of a new semiconductor plant must determine the initial quality of the production line as a basis for evaluating steps to enhance productivity. During the initial production runs, the yield rate—or the fraction of products that pass inspection—could be dismally low. For the initial estimate, the manager is contemplating using one of two decision functions. The first is to assume the value of 1/2 as the yield rate, regardless of the observed quality of the products. The second is to take the average of the first two samples. In this case $X_1 = 1$ if the first item passes inspection and $X_1 = 0$ otherwise; the approach for the second sample is similar. The manager assesses the loss function to be directly proportional to the error in the estimated value in relation to the actual yield rate. For a given value θ of the yield rate, what is the risk for each decision procedure? Further, if the actual yield were 0.8, which procedure would be superior?

The first decision function is independent of any observed values, and is fixed at 1/2; in other words, $\delta_1(X) = 1/2$. The second procedure is the average of two observations, corresponding to

$$\delta_2(X) = \delta_2(X_1, X_2) = (X_1 + X_2)/2$$

The loss function is the discrepancy between the estimated and actual values of the yield, given by $L(\theta, \delta(X)) = |\delta(X) - \theta|$. The risk is the average loss resulting from a particular decision procedure. For the first procedure, the risk is

$$R_1 \equiv R(\theta, \delta_1) = EL(\theta, \delta_1(X)) = |\tfrac{1}{2} - \theta|$$

The risk for the second procedure is

$$R_2 \equiv R(\theta, \delta_2) = EL(\theta, \delta_2(X_1, X_2))$$
$$= p_{0,0}L(\theta, \delta_2(0,0)) + 2p_{0,1}L(\theta, \delta_2(0,1)) + p_{1,1}L(\theta, \delta_2(1,1))$$
$$= (1 - \theta)^2 \left|\frac{0+0}{2} - \theta\right| + 2\theta(1 - \theta)\left|\frac{0+1}{2} - \theta\right| + \theta^2\left|\frac{1+1}{2} - \theta\right|$$
$$= (1 - \theta)^2\theta + 2\theta(1 - \theta)|\tfrac{1}{2} - \theta| + \theta^2|1 - \theta|$$

The notation $p_{i,j}$ refers to the probability that $x_1 = i$ and $x_2 = j$.

When the actual yield is $\theta = 0.8$, the risks are given by $R_1 = 0.3$ and $R_2 = 0.256$. Since the second risk is smaller than the first, the second decision procedure is superior.   ◆

In the preceding example, the second decision function is seen to perform better than the first procedure when the value of the parameter is $\theta = 0.8$. For other values of the parameter, however, this evaluation could be reversed. In fact, this happens when the parameter is only slightly smaller. For instance, when $\theta = 0.75$, the risks are $R_1 = 0.25$ and $R_2 \cong 0.28$, in which case the first procedure is preferable.

Comparing Procedures

We have seen how the risk can be used as a measure of the performance of a decision rule. Given two procedures δ and ε, the first procedure is better than

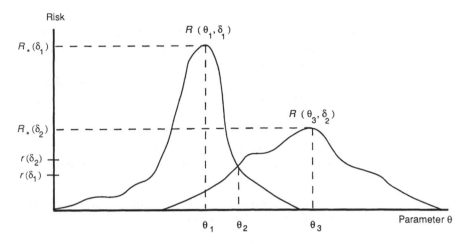

FIGURE 5–3. Risk as a function of the parameter and decision rule. For small values of θ below θ_2, the risk due to procedure δ_1 exceeds that for procedure δ_2; hence, δ_2 is preferred in this region. The converse applies for θ exceeding θ_2. The maximum risk for δ_1 is $R_*(\delta_1)$ which ocurs at $\theta = \theta_1$; the worst for δ_2 is $R_*(\delta_2)$ which arises at $\theta = \theta_3$. The average risk for δ_1 over all values of θ is $r(\delta_1)$; and likewise for $r(\delta_2)$ as the mean risk for δ_2.

the second for a given value of θ if its risk is less; in other words,

$$R(\theta, \delta) < R(\theta, \varepsilon)$$

But there is a catch here lurking within the phrase "for a given value of θ." Since the risk depends on the parameter, one decision procedure could be favored over another for one value of θ, and the preference reversed for a different value. In fact, we have seen this reversal illustrated in connection with the last example. Figure 5–3 depicts two risk functions as they vary with the value of the parameter. The risk $R(\theta, \delta_1)$ due to decision procedure δ_1 increases with θ, reaching a maximum at θ_1 and decreasing thereafter. The risk $R(\theta, \delta_2)$ resulting from procedure δ_2 increases with θ until it peaks at θ_3 and declines thereafter. Both risks have the same magnitude at θ_3.

For small values of θ below θ_2, the risk for procedure δ_2 is smaller; hence, δ_2 is preferred to δ_1. The converse applies for large values of θ in excess of θ_2; and the procedures are equally favorable at the crossover point θ_2.

If the value of the parameter θ is known, it is a simple matter to select one procedure over the other. However, life is not always so simple. As discussed earlier, the parameter may itself be viewed as a random variable.

Since the value of θ is not known in advance, how is one decision procedure to be selected over another? The two most popular approaches are based on the maximum and average values of the risk.

Minimax Criterion

One approach to selecting a decision procedure is to favor the one which limits the maximum penalty. In Figure 5–3, the maximum risk for procedure δ is $R_*(\theta, \delta_1)$, which occurs at $\theta = \delta_1$. The worst risk for procedure δ_2 is $R_*(\theta, \delta_2)$, which obtains at $\theta = \theta_3$. Since the worst risk for the second procedure is smaller than that for the first, procedure δ_2 is favored.

More generally, let Δ be a collection of decision functions: $\Delta = \{d_1, d_2, \ldots, d_n\}$. Among these functions δ_i, the minimax procedure δ^* is the one for which the worst risk is smallest. In other words, the minimax procedure minimizes the maximum risk. We encapsulate this approach in the form of a definition.

• Let Δ be a class of decision functions. The procedure δ^* in Δ is called a *minimax* procedure if it satisfies the relation

$$\max_{\theta} R(\theta, \delta^*) = \min_{\delta \in \Delta} \max_{\theta} R(\theta, \delta)$$

In the context of Figure 5–3, the class Δ of candidate decision functions contains only two members; that is, $\Delta = \{d_1, d_2\}$. Within this class, procedure $\delta^* = \delta_2$ is the favored decision function by the minimax criterion.

We illustrate the notion of a minimax strategy with another example.

Example

Let X be a random variable from a Poisson mass function with parameter θ. The parameter is to be estimated from a single observation X. Consider the loss function $L(\theta, \delta) \equiv (\delta - \theta)^2 / \theta$ and a decision function which is linear in the observation. In other words, $\delta(X) = cX$ for some positive constant c. What value of the scaling constant c will minimize the risk?

The Poisson distribution has the form

$$f(x|\theta) = \frac{e^{-\theta} \theta^x}{x!}$$

where $x = 0, 1, 2, \ldots$. We note from Table 3–1 that the mean and variance of the Poisson distribution are both equal to θ.

The risk function is given by

$$R(\theta, \delta) = EL(\theta, \delta(X)) = E\left(\frac{(cX - \theta)^2}{\theta}\right)$$

The second equality makes use of the explicit form of the decision procedure, namely, $\delta(X) = cX$. The constant c as well as the parameter θ are regarded as fixed. We can, therefore, extract them from the expectation operator:

$$R(\theta, \delta) = \frac{c^2}{\theta} E\left[\left(X - \frac{\theta}{c}\right)^2\right]$$

$$= \frac{c^2}{\theta} E\left[(X - \theta) + \left(\theta - \frac{\theta}{c}\right)\right]^2$$

$$= \frac{c^2}{\theta}\left[E(X - \theta)^2 + 2\left(\theta - \frac{\theta}{c}\right)E(X - \theta) + \left(\theta - \frac{\theta}{c}\right)^2\right]$$

$$= \frac{c^2}{\theta}\left[\theta + 0 + \left(\theta - \frac{\theta}{c}\right)^2\right] = c^2 + \theta(c - 1)^2$$

The second relation follows from adding and subtracting θ. The third equality springs from completing the square, then applying the expectation operator. The fourth transformation depends on noting that $E(X - \theta)^2$ is the variance of X, which equals θ, and that the expectation $E(X)$ also equals θ.

Which value of c will minimize the risk? If $c = 1$, the risk is fixed at $R = 1$. For any other value of c, the risk is a linear function of the parameter θ; as θ increases, the risk rises without bound. Since $c = 1$ is the only value that constrains the risk for all values of θ, the minimax criterion call for a decision function $\delta(X) = X$. In other words, $\delta(X) = X$ is the minimax procedure within the class of decision function given by $\Delta = \{cX : c > 0\}$.   ◆

The minimax criterion is a conservative approach that prepares for the worst possible outcome. But this might not always be the best strategy.

Referring to Figure 5–3, suppose that the risk for procedure δ_1 is more accentuated than that shown. In particular, assume that the risk for δ_1 is 0 everywhere except at $\theta = \theta_1$, for which the risk is a large value $R_*(\delta_1)$. Then decision procedure δ_2, having a smaller worst risk, will still be favored.

However, if θ is a continuous parameter, the probability that it will take on any particular value, including $\theta = \theta_1$, is 0. Consequently, the chance of observing risk $R_*(\delta_1)$ is nil. Even so, the minimax criterion would preclude the use of procedure δ_1.

In many applications, the decision maker's attitude would be more optimistic than that reflected in the minimax approach. A plausible strategy along these lines is to consider the average risk.

Bayes' Criterion

The parameter θ can be regarded as a random variable from a parameter space Θ with a probability density function $\pi(\theta)$. (As usual, if θ is a discrete rather than continuous variable, then the term "mass function" would be more appropriate than "density function.")

When the parameter θ is a random variable, so is the risk $R(\theta, \delta_i)$. As usual, we can obtain a summarizing quantity for the risk in terms of its average value. The mean value of the risk is called the *Bayes' risk*.

- The Bayes' risk for a procedure δ is its mean risk given by

$$r(\delta) = ER(\theta, \delta)$$

Here the expectation operator applies with respect to the parameter θ. More specifically, let θ be a continuous random variable with a density function $\pi(\theta)$. Then the Bayes' risk for procedure δ is

$$r(\delta) = \int r(\theta, \delta)\pi(\theta)\, d\theta$$

If θ is a discrete rather than continuous variable, then the integral is replaced by a corresponding sum.

The parameter θ has been "averaged out" from the risk $R(\theta, \delta)$ to obtain the Bayes' risk $r(\delta)$. The quantity $r(\delta)$ indicates the average penalty incurred by use of the procedure δ.

A procedure with lower average risk than another ought to be preferable. The use of the average risk as a way to select among decision functions is called the Bayes' criterion. The procedure thus selected is the *Bayes' procedure*.

- Let Δ be a class of decision procedures. A procedure δ^* from Δ is called a Bayes' procedure if it satisfies the relation

$$r(\delta^*) = \min_{\delta \in \Delta} r(\delta)$$

In other words, suppose procedure δ^* is a member of a collection of decision function $\Delta = \{d_1, d_2, \ldots, d_n\}$. If the mean risk for δ^* is minimal among all the procedures δ_i, then it is known as the Bayes' rule. According to the Bayes' criterion, δ^* should be selected over the other procedures in Δ.

To illustrate, let us return to Figure 5–3. Suppose the average risk $r(\delta_1)$ for procedure δ_1 is smaller than the corresponding value $r(\delta_2)$ for δ_2. Then the Bayes' procedure is $\delta^* = \delta_1$, from the class of decision functions $\Delta = \{\delta_1, \delta_2\}$. By the Bayes' criterion, δ_1 is the preferred procedure. This contrasts with the minimax approach, for which procedure δ_2 was favored over δ_1.

Example

Consider the previous example of estimating the mean of a Poisson distribution. If the prior distribution for θ is a uniform function on the interval $[2, 6]$, what is the Bayes' procedure and its risk?

The prior density is $\pi(\theta) = 1/4$ on the interval $[2, 6]$. We recall from the previous example that the risk for procedure $\delta(X) = cX$ is equal to $R(\theta, \delta) = c^2 + \theta(c - 1)^2$. The Bayes' risk is, therefore,

$$r(\delta) = \int_2^6 [c^2 + \theta(c - 1)^2]\tfrac{1}{4}\,d\theta$$

$$= 5c^2 - 8c + 4$$

By differentiating with respect to c, we infer that the minimum risk occurs for $c = 4/5$. For this value of c, the risk $r(\delta)$ also happens to be $4/5$.

The Bayes' procedure $\delta(X) = 0.8X$ is optimal among the class of decision functions $\Delta = \{cX: c > 0\}$. The scaling constant $c = 0.8$ for the Bayes' procedure is smaller than the corresponding value of 1 for the minimax approach. This results from the fact that the prior density gives weight only to small values, namely, those in the interval $[2, 6]$. The minimax criterion, in contrast, ignores such information.   ◆

5.3   CLOSURE

The basic problem of statistics is that of inference, relating to the tasks of estimation, localization, and testing. Statistical decision theory, however, extends much beyond the confines of these basic problems. Statistical methods deal also with the ways in which experiments should be conducted, analytic models constructed, information utilized, and actions selected.

We will explore a number of these topics in the subsequent chapters.

PROBLEMS

The problems below are numbered according to the sections to which they correspond. For instance, Problem 5.1A is the first ("A") exercise pertaining to Section 5.1.

5.1A   In Frank Einstein's laboratory is a digital scale. The scale is accurate to ± 1 gram. When a specimen weighs x grams, the scale will read x grams with probability 1/2, and indicate $(x + 1)$ or $(x + 2)$ with even odds. The samples for the current experiment are uniformly distributed between 50 and 60 grams. Suppose Frank takes a single reading of the weight of each sample. Define the following quantities: (a) the outcome space Ω; (b) the action space A; (c) the decision procedure $\delta(X)$; (d) the parameter θ; (e) the prior density $\pi(\theta)$.

5.2A A machine is known to produce defective parts with probability $\theta = .1$ or $\theta = .2$. The production supervisor decides to assess the nature of the machine based on three trials. Let X be the number of defective items. Suppose the loss is proportional to the error: $L(\theta, \delta) = |\delta - \theta|$. Consider the following decision rules:

X	0	1	2	3
δ_1	.1	.1	.1	.1
δ_2	.2	.2	.2	.2
δ_3	.1	.2	.2	.2

(a) Tabulate the loss $L(\theta_i, \delta_j)$ for each value of θ_i, procedure δ_j, and observation x_k. (b) Calculate the risk $R(\theta_i, \delta_j)$ for each procedure δ_j. (c) What is the minimax procedure? (d) Suppose the prior distribution for θ is given by $\pi(.1) = .6$ and $\pi(.2) = .4$. What is the Bayes' procedure?

5.2B Consider the normal density with mean θ and variance 1: $f(x) = \exp[-(x - \theta)^2/2]/\sqrt{2\pi}$. The mean is to be estimated as $\delta(X) = cX$ based on a single observation X. Assume the scaling factor c is a constant, and the loss function is $L(\theta, \delta) = (\delta - \theta)^2$. (a) Calculate the risk $R(\theta, \delta)$. (b) What choice of c is dictated by the minimax criterion? (c) Suppose the prior density for θ is exponential with parameter 1: $\pi(\theta) = e^{-\theta}$ for $\theta \geq 0$. What is the Bayes' procedure?

FURTHER READING

Hoel, Paul G., Sidney C. Port, and Charles J. Stone. 1971. *Introduction to Statistical Theory*. New York: Houghton Mifflin.

Kendall, M.G., and A. Stuart. 1967. *The Advanced Theory of Statistics*, Vol. II, 2nd ed. New York: Hafner.

Savage, Leonard J. 1972. *The Foundations of Statistics*, 2nd ed. New York: Dover.

Wilks, Samuel S. 1962. *Mathematical Statistics*. New York: Wiley.

6

Estimation and Localization

Many problems in decision making involve, in whole or in part, the estimation of an unknown parameter or its localization within a narrow range. In this chapter, we will assume that a random variable X possesses a density function $f(x|\theta)$. For the sake of simplicity, we will use the term "density function" in referring to both continuous and discrete random variables, although in the discrete case the term "mass function" would be more descriptive.

The quantity θ in the density function $f(x|\theta)$ indicates that f belongs to a family of density functions, and its particular identity is given by the value of the parameter. As an illustration, the exponential family is given by $\Delta = \{\theta e^{-\theta x}: \theta > 0 \text{ \& } x \geq 0\}$. Selecting a value for the parameter, such as $\theta = 3$, yields the exponential function $f(x|3) = 3e^{-3x}$ for $x \geq 0$.

The parameter θ characterizes some process such as the rate of decay of a radioactive sample. The parameter is not given directly to a decision maker, but rather must be inferred by observing the process under study.

The procedure in estimation, and in its sibling task of localization, is to observe a number of trials of the random variable X and determine the location of the parameter. In other words, a sample $X_1, \ldots, X_n$ is taken of the random variable and the value of θ is estimated through some decision function $\delta(X_1, \ldots, X_n)$. The key step is to choose an effective decision procedure δ to transform the data into a good estimate of θ.

The decision function $\delta(X_1, \ldots, X_n)$ is also called an *estimator* of the parameter θ. A specific value of δ resulting from an instantiation of each argument is called an *estimate* of the parameter. In other words, action $a = \delta(X_1, \ldots, X_n)$ is also known as an estimate of θ.

In estimation problems, a popular loss function is the squared error given by $L(\theta, \delta) = (\delta - \theta)^2$. The risk function, defined by the average loss, is one way to evaluate the performance of a decision procedure:

$$R(\theta, \delta) \equiv EL(\theta, \delta) = E[\delta(X_1, \ldots, X_n) - \theta]^2$$

In other words, the mean squared error resulting from a decision procedure δ is a measure of its performance—or more accurately, of its incompetence in estimating θ. A procedure with high risk is to be rejected in comparison to one with low risk.

As explained in the previous chapter, the risk $R(\theta, \delta)$ for a decision procedure δ depends on the value of θ. For most problems, no single procedure possesses minimal risk for all values of the parameter.

Consequently, a further criterion is needed to select one procedure over another. One such criterion is the minimax principle, which favors the procedure whose worst risk is the smallest. As we have seen in the last chapter, the minimax criterion is a conservative approach. In fact, the approach springs from the theory of games, where two opponents are actively attempting to undermine the performance of the other. In this case, it becomes imperative to guard against the worst possible scenario since it may well arise. For the task of estimation or localization, on the other hand, the two "players" are the decider and nature. Since nature takes a neutral stance, seeking neither to help nor hinder the decider, it may not be necessary to give so much emphasis to the worst scenario.

In many situations, a more appropriate criterion is the average risk over all possible values of the parameter θ. The Bayes' criterion favors the procedure δ with minimal mean risk. In other words, The Bayes' procedure δ is one which yields the least loss on average.

Both the minimax and Bayes' criteria characterize a decision rule in terms of risk—that is, in terms of the expected loss due to selecting one action over another. It is reasonable, however, to consider other criteria for selecting decision rules. Among these are the accuracy and precision of a decision procedure.

The next section presents alternative criteria for evaluation procedures. This is followed by a decision procedure known as the maximum likelihood estimator and its evaluation in terms of desirable properties of procedures. The last section presents the topic of localization, which is closely linked to the problem of estimation.

6.1   PROPERTIES OF ESTIMATORS

One desirable property of a decision procedure is its accuracy in identifying the target parameter. We prefer a procedure that tends to estimate accurately to one that does not. When the mean value of an estimator for parameter θ exactly matches the value of θ, it is said to be unbiased.

A second criterion is the precision of an estimator. We would like a procedure to cluster tightly around its estimated value for parameter θ rather than scatter loosely around this value. This property is also called *efficiency*.

A third desirable characteristic is an increase in precision with expanding sample size. The precision of an estimator should increase as the number of observations rises. This property is also called *asymptotic efficiency*.

Unbiasedness

Given the probabilistic nature of statistical problems, no procedure will yield perfect results. In the context of estimation, we cannot expect any procedure to perform without error on every occasion. It is reasonable, however, to expect accurate performance *on average*. In other words, we can stipulate that the average value of a procedure match its target value.

The discrepancy between the mean value of an estimator δ and the underlying parameter θ is called the *bias*. In other words, the bias is given by $(E\delta - \theta)$, where E is the expectation operator. If the mean value of procedure δ is θ, then the bias is zero, and δ is called an *unbiased* estimator.

- Let $\delta(X)$ be an estimator of a parameter θ. Then δ is said to be *unbiased* if its mean equals the parameter for all values of θ: $E\delta(X) = \theta$.

As usual, the notation X in the definition may refer to a simple variable or to the vector quantity $X = (X_1, \ldots, X_n)$.

As an illustration, consider the sample average

$$\delta(X_1, \ldots, X_n) \equiv \overline{X} = \frac{\sum_{i=1}^{n} X_i}{n}$$

as an estimator for the mean of a normal density. In particular, let X_i be the ith observation of a random variable X from a normal density with mean θ. The mean value of procedure δ is given by

$$E\delta(X_1, \ldots, X_n) = E(\overline{X}) = E\left(\frac{\sum_{i=1}^{n} X_i}{n}\right)$$

$$= \frac{1}{n} E\left(\sum_{i=1}^{n} X_i\right) = \frac{1}{n} \sum_{i=1}^{n} E(X_i)$$

$$= \frac{1}{n} n\theta = \theta$$

The third and fourth equalities spring from the linearity property of the expectation operator: constants can be extracted, and the expectation of a sum is the sum of expectations. The fifth equality depends on the fact that the mean of each X_i is θ. Since $E(\overline{X}) = \theta$, we have shown that the sample average $\overline{X}$ is an unbiased estimator of the parameter θ.

When the loss function takes the form of the squared error, the risk for an unbiased procedure reduces to a familiar expression. This is the variance of the random variable. We can show this as

$$R(\theta, \delta) = EL(\theta, \delta) = E[\delta(X) - \theta]^2 = \text{Var}(X)$$

The second equality relies on the choice of the squared error for the loss function L. The last conversion depends on the unbiased property of the decision procedure, namely, $E\delta(X) = \theta$.

To illustrate, let $\delta(X) \equiv X$ be the estimator for the mean of a normal density based on a single observation. Suppose the density function has mean θ and variance σ^2. The procedure is unbiased since $E\delta(X) = E(X) = \theta$. The risk for a squared error loss is given by

$$R(\theta, \delta) = E[\delta(X) - \theta]^2 = E(X - \theta)^2 = \text{Var}(X) = \sigma^2$$

In other words, the risk for decision function $\delta(X) \equiv X$ is the variance of the underlying variable.

A similar result obtains for the case of the sample average $\overline{X}$. The risk is given by

$$R(\theta, \delta) = E(\overline{X} - \theta)^2 = \text{Var}(\overline{X}) = \text{Var}\left(\frac{\sum_{i=1}^{n} X_i}{n}\right)$$

$$= \frac{1}{n^2} \text{Var}\left(\sum_{i=1}^{n} X_i\right) = \frac{1}{n^2} n \,\text{Var}(X_i) = \frac{\sigma^2}{n}$$

The second equality depends on the definition of the variance and our earlier result that $E(\overline{X}) = \theta$. The fourth results from the squaring property of the variance: $\text{Var}(cY) = c^2 \,\text{Var}(Y)$ for any constant c and variable Y. The next transformation relies on the independence of the variables X_i. As before, we see that the risk for the decision function X is given by its variance.

In summary, the risk for an unbiased procedure under quadratic loss is simply its variance.

Efficiency

A decision procedure $\delta(X)$ is a function of the random variable X. Consequently, the procedure δ is itself a random variable. When the procedure is unbiased, it is accurate on average. However, the procedure may still be of little use if it is imprecise, tending to scatter widely about its target value. In other words, a good procedure should take values that cluster tightly around its mean value.

Since the procedure δ is a random variable, high precision is tantamount to low variance. More specifically, a procedure δ is more precise than another procedure δ' if its variance is less: $\text{Var}(\delta) < \text{Var}(\delta')$.

Consider a set of procedures $\Delta = \{\delta_i\}$ for estimating a parameter θ. Suppose that a lower bound $\mathcal{M}$ can be found for the variance of all the procedures within the set Δ. In algebraic terms, there is some $\mathcal{M}$ such that $\mathcal{M} \le \text{Var}(\delta_i)$ for all i.

Since $\mathcal{M}$ is no greater than $\text{Var}(\delta_i)$, the ratio

$$\eta(\delta_i) \equiv \frac{\mathcal{M}}{\text{Var}(\delta_i)}$$

is a measure of the relative precision of procedure δ_i. The quantity $\eta(\delta_i)$ is called the *efficiency* of procedure δ_i. When δ_i is loosely scattered, $\text{Var}(\delta_i)$ is large and the efficiency $\eta(\delta_i)$ is low. In the best possible case, δ_i has maximum precision; then $\text{Var}(\delta_i) = \mathcal{M}$ and the efficiency η equals 1.

In this section, we will derive a formula for the lower bound $\mathcal{M}$. The quantity $\mathcal{M}$ actually depends on the value of the parameter θ; that is, $\mathcal{M} = \mathcal{M}(\theta)$. These concepts are also illustrated through a number of examples.

Minimal Dispersion

We consider the situation where X is a continuous random variable having the density function $f(x|\theta)$. The argument for the discrete case is similar.

The density function is a member of the family parameterized by θ, namely, $\{f(x|\theta): \theta \in \Theta\}$. The parameter space Θ is considered to be some subset of the real line. We require two assumptions concerning the regularity of the family of density functions:

A. The state space is independent of the parameter; that is, $\Omega = \{x: f(x|\theta) > 0\}$ is unaffected by the value of θ. Further, for all $x \in \Omega$ and $\theta \in \Theta$, the derivative

$$\frac{\partial}{\partial \theta} \ln f(x|\theta)$$

exists and is finite.

B. For a decision procedure δ whose expected absolute value is finite, the operations of integration and differentiation with respect to θ can be exchanged in taking the expected value. More specifically, suppose $E_\theta(|\delta|) < \infty$ for all θ in Θ. Then

$$\frac{\partial}{\partial\theta}\int_{-\infty}^{\infty}\delta(x)f(x|\theta)\,dx = \int_{-\infty}^{\infty}\delta(x)\frac{\partial}{\partial\theta}f(x|\theta)\,dx$$

When the random variable is a vector quantity $X = (X_1,\ldots,X_n)$, then the integral is a multiple operation taken over each dimension X_i.

We note that $\delta(x) = 1$ is a valid decision procedure for which the expectation is finite: $E_\theta(|1|) = 1$. As a result, condition B also applies when the density function by itself is integrated. In other words,

$$\frac{\partial}{\partial\theta}\int_{-\infty}^{\infty}\delta(x|\theta)\,d\theta = \int_{-\infty}^{\infty}\frac{\partial}{\partial\theta}f(x|\theta)\,d\theta$$

Regularity conditions A and B are satisfied by many familiar distributions, including the exponential and normal density functions.

When assumption A holds, we can define a characteristic quantity for a family $\{f(x|\theta)\}$ of density functions. This is the *Fisher information number* defined as

$$I(\theta) = E\left(\frac{\partial}{\partial\theta}\ln f(x|\theta)\right)^2$$

$$= \int_{-\infty}^{\infty}\left(\frac{\partial}{\partial\theta}\ln f(x|\theta)\right)^2 f(x|\theta)\,dx$$

$$= \int_{-\infty}^{\infty}\frac{1}{f(x|\theta)}\left(\frac{\partial}{\partial\theta}f(x|\theta)\right)^2 dx$$

The last equality depends on the differentiation property of logarithms, for which $(\partial/\partial u)\ln f(u) = (1/u)(\partial f/\partial u)$. Since each factor under the integral is non-negative, the Fisher information number is also non-negative.

As an illustration, suppose $X = (X_1,\ldots,X_n)$ is a vector of observations from an exponential density. The density of X is given by

$$f(X_1,\ldots,X_n|\theta) = \prod_{i=1}^{n}\theta e^{-\theta x_1} = \theta^n e^{-\theta\sum_{i=1}^{n}X_i}$$

The information number is

$$I(\theta) = E\left(\frac{\partial}{\partial \theta}(n \ln \theta - \theta \sum_{i=1}^{n} X_i)\right)^2$$

$$= E\left(\frac{n}{\theta} - \sum_{i=1}^{n} X_i\right)^2 = \text{Var}\left(\sum_{i=1}^{n} X_i\right)$$

$$= n \, \text{Var}(X_i) = \frac{n}{\theta^2}$$

The third equality depends on the fact that the expectation of $\sum_i X_i$ is n/θ. The information number n/θ is a general characteristic of the exponential family of densities.

From the regularity assumptions A and B, we can show that the variance of any decision procedure has a lower bound. This bound is given by the ratio of $(\partial/\partial\theta)E_\theta(\delta)$ and the information number $I(\theta)$.

More specifically, let Δ be a collection of procedures for estimating the parameter θ from a density function $f(x|\theta)$. A lower bound exists on the precision of any procedure in the set. The *minimal dispersion* of any procedure δ in Δ is given by

$$\mathcal{M}(\theta) \equiv \frac{[\partial E_\theta(\delta)/\partial\theta]^2}{I(\theta)}$$

where $I(\theta)$ is the Fisher information number. We state this result in the form of a theorem and its proof.

Theorem (Minimal Dispersion). Let $\delta(X)$ be a decision procedure with a finite variance, namely, $\text{Var}_\theta[\delta(X)] < \infty$. Suppose the Fisher information number for the underlying parameter is a finite positive number: $0 < I(\theta) < \infty$. Under regularity conditions A and B, the quantity $E_\theta[\delta(X)]$ is differentiable for all θ. The quantity

$$\mathcal{M}(\theta) \equiv \frac{[\partial E_\theta(\delta)/\partial\theta]^2}{I(\theta)}$$

is a lower bound for the dispersion of procedure δ. In other words, $\text{Var}_\theta[\delta(X)] \geq \mathcal{M}(\theta)$.

Proof. The following relations hold for any value of θ:

$$\int_{-\infty}^{\infty} f(x|\theta)\,\delta x = 1$$

$$\int_{-\infty}^{\infty} \delta(x)f(x|\theta)\,dx = E_\theta \delta(x)$$

Under regularity conditions A and B, these relations can be transformed into

$$\int_{-\infty}^{\infty} \frac{\partial}{\partial\theta} f(x|\theta)\,dx = \frac{\partial}{\partial\theta}\int_{-\infty}^{\infty} f(x|\theta)\,dx = 0$$

$$\int_{-\infty}^{\infty} \delta(x)\frac{\partial}{\partial\theta} f(x|\theta)\,dx = \frac{\partial}{\partial\theta}\int_{-\infty}^{\infty} \delta(x)f(x|\theta)\,dx = \frac{\partial}{\partial\theta} E_\theta \delta(x)$$

At this point, we multiply and divide by $f(x|\theta)$ within the integrals. Using the identity

$$\frac{d}{dx}\ln g(x) = \frac{1}{g(x)}\frac{d}{dx} g(x)$$

the last two relations can be converted into the following:

$$E\left(\frac{\partial}{\partial\theta}\ln f(x|\theta)\right) = \int_{-\infty}^{\infty} \left(\frac{\partial}{\partial\theta}\ln f(x|\theta)\right) f(x|\theta)\,dx = 0$$

$$E\left(\delta(x)\frac{\partial}{\partial\theta}\ln f(x|\theta)\right) = \int_{-\infty}^{\infty} \delta(x)\left(\frac{\partial}{\partial\theta}\ln f(x|\theta)\right) f(x|\theta)\,dx = \frac{\partial}{\partial\theta} E_\theta \delta(x)$$

To simplify matters, we make the substitutions $U(x) \equiv \partial \ln f(x|\theta)/\partial\theta$ and $V(x) \equiv \delta(x)$. The last two results then transform into

$$E[U(x)] = 0 \tag{6-1}$$

$$E[U(x)V(x)] = \frac{\partial}{\partial\theta} E_\theta V(x) \tag{6-2}$$

The variance of U is given by

$$\mathrm{Var}(U) = E[U(x)]^2 - [EU(x)]^2$$

$$= E\left(\frac{\partial}{\partial\theta}\ln f(x|\theta)\right)^2 - 0 = I(\theta) \tag{6-3}$$

The second equality relies on the description of U and on Eq. (6–1). The last equality depends on the definition of the Fisher information number.

The covariance between any two variables U and V is given by

$$\text{Cov}(U, V) = E(UV) - E(U)E(V)$$

$$= \frac{\partial}{\partial \theta} E_\theta V(X) - 0 = \frac{\partial}{\partial \theta} E_\theta \delta(X) \tag{6–4}$$

The previous line depends on Eqs. (6–1) and (6–2).

Next we note that the correlation between any two variables U and V is given by

$$\rho \equiv \frac{\text{Cov}(U, V)}{[\text{Var}(U)\,\text{Var}(V)]^{1/2}} = \frac{\partial E_\theta \delta(X)/\partial \theta}{\{I(\theta)\,\text{Var}[\delta(X)]\}^{1/2}}$$

The second equality relies on Eqs. (6–3) and (6–4) as well as the definition of V.

Since any correlation coefficient has an upper bound of unity, we obtain

$$\text{Var}[\delta(X)] \geq \frac{[\partial E_\theta \delta(X)/\partial \theta]^2}{I(\theta)}$$

Since the lower bound on the right-hand side is the minimal dispersion $\mathcal{M}(\theta)$, we have validated the theorem.   ∎

The minimal dispersion may be written in the following from:

$$\mathcal{M}(\theta) = \frac{[\partial E_\theta(\delta)/\partial \theta]^2}{E[\partial \ln f(x|\theta)/\partial \theta]^2} = \frac{[\partial E_\theta(\delta)/\partial \theta]^2}{E[(\partial f/f)/\partial \theta]^2}$$

The second equality results from taking the derivative of the logarithm in the denominator. The quantity $(\partial f/f)/\partial \theta$ is approximately equal to $(\Delta f/f)/\Delta \theta$ for small changes in f and θ. This latter expression is the fractional change in the magnitude of the density f as the parameter θ is varied.

In the numerator, the expression $\partial E_\theta(\delta)/\partial \theta$ refers to the change in the mean value of procedure δ as the parameter θ varies. Consequently, the minimal dispersion is given by the square of the mean change in δ induced by θ, normalized by the mean of the squared fractional change in f as θ varies.

The lower bound $\mathcal{M}(\theta)$ depends on $\delta(X)$ through the parametric derivative of the latter's mean:

$$\frac{[\partial E_\theta \delta(X)/\partial \theta]^2}{I(\theta)}$$

When the procedure $\delta(X)$ is unbiased, its expectation is $E_\theta \delta(X) = \theta$. The resulting derivative with respect to θ is 1, and the lower bound is simply the reciprocal of the information number, namely, $1/I(\theta)$.

We state this result as a corollary.

Corollary. Suppose procedure $\delta(X)$ is an unbiased estimator of parameter θ. Under the assumptions of the Minimal Dispersion Theorem, the precision of $\delta(X)$ is bounded below by the reciprocal of the Fisher information number:

$$\text{Var}_\theta[\delta(X)] > \frac{1}{I(\theta)}$$

The best precision $1/I(\theta)$ achievable by an unbiased estimator is called the Cramer–Rao lower bound.

When multiple observations are involved, we may state the information number for the set in terms of that for a simple observation. More specifically, we obtain

$$I(\theta) = nI_1(\theta)$$

where n is the number of observations and $I_1(\theta)$ is the information number for a single trial. We can show this as follows.

Let $X = (X_1, \ldots, X_n)$ be a vector representing n independent trials of a random variable from a density function $f(x|\theta)$. The information number for X is

$$I(\theta) = \text{Var}\left(\frac{\partial}{\partial\theta}\ln f(X|\theta)\right) = \text{Var}\left(\frac{\partial}{\partial\theta}\ln \prod_{i=1}^{n} f(X_i|\theta)\right)$$

$$= \text{Var}\left(\sum_{i=1}^{n}\frac{\partial}{\partial\theta}\ln f(X_i|\theta)\right)$$

$$= \sum_{i=1}^{n} \text{Var}\left(\frac{\partial}{\partial\theta}\ln f(X_i|\theta)\right) = nI_1(\theta)$$

The first equation relies on Eq. (6–3). The second and fourth equalities spring from the independence of the variables X_i. We have just verified the following.

Corollary. Let $X = (X_1, \ldots, X_n)$ be a sample from a density function $f(x|\theta)$. Under the assumptions of the Minimal Dispersion Theorem, the relation

$$\text{Var}_\theta[\delta(X)] \geq \frac{[\partial E_\theta \ln f(X|\theta)/\partial\theta]^2}{nI_1(\theta)}$$

holds.

The quantity $I_1(\theta)$ is the information conveyed by one observation. It is intuitively satisfying that the information in a sample size of n is proportionately larger, namely, $nI_1(\theta)$.

Efficiency

The existence of a lower bound on the dispersion of a procedure suggests the use of a relative measure of precision. The ratio of the lower bound to the actual variance of a procedure is called the efficiency.

- Let $\mathcal{M}(\theta)$ denote the minimal dispersion for any procedure δ designed to estimate a parameter θ. The efficiency of δ is given by

$$\eta(\delta) \equiv \frac{\mathcal{M}(\theta)}{\text{Var}(\delta)}$$

The efficiency η ranges between 0 and 1. When the procedure δ is precise, its variance takes the minimal value of $\mathcal{M}(\theta)$, and the efficiency η equals 1. Conversely, when δ is highly scattered, its variance is large and η falls toward 0.

Example

Let $X_1, \ldots, X_n$ be a set of observations of a binary random variable with parameter θ. In other words, X_i equals 1 and 0 with probability θ and $1 - \theta$, respectively. If θ is to be estimated by the sample average, what is the efficiency of this procedure?

The decision procedure is

$$\delta(X_1, \ldots, X_n) = \frac{\sum_{i=1}^{n} X_i}{n}$$

This procedure is unbiased because

$$E(\delta) = E\left(\frac{\sum_{i=1}^{n} X_i}{n}\right) = \frac{1}{n} E\left(\sum_{i=1}^{n} X_i\right)$$

$$= \frac{1}{n} \sum_{i=1}^{n} E(X_i) = \frac{1}{n}(n\theta) = \theta$$

We know that the lower bound on the dispersion of an unbiased estimator is

$\mathcal{M}(\theta) = 1/I(\theta)$. The Fisher information number is

$$I(\theta) = E\left(\frac{\partial}{\partial\theta}\ln f(X_1,\ldots,X_n|\theta)\right)^2$$

$$= E\left(\frac{\partial}{\partial\theta}\ln\prod_{i=1}^{n}\theta^{X_i}(1-\theta)^{1-X_i}\right)^2$$

$$= E\left(\frac{\sum_{i=1}^{n}X_i}{\theta} - \frac{n-\sum_{i=1}^{n}X_i}{1-\theta}\right)^2$$

$$= \frac{E(\sum_{i=1}^{n}X_i - n\theta)^2}{\theta^2(1-\theta)^2}$$

$$= \frac{\text{Var}(\sum_{i=1}^{n}X_i)}{\theta^2(1-\theta)^2} = \frac{\sum_{i=1}^{n}\text{Var}(X_i)}{\theta^2(1-\theta)^2} = \frac{n[\theta(1-\theta)]}{\theta^2(1-\theta)^2} = \frac{n}{\theta(1-\theta)}$$

The second equality depends on the independence of the X_i variables and the fact that each has the mass function $\theta^{X_i}(1-\theta)^{1-X_i}$ for X_i equal to 1 or 0. The fourth equality springs from the linearity property of expectation. The next transformation depends on the fact that the expectation of $\sum_i X_i$ is simply $n\theta$. The sixth equality relies once more on the independence of variables. We deduce from the results that the lower bound is $\mathcal{M}(\theta) = \theta(1-\theta)/n$.

The actual variance of the sample average is

$$\text{Var}(\delta) = \text{Var}\left(\frac{\sum_{i=1}^{n}X_i}{n}\right) = \frac{1}{n^2}\sum_{i=1}^{n}\text{Var}(X_i)$$

$$= \frac{1}{n^2}[n\theta(1-\theta)] = \frac{\theta(1-\theta)}{n}$$

which happens to match the lower bound $\mathcal{M}(\theta)$. Consequently, the efficiency of the procedure is unity: $\eta(\delta) = 1$.   ◆

A decision function δ with the maximal efficiency of 1 corresponds to a procedure with minimal variance. When a procedure has minimal variance for each value of the parameter θ, it is said to be a *uniformly minimum variance* estimator. If the procedure is unbiased as well, it is called a *uniformly minimum variance unbiased* (UMVU) estimator.

Consider the task of estimating the mean of a binary mass function. The constant decision function $\delta_*(X) = 0.5$ has perfect precision and, therefore, possesses the minimal possible variance of 0. On the other hand, δ_* is accurate only when the value of the parameter is $\theta = 0.5$. For all other values of θ, the procedure is biased. In contrast, we have seen that $\delta^*(X) = \sum_i X_i/n$ is unbiased. Further, it has minimal variance among unbiased estimators. Consequently, procedure δ^* is a minimum variance unbiased estimator.

To summarize, efficiency is a measure of the precision of a decision procedure. The precision, in turn, is a special case of the attribute of risk as an index of a procedure's performance.

The risk is tantamount to the variance when a procedure is unbiased and the loss function is quadratic. In this case, the risk for a procedure δ at a specific value of parameter θ is given by

$$R(\theta, \delta) \equiv EL(\theta, \delta) = E(\delta - \theta)^2 = \text{Var}(\delta)$$

That is, the risk for an unbiased procedure under quadratic loss is simply the variance of the procedure.

Asymptotic Properties

The performance of a decision function should improve as the amount of data increases. In particular, the decision procedure should become more accurate and precise as the number of observations rises. The increase in accuracy and precision is referred to as asymptotic unbiasedness and efficiency. A number of methodologies for constructing decision procedures exhibit these desirable properties.

Many decision procedures, however, offer an additional bonus. In particular, the procedure tends toward a normal density as the sample size increases. This phenomenon springs from the Strong Law of Large Numbers presented in Chapter 3.

Asymptotic Unbiasedness

A desirable characteristic of a procedure is that it should approach its target more closely as the available data increases. Suppose $\delta_n = \delta_n(X_1, \ldots, X_n)$ is a procedure for estimating some function $g(\theta)$ of the parameter θ. If the sequence $\{\delta_1, \delta_2, \delta_3, \ldots\}$ approaches $g(\theta)$ with high probability, we say that the procedure δ_n is *consistent* or *asymptotically unbiased*.

- Suppose $\delta_n = \delta_n(X_1, \ldots, X_n)$ is a procedure for estimating some function $g(\theta)$ of a parameter θ. Then procedure δ_n is said to be *consistent* or *asymptotically unbiased* if it converges to $g(\theta)$ in probability. In other words, for any $\varepsilon > 0$:

$$P[|\delta_n - g(\theta)| < \varepsilon] \to 1 \quad \text{as } n \to \infty$$

Convergence in probability implies that δ_n will be arbitrarily close to $g(\theta)$ with unit probability as the sample size n increases.

The concept of consistency is relevant for procedures that are biased but asymptotically unbiased. In other words, the mean value of δ_n does not exactly match $g(\theta)$ for any finite value of n; but the mean does coincide with its target value for indefinitely large sample sizes.

When the function g is the identity function, the procedure δ_n is simply the estimator of the underlying parameter. That is, δ_n estimates the parameter θ when $g(\theta) \equiv \theta$.

An example of a procedure that is biased but asymptotically unbiased is the popular estimator of the sample variance. Suppose X_i is an observation from a normal density with mean μ and variance σ^2, where neither parameter is known beforehand. The procedure $\gamma_n = \sum_i X_i/n$ is generally used to estimate μ, and the function $\delta_n = \sum_i (X_i - \gamma_n)/n$ to estimate the variance σ^2.

The decision function δ_n, however, is biased. Its expected value is not σ^2, but rather $\sigma^2(n - 1)/n$, as we will verify in the next section. On the other hand, δ_n is asymptotically unbiased because its expected value converges to σ^2 as n increases.

The desirable properties of a procedure do not always carry across transformations. Unbiasedness is one such property. More specifically, γ_n may be an unbiased estimator of a parameter θ, but the procedure $\delta_n \equiv g(\gamma_n)$ may be biased in estimating $g(\theta)$.

As an illustration, consider the task of estimating the square of an unknown mean μ from a density function with variance σ^2. We have seen that the sample average given by $\gamma_n \equiv \overline{X} = \sum_i X_i/n$ is an unbiased estimator for the mean μ. Suppose we try to estimate μ^2 with the function $\delta_n \equiv (\gamma_n)^2 = (\sum_i X_i/n)^2$. We can show that δ_n is a biased estimator for μ^2:

$$E(\delta_n) = E\left(\frac{\sum X_i}{n}\right)^2 = \frac{1}{n^2} E[(\sum X_i - n\mu) + n\mu]^2$$

$$= \frac{1}{n^2}[\text{Var}(\sum X_i) + n^2\mu^2] = \frac{1}{n^2}(n\sigma^2 + n^2\mu^2)$$

$$= \frac{\sigma^2}{n} + \mu^2$$

The second equality results from adding and subtracting $n\mu$. The third equality springs from completing the square, using the definition of the variance, and applying the expectation operator.

We see that δ_n has a bias of σ^2/n in estimating μ^2. On the other hand, the bias vanishes as n increases. Consequently, δ_n is biased but asymptotically unbiased.

Asymptotic Efficiency
A procedure should become more precise with increasing sample size. In other words, the procedure should be asymptotically efficient.

In parallel with the previous section, we may define asymptotic efficiency of a procedure as the ratio of the minimal dispersion $\mathcal{M}(\theta)$ to the asymptotic variance $\mathrm{Var}(\delta_n)$.

• Let δ_n be a procedure for estimating parameter θ from a density function $f(x|\theta)$. The *asymptotic efficiency* of the procedure is given by

$$\eta(\delta_n) \equiv \frac{\mathcal{M}(\theta)}{\mathrm{Var}(\delta_n)}$$

where $\mathcal{M}(\theta)$ is the minimal dispersion of δ_n and $\mathrm{Var}(\delta_n)$ is its variance.

As usual, when δ_n is loosely scattered about its mean value, the variance is high and the efficiency is low. When a procedure attains the minimal level of dispersion, its efficiency is 1, and we say that the procedure is *asymptotically efficient*.

The variance for many procedures decreases directly with the sample size. In other words, the variance is given by $c(\theta)/n$ where c is some function of the underlying parameter. In this case the efficiency may be written as

$$\eta(\delta_n) = \frac{\mathcal{M}(\theta)}{c(\theta)/n} = \frac{[\partial E(\delta_n)/\partial\theta]^2}{c(\theta)I_1(\theta)}$$

The second equality depends on the definition of the minimal dispersion $\mathcal{M}(\theta)$ and of $I_1(\theta)$ as the Fisher information number for a single observation.

Example

Consider the sample average $\overline{X}$ as an estimator for the mean θ of a normal density function with known variance σ^2. What is the asymptotic efficiency of the procedure?

The variance of $\delta_n \equiv \overline{X}$ is given as

$$\mathrm{Var}(\overline{X}) = \mathrm{Var}\left(\frac{\sum_i X_i}{n}\right) = \frac{1}{n^2}\mathrm{Var}(\sum_i X_i) = \frac{1}{n^2}(n\sigma^2) = \frac{\sigma^2}{n}$$

In other words, the dispersion of δ_n is $c(\theta)/n$ where the function $c(\theta) \equiv \sigma^2$ is simply a constant.

The Fisher information contained in a single observation is

$$I_1(\theta) \equiv E\left(\frac{\partial}{\partial\theta}\ln f(x|\theta)\right)^2 = E\left(\frac{\partial}{\partial\theta}\ln\frac{\exp[-(x-\theta)^2/2\sigma^2]}{\sigma\sqrt{2\pi}}\right)^2 = E\left(\frac{X-\theta}{\sigma^2}\right)^2$$

$$= \frac{1}{\sigma^4}\text{Var}(X) = \frac{1}{\sigma^4}\sigma^2 = \frac{1}{\sigma^2}$$

We know that δ_n is unbiased: $E(\overline{X}) = \sum_i(X_i/n) = \theta$. Hence, the asymptotic efficiency is

$$\eta(\delta_n) = \left(\frac{\partial\theta/\partial\theta}{c(\theta)I_1(\theta)}\right) = \frac{1}{\sigma^2(1/\sigma^2)} = 1$$

We conclude that the sample average is an asymptotically efficient estimator for the mean of a normal density function.   ◆

Often it is convenient to compare directly the relative efficiency of two procedures. A procedure is more useful than another if its precision increases at a faster rate. The *relative efficiency* of two procedures is given by the inverse ratio of their variances.

- Consider the task of estimating a function $g(\theta)$ of a parameter θ based on observations $X_1, X_2, \ldots, X_n$. The relative efficiency of the two estimators γ_n and δ_n is defined as follows:

$$\eta_r(\gamma_n, \delta_n) \equiv \frac{\text{Var}(\delta_n)}{\text{Var}(\gamma_n)}$$

For instance, suppose γ_n is relatively imprecise for each sample size n. Then its variance would be large and, therefore, its relative efficiency η_r would be small. In this case, the procedure γ_n would be relatively inefficient.

Example

The upper bound θ is to be estimated for a uniform density on the interval $[0, \theta]$. One procedure is to double the sample average: $\delta_n = 2\sum_i X_i/n$, where $X_1, \ldots, X_n$ is the set of observations. An alternative procedure is to use the largest observation as the estimate of the parameter; in other words, $\gamma_n = \max_i\{X_i\}$. What is the relative efficiency of these procedures?

The procedure δ_n happens to be unbiased:

$$E(\delta_n) = E\left(2\frac{\sum_i X_i}{n}\right) = \frac{2}{n}E(\sum_i X_i) = \frac{2}{n}\left(n\frac{\theta}{2}\right) = \theta$$

The third equality relies on the linearity of expectation operator and the fact that $E(X_i) = \theta/2$. The variance of δ_n is as follows:

$$\text{Var}(\delta_n) = \text{Var}\left(2\frac{\sum_i X_i}{n}\right) = \frac{4}{n^2}\text{Var}(\sum_i X_i) = \frac{4}{n^2}\left(n\frac{\theta^2}{12}\right) = \frac{\theta^2}{3n}$$

The third relation springs from the independence of the X_i and the fact of their individual variances being equal to $\theta^2/12$.

To determine the characteristics for procedure γ_n, we must first derive its density function. Let $X_{(1)}, X_{(2)}, \ldots, X_{(n)}$ denote the *order statistics* of the observations $X_1, X_2, \ldots, X_n$; in other words, $X_{(1)}$ is the smallest value among the set $\{X_i\}$, $X_{(2)}$ is the next smallest value, and so on. The decision procedure is equal to the nth-order statistic, namely, $\gamma_n = X_{(n)} \equiv \max_i\{X_i\}$. The distribution function for γ_n takes the form

$$F_{\gamma_n}(x) = P\{\gamma_n \leq x\} = P\{X_{(n)} \leq x\}$$

$$= P\{X_1 \leq x, \ldots, X_n \leq x\} = \left(\frac{x}{\theta}\right)^n$$

The event "$\gamma_n < X$" corresponds to the event where the largest observation $X_{(n)}$ is bounded by x. This is equivalent to the case where each X_i is less than or equal to x. The corresponding probability is x/θ, and the last equality springs from the independence of the X_i variables. We note that the order statistics $X_{(j)}$ constitute a *dependent* set because the relative rank of an observation depends on the values of the other random variables; but each of the raw observations X_i arises independently of the others.

Differentiating the distribution function with respect to x yields the density function for γ_n. The result is

$$f_{\gamma_n}(x) = \frac{nx^{n-1}}{\theta^n}$$

for x lying in the interval between 0 and θ. We can discern that procedure γ_n is biased:

$$E(\gamma_n) = \int_0^\theta x\left(\frac{n}{\theta^n}x^{n-1}\right)dx = \frac{n}{n+1}\theta$$

On the other hand, the procedure is asymptotically unbiased because $E(\gamma_n) \to \theta$ as $n \to \infty$.

The variance of γ_n can be obtained as

$$
\begin{aligned}
\mathrm{Var}(\gamma_n) &= E(\gamma_n^2) - E^2(\gamma_n) \\
&= \int_0^\theta x^2 \left(\frac{n}{\theta^n} x^{n-1} \right) dx - \left(\frac{n\theta}{n+1} \right)^2 \\
&= \frac{n\theta^2}{n+2} - \left(\frac{n\theta}{n+1} \right)^2 = \frac{n\theta^2}{(n+1)^2(n+2)}
\end{aligned}
$$

Finally, the relative efficiency of δ_n compared to γ_n is

$$
\eta_r(\delta_n, \gamma_n) = \frac{\mathrm{Var}(\gamma_n)}{\mathrm{Var}(\delta_n)} = \frac{3n^2}{(n+1)^2(n+2)}
$$

The relative efficiency is $\eta_r = 1/4$ for $n = 1$ and falls toward 0 as the sample size n increases. We conclude that procedure γ_n is a biased but consistent estimator, and it is asymptotically efficient compared to procedure δ_n.   ◆

Asymptotic Normality

Many procedures for estimation tend toward a normal distribution with increasing sample size. For these procedures, the cumulative distribution function approaches that for a normal density function.

More specifically, let procedure $\delta_n(X_1, \ldots, X_n)$ be an estimator for parameter $g(\theta)$ based on observations $X_1, \ldots, X_n$. Suppose the mean of procedure δ_n is $\mu_n(\theta)$, whereas the standard deviation is $\sigma_n(\theta)$. The cumulative distribution function for δ_n is given by

$$
F_{\delta_n}(x) \equiv P\{\delta_n \leq x\} \approx \Phi\left(\frac{x - \mu_n(\theta)}{\sigma_n(\theta)} \right)
$$

where Φ denotes the distribution function for the standard normal density. The approximation converges to an equality as the sample size increases. In algebraic terms, we can write

$$
\lim_{n \to \infty} P\left(\frac{\delta_n(X_1, \ldots, X_n) - \mu_n(\theta)}{\sigma_n(\theta)} \leq z \right) = \Phi(z)
$$

for all values of z. When this convergence holds, the quantity $\mu_n(\theta)$ is called the *asymptotic mean* of δ_n and $\sigma_n(\theta)$ the *asymptotic standard deviation*. The *asymptotic bias* is defined as $\mu_n(\theta) - g(\theta)$; when this difference is 0, we say that δ_n is *asymptotically unbiased*.

For many plausible procedures, the asymptotic variance is on the order of $1/n$. In other words, $\sigma_n^2(\theta)$ converges to $\sigma^2(\theta)/n$ as n increases; or equivalently, $n\sigma_n^2(\theta)$ converges to $\sigma^2(\theta)$.

A second property of many procedures is that of asymptotic unbiasedness. More precisely, the "standardized" value of the decision function vanishes with increasing sample size:

$$\frac{\mu_n(\theta) - g(\theta)}{\sigma_n(\theta)} = \frac{\mu_n(\theta) - g(\theta)}{\sigma(\theta)/\sqrt{n}} \to 0$$

Since the actual standard deviation $\sigma(\theta)$ is a fixed positive number, we can state equivalently that $\sqrt{n}[\mu_n(\theta) - g(\theta)]$ converges to 0. Consequently, the corresponding decision procedure δ_n is a consistent estimator.

6.2   MAXIMUM LIKELIHOOD ESTIMATORS

A popular method of constructing procedures for estimation is based on the concept of maximum likelihood. The *maximum likelihood* principle calls for an estimate of a parameter which has the greatest chance of yielding the observed sample.

More precisely, let $X_1, \ldots, X_n$ be a set of random variables from a common density function $f(x|\theta)$. Suppose that x_i is the actual outcome for variable X_i. The *likelihood function* $\mathscr{L}(\theta)$ for the parameter θ is the density function defined by the observations x_i:

$$\mathscr{L}(\theta) \equiv f(x_i, \ldots, x_n|\theta)$$

Since the x_i are outcomes for independent observations X_i, the joint density $f(x_1, \ldots, x_n|\theta)$ can be factored into the product of individual densities. The likelihood function then becomes

$$\mathscr{L}(\theta) = \prod_{i=1}^{n} f(x_i|\theta)$$

When the x_i are regarded as fixed values, the likelihood function $\mathscr{L}(\theta)$ depends solely on the parameter θ. The *maximum likelihood estimator* is the value of θ which maximizes the likelihood $\mathscr{L}(\theta)$; this decision procedure is also called the MLE for short, and indicated by the notation $\hat{\theta}$.

Since the likelihood is based on the joint density $f(x_1, \ldots, x_n|\theta)$, the actual value of $\hat{\theta}$ will, of course, depend on the values for $x_1, \ldots, x_n$. Sometimes it will be convenient to underscore this dependence by writing $\hat{\theta} = \hat{\theta}(x_1, \ldots, x_n)$.

Although the particular value of $\hat{\theta}$ will depend on the x_i, our concern focuses on the *procedure* for obtaining $\hat{\theta}$ regardless of the actual outcomes for the X_i. The general dependence of $\hat{\theta}$ on the X_i can be highlighted by writing $\hat{\theta} = \hat{\theta}(X_1, \ldots, X_n)$.

To clarify these ideas, we return to the binomial example given in Chapter 4.

Example

Let $X_1, \ldots, X_n$ be a set of observations from a binomial distribution with parameter θ. The individual density for X_i is given by

$$f(x_i|\theta) = \theta^{x_i}(1 - \theta)^{1-x_i} \qquad x_i = 0, 1 \quad \text{and} \quad 0 < \theta < 1$$

In other words, $f(1|\theta) = \theta$ and $f(0|\theta) = 1 - \theta$.

We can write the likelihood function as

$$\mathscr{L}(\theta) = \prod_{i=1}^{n} f(x_i|\theta) = \prod_{i=1}^{n} \theta^{x_i}(1 - \theta)^{1-x_i}$$

$$= \theta^{\sum_i x_i}(1 - \theta)^{n - \sum_i x_i}$$

Our task is to find the value of θ which maximizes $\mathscr{L}(\theta)$. The likelihood function has value 0 at $\theta = 0$, rises with increasing θ, and eventually vanishes for $\theta = 1$. The peak of $\mathscr{L}(\theta)$, therefore, lies somewhere between 0 and 1. Since the logarithm of $\mathscr{L}(\theta)$ increases monotonically with $\mathscr{L}(\theta)$ itself, it suffices to maximize the quantity $\ln \mathscr{L}(\theta)$. By differentiating with respect to θ, we obtain

$$\frac{\partial \ln \mathscr{L}(\theta)}{\partial \theta} = \frac{\sum_i x_i}{\theta} - \frac{n - \sum_i x_i}{1 - \theta}$$

By equating the right-hand side to 0 and simplifying, we obtain the maximum likelihood estimator $\hat{\theta}$ for parameter θ:

$$\hat{\theta} = \frac{\sum_{i=1}^{n} x_i}{n}$$

Suppose that a sequence of five binomial trials results in three successes. Then the maximum likelihood *estimate* is given by $\hat{\theta} = \sum_i x_i/n = 3/5$. On the other hand, the *estimator* is a procedure valid for all values of $X_1, \ldots, X_n$. We, therefore, refer to the estimator as $\hat{\theta}(X_1, \ldots, X_n) = \sum_i X_i/n$. ◆

Properties of Maximum Likelihood Estimators

Procedures based on the maximum likelihood principle exhibit a number of desirable properties. Among these are efficiency among unbiased estimators, as well as asymptotic efficiency for increasing sample sizes.

Maximum likelihood estimators are often unbiased, but this need not be the case. In the previous example, the estimator for the binomial parameter θ happens to be unbiased. We can verify this as follows:

$$E(\hat{\theta}) = E\left(\frac{\sum_{i=1}^{n} X_i}{n}\right) = \frac{1}{n} \sum_{i=1}^{n} E(X_i) = \frac{1}{n} n(\theta) = \theta$$

In the next section, however, we will see an example of an MLE which is, in fact, biased.

When a maximum likelihood procedure is unbiased, it is also efficient. Consequently, the maximum likelihood approach can be used to generate the most precise procedures among those that are unbiased.

More precisely, suppose that an efficient unbiased estimator $\hat{\theta}$ exists. Then the maximum likelihood method will produce it when certain regularity conditions are fulfilled.

Another important feature of MLEs lies in their asymptotic efficiency. When certain regularity conditions are satisfied, maximum likelihood estimators are asymptotically efficient.

The operational consequence of this result lies in a straightforward approach to obtaining asymptotically efficient procedures. We need not seek efficient procedures by trial and error, but simply pursue the maximum likelihood methodology.

Under regularity assumptions, the maximum likelihood procedure will tend toward a normal density. From Section 6.1, we know that the maximum likelihood estimator $\hat{\theta}$ will converge asymptotically to a normal density with mean θ and a variance given by

$$\text{Var}(\hat{\theta}) = \frac{1}{I(\theta)} = \frac{1}{nI_1(\theta)}$$

Here n is the sample size and $I_1(\theta)$ is the Fisher information number for a single observation. We see that the variance of the estimator $\hat{\theta}$ falls directly with the sample size n.

Example

We saw earlier that the maximum likelihood estimator for the binomial parameter θ is given by the sample average: $\hat{\theta} = \sum_i X_i/n$. The Fisher information for a single observation X_i is

$$I_1(\theta) = E\left(\frac{\partial}{\partial \theta} \ln f(X_i|\theta)\right)^2 = E\left(\frac{\partial}{\partial \theta} \ln[\theta^{X_i}(1-\theta)^{1-X_i}]\right)^2$$

$$= E\left(\frac{X_i - \theta}{\theta(1-\theta)}\right)^2 = \frac{1}{[\theta(1-\theta)]^2} \text{Var}(X_i) = \frac{1}{\theta(1-\theta)}$$

The last equality depends on the variance for X_i being equal to $\theta(1 - \theta)$.
The asymptotic variance for the maximum likelihood estimator is

$$\text{Var}(\hat{\theta}) = \frac{1}{nI_1(\theta)} = \frac{\theta(1-\theta)}{n}$$

This value represents the best possible precision for each value of n.   ◆

To summarize, maximum likelihood estimators possess a number of desirable properties. Maximum likelihood procedures are usually asymptotically normal as well as efficient. They may be biased, but are often asymptotically unbiased.

Further, maximum likelihood estimators exhibit the property of functional independence. Suppose a parameter θ is estimated by the procedure $\hat{\theta}$; then a transformation of $\hat{\theta}$ is estimated by the same transformation of $\hat{\theta}$. In algebraic terms, $\hat{g}(\theta) = g(\hat{\theta})$, where g is some function of θ, and the caret (^) denotes the corresponding maximum likelihood estimator. In other words, the MLE of the transform is the transform of the MLE.

The MLE is not all-wonderful, however. In particular, the maximum likelihood method is not robust: the estimate can be distorted by extreme values in the data sample. As an illustration, consider the sample average as an estimate of the mean: $\hat{\theta} = \sum_i X_i/n$. One extreme value X_i can wreak havoc with the estimated value of θ.

This is in contrast to, say, the median as an estimator of the mean θ. In estimating the average income of individuals in the United States, for example, the presence of millionaires and billionaires can severely distort the estimate when the sample average is used. Although wealthy individuals comprise only several percent of the population, a single billionaire may carry the financial weight of thousands of individuals earning the average wage. On the other hand, the handful of wealthy individuals will have little or no impact on the estimated average income when the sample median is used.

The maximum likelihood approach may also perform poorly against global criteria of risk. More specifically, the minimax and Bayes' criteria may prohibit the use of MLE procedures. We will return to this topic in Chapter 8 on Decision Theory.

6.3 MULTIDIMENSIONAL PARAMETERS

Thus far, our discussion has focused on the estimation of a single parameter. As an example, we developed decision procedures for estimating the parameter for the binomial distribution. Even when dealing with distributions possessing more than one parameter, we addressed only one at a time, while assuming that any other parameters were already known. For instance, we developed procedures for the mean or variance of a normal density function —but not both, since the other was assumed to be known.

Loss Function

For the estimation of a single parameter θ, our staple loss function has taken the form of a quadratic. More specifically, if δ is the procedure for estimating parameter θ, the standard loss function has been $L(\theta) = a(\delta - \theta)^2$, where a is some constant usually taken to equal 1.

A generalization of the quadratic loss function would involve not only the squared error for each decision function, but the cross-products of the errors as well. For the case of two parameters θ_1 and θ_2, suppose δ_1 and δ_2 are their respective estimators. We can denote these more compactly as a *vector*, namely, a list of objects $\theta \equiv (\theta_1, \theta_2)$ for the parameters and $\delta \equiv (\delta_1, \delta_2)$ for the procedures. The general quadratic loss function can be expressed as

$$L(\theta, \delta) = c_{11}(\delta_1 - \theta_1)^2 + c_{12}(\delta_1 - \theta_1)(\delta_2 - \theta_2) + c_{22}(\delta_2 - \theta_2)^2$$

Each c_{ij} is an arbitrary constant that serves as a weight for the relative contribution due to the corresponding term. The cross-term $(\delta_1 - \theta_1) \cdot (\delta_2 - \theta_2)$ accounts for the loss due to interaction effects between the errors in δ_1 and δ_2.

Unfortunately, the task of estimation becomes unwieldy when the interaction effects are taken into consideration. To maintain the estimation problems at manageable levels, the interaction effects are usually ignored. The elimination of the cross-product in the above equation corresponds to setting its multiplicative constant to zero; that is, $c_{12} \equiv 0$.

As a further simplification, the contribution of the loss $(\delta_i - \theta_i)^2$ due to each procedure δ_i is weighted equally in determining the overall loss L. In algebraic terms, this corresponds to setting to unity the multiplicative constants for the squared errors; that is, $c_{ii} \equiv 1$. The loss function for the case of a two-dimensional parameter is then

$$L(\theta, \delta) = (\delta_1 - \theta_1)^2 + (\delta_2 - \theta_2)^2$$

When interaction effects are ignored and the individual loss due to each δ_i is weighted equally, the overall loss is the simple sum of the individual losses. In other words, suppose $\delta \equiv (\delta_1, \ldots, \delta_k)$ is a vector of procedures for estimating the multidimensional parameter $\theta \equiv (\theta_1, \ldots, \theta_k)$. The loss function is

$$L(\theta, \delta) = \sum_{i=1}^{k} (\delta_i - \theta_i)^2$$

The risk is the average value of loss as a function of θ: $R(\theta) = EL(\theta, \delta)$. When the decision procedures are unbiased, we can write $E(\delta_i) = \theta_i$. The risk then becomes the sum of the variances:

$$R(\theta) = E_\theta L(\theta, \delta) = E_\theta \sum_{i=1}^{k} [\delta_i - E(\delta_i)]^2 = \sum_{i=1}^{k} \mathrm{Var}(\delta_i)$$

As in the single-parameter case, the method of maximum likelihood yields procedures with a number of notable properties.

Maximum Likelihood Estimators

The maximum likelihood approach to the generation of decision procedures can be applied to the estimation of multiple parameters as well as solitary ones. The procedures produced by this method exhibit a number of desirable characteristics. For instance, a maximum likelihood procedure is asymptotically efficient among estimators that follow an asymptotically normal density.

The maximum likelihood method for multiple parameters can be specified as follows. Let X denote a random variable from a density function $f(x|\theta)$. The parameter θ is a vector of k components: $\theta \equiv (\theta_1, \ldots, \theta_k)$. The likelihood function for the set of observations $x_1, \ldots, x_n$ is given by

$$\mathscr{L}(\theta) \equiv f(x_1, \ldots, x_k | \theta_1, \ldots, \theta_k) = \prod_{i=1}^{n} f(x_i | \theta_1, \ldots, \theta_k)$$

The second equality depends on the fact that the data x_i are manifestations of independent variables X_i. *The maximum likelihood estimator $\hat{\theta}$ for the multidimensional parameter θ is the list of procedures $\hat{\theta}_i$ that maximize the likelihood function $\mathscr{L}$.* In other words, the vector estimator is $\hat{\theta} = (\hat{\theta}_1, \ldots, \hat{\theta}_k)$, where the procedures $\hat{\theta}_i$ maximize the value of $\mathscr{L}$.

The value of each $\hat{\theta}_i$ will depend on the observations $X_1, \ldots, X_n$. We may underscore this dependence by writing $\theta_i = \theta_i(X_1, \ldots, X_n)$.

Example

Consider the normal density function. Determine the maximum likelihood estimators for the mean and variance, where both parameters are assumed to be unknown.

For notational convenience, let τ denote the variance σ^2. The likelihood function based on observations $x_1, \ldots, x_n$ is

$$\mathscr{L}(\theta) = f(x_1, \ldots, x_n | \theta) = \frac{1}{(2\pi\tau)^{n/2}} \exp\left(-\frac{1}{2\tau} \sum_{i=1}^{n} (x_i - \mu)^2\right)$$

where $\theta \equiv (\mu, \tau)$ is the vector parameter to be estimated.

Rather than maximizing $\mathscr{L}$ directly, we first take logarithms to obtain

$$\ln \mathscr{L}(\theta) = -\frac{n}{2} \ln(2\pi) - \frac{n}{2} \ln \tau - \frac{1}{2\tau} \sum_{i=1}^{n} (x_i - \mu)^2$$

Taking the partial derivatives with respect to μ and τ yields

$$\frac{\partial}{\partial \mu} \ln \mathscr{L} = \frac{1}{\tau} \sum_{i=1}^{n} (x_i - \mu)$$

$$\frac{\partial}{\partial \tau} \ln \mathscr{L} = -\frac{n}{2\tau} + \frac{1}{2\tau^2} \sum_{i=1}^{n} (x_i - \mu)^2$$

Equating these derivatives to zero and solving the equations simultaneously, we have

$$\hat{\mu} = \bar{x}$$

$$\hat{\tau} = \frac{1}{n} \sum_{i=1}^{n} (x_i - \bar{x})^2$$

where $\bar{x} = \sum_i x_i / n$ is the sample average. These are the maximum likelihood estimators for μ and σ^2, respectively.

We can show that these values maximize $\ln \mathscr{L}$ in the following way. The expression $\ln \mathscr{L}$ is maximized when the quantity

$$\mathscr{L}_0 \equiv n \ln \tau + \frac{1}{\tau} \sum_{i=1}^{n} (x_i - \bar{x})^2$$

is minimized. For any value of $\tau > 0$, $\mathscr{L}_0$ is minimized when $\mu = \bar{x}$.

Our remaining task is to show that $\tau = c^2/n$ minimizes $\mathscr{L}_0$, where $c^2 \equiv \sum(x_i - \bar{x})^2$. The first and second partial derivatives of $\mathscr{L}_0$ at $\tau = c^2/n$ are as follows:

$$\frac{\partial}{\partial \tau} \mathscr{L}_0 = \frac{n}{\tau} - \frac{c^2}{\tau^2} = \frac{1}{\tau^2}(n\tau - c^2) = 0$$

$$\frac{\partial^2}{\partial \tau^2} \mathscr{L}_0 = \frac{n}{\tau} - \frac{c^2}{\tau^2} = \frac{1}{\tau^2}(n\tau - c^2) = 0$$

Therefore, the functions $\hat{\mu} = \bar{x}$ and $\hat{\tau} = \sum_i (x_i - \bar{x})^2/n$ minimize $\mathscr{L}_0$, or equivalently maximize the likelihood function $\mathscr{L}$. They are, therefore, maximum likelihood estimators for μ and σ^2, respectively.   ◆

Unbiasedness and Efficiency

A maximum likelihood estimator can be both unbiased and efficient. For instance, the sample average is an unbiased estimate for the mean of a normal density. This is verified easily:

$$E(\hat{\mu}) = E\left(\frac{\sum_{i=1}^{n} X_i}{n}\right) = \frac{1}{n}\sum_{i=1}^{n}(EX_i) = \frac{1}{n}n(\mu) = \mu$$

The sample average is unbiased whether or not the variance σ^2 is known. We saw in Section 6.1 that the sample mean is also efficient. Consequently, the procedure $\hat{\mu} = \bar{x}$ is an unbiased, efficient estimator regardless of knowledge of the variance.

On the other hand, the complementary statement for the variance is false. In particular, the estimator $\hat{\tau} = \sum_i (X_i - \mu)^2/n$ happens to be unbiased when μ is known:

$$E(\hat{\tau}) = E\left(\frac{\sum_{i=1}^{n}(X_i - \mu)^2}{n}\right) = \frac{1}{n}\sum_{i=1}^{n} E(X_i - \mu)^2$$

$$= \frac{1}{n}n \operatorname{Var}(X_i) = \frac{1}{n}n\sigma^2 = \sigma^2$$

In contrast, $\hat{\tau}$ is biased when the mean μ is unknown. In this case, the mean must be estimated by the sample average $\overline{X}$. The expected value of $\hat{t}$ is then

$$E(\hat{t}) = E\left(\frac{1}{n}\sum_i (X_i - \overline{X})^2\right)$$

$$= E\left(\frac{1}{n}\sum_i [(X_i - \mu) - (\overline{X} - \mu)]^2\right)$$

$$= E\left(\frac{1}{n}\sum_i [(X_i - \mu)^2 - 2(X_i - \mu)(\overline{X} - \mu) + (\overline{X} - \mu)^2]\right)$$

$$= E\left(\frac{1}{n}\left[\sum_i (X_i - \mu)^2 - n(\overline{X} - \mu)^2\right]\right) = \frac{1}{n}\sum_i E(X_i - \mu)^2 - E(\overline{X} - \mu)^2$$

$$= \frac{1}{n}\sum_i \tau - \frac{\tau}{n} = \frac{n-1}{n}\tau$$

The second equality follows from adding and subtracting μ. The third relation involves completing the square. The fourth equality results from distributing the summation operator, and the fifth results on applying the expectation operator. The sixth relation springs from the properties of the variance: $\mathrm{Var}(X_i) = \tau$ and $\mathrm{Var}(\overline{X}) = \tau/n$.

Although $\hat{t}$ is biased, we can easily generate an unbiased estimator for τ. The procedure

$$\delta' = \frac{n}{n-1}\hat{t} = \frac{1}{n-1}\sum_i (X_i - \overline{X})^2$$

provides an unbiased estimate[1] for the variance τ.

The square of a standard normal variable follows a χ^2-distribution having one degree of freedom. Moreover, the sum of independent χ^2-variables also follows the χ^2-distribution whose degree of freedom is the sum of the individual degrees of freedom.

Let $X_1, \ldots, X_n$ be a set of observations from a normal density with mean μ and variance σ^2. The quantity $(X_i - \mu)/\sigma$ is a standard normal variable. The

---

[1] The sample standard deviation is the square root of the sample variance. Hence, it may be defined as $[\sum_i (X_i - \overline{X})^2/n]^{1/2}$ or $[\sum_i (X_i - \overline{X})/(n-1)]^{1/2}$ when the mean μ is unknown. Unfortunately, neither of these procedures represents an unbiased estimator of the standard deviation σ. The procedure $\sum_i (X_i - \overline{X})/(n-1)$ is unbiased only when the variance σ^2 is to be estimated.

sum of square given by

$$W = \sum_{i=1}^{n} \left(\frac{X_i - \mu}{\sigma} \right)^2 = \frac{1}{\sigma^2} \sum_{i=1}^{n} (X_i - \mu)^2$$

follows a χ^2-distribution with degree of freedom equal to n.

For a χ^2-variable with n degrees of freedom, the mean has value n, whereas the variance equals $2n$. When the parameter μ is known, we have seen that the procedure $\delta = \sum_i (X_i - \mu)^2/n$ is an unbiased estimator for the variance σ^2. The efficiency of this procedure can be shown to be 100%.

When the mean μ is unknown, the procedure $\delta' = \sum_i (X_i - \overline{X})^2/(n-1)$ is a good estimator for the variance σ^2. We know that this estimator is unbiased. In this case, the parameter μ must be estimated by the relation $\overline{X} = \sum_i X_i/n$. The relation imposes one constraint on the freedom of movement of the X_i variables, leading to the loss of one degree of freedom. Hence, the procedure δ' follows a χ^2-distribution with $n-1$ degrees of freedom. The efficiency of δ' can be shown to equal $(n-1)/n$.

The efficiency calculations for both δ and δ' are left as exercises. Efficiency is a measure of the speed with which an estimator converges to its target. As we might expect, the efficiency suffers somewhat when the data must do double duty by attempting to identity the mean as well as the variance of the underlying distribution.

6.4   LOCALIZATION

The goal of estimation is to identify the value of some unknown quantity. Since the ascertained value is based on a set of random observations, it is possible for the estimate to lie near or far from its target value. The estimated value, by itself, provides no information on the accuracy of the conclusion.

For instance, suppose that the average height of teenagers is to be estimated from a sample of 63 individuals encountered in a shopping mall. If the sample average is 1.637 meters, is this an accurate estimate of the mean height of teenagers in the population at large? Or could it merely be a statistical aberration due to the uncertain nature of random observations?

The goal of localization is to produce a range of estimates rather than a single value and to attribute a level of belief to that range. The degree of belief is called the confidence level α. If, say, α equals 0.90, then we would expect the real parameter to lie in the estimated band in 9 out of 10 instances where a confidence level of α was given. Given the inherently chancy nature of the observations, the conclusion could be incorrect. For $\alpha = 0.90$, we would expect to be wrong roughly once in every 10 cases.

In the teenager example, we might infer that the mean height of teenagers in the population at large lies in the range from 1.53 to 1.73 meters with a confidence level of 95%. In this case, we would be prepared to face a 5% probability of being mistaken.

The basic idea behind localization is to obtain a good estimate for the target parameter, determine the distribution of the estimation procedure, then construct a small interval around the estimated value which spans a large fraction of the distribution. This strategy was explained and demonstrated through a number of examples in Chapter 4.

PROBLEMS

The problems below are numbered according to the sections to which they correspond. For instance, Problem 6.1A is the first ("A") exercise pertaining to Section 6.1.

6.1A   Let $X_1, \ldots, X_n$ be a set of observations from a normal density with mean 0 and variance θ. The variance is to be estimated through the procedure $\delta(X) = \sum_{i=1}^{n} X_i^2/n$. Show that the procedure is unbiased.

6.1B   Let X be a random variable from a uniform density on the interval $[0, \theta]$. The value of θ is to be estimated from the single observation $\delta(X) = cX$. What value of c yields an unbiased estimator for θ?

6.1C   A zoological expedition is to study gorillas in the wild. Suppose p is the proportion of gorillas of a particular type. Let k denote the number of gorillas of the relevant type among the total number X of gorillas sighted. (a) Derive the density function for X:

$$\binom{x-1}{k-1} p^k (1-p)^{x-k}$$

where $k = 0, 1, 2, \ldots$ and $x = k, k+1, \ldots$. (b) A straightforward, intuitive estimator for p is $\delta'(X) = k/X$; show that δ' is biased. (c) A better estimator for p is $\delta(X) = (k-1)/(X-1)$; prove that it is unbiased.

6.1D   Let $X_1, \ldots, X_n$ be a random sample from a Poisson density $f(x|\theta) = e^{-\theta}\theta^x/x!$ where $x = 0, 1, \ldots$. The sample mean $\overline{X} = \sum_i X_i/n$ is to be used as an estimator for θ; show that it is unbiased and efficient.

6.1E   Let $X_1, \ldots, X_n$ be a sample from a normal density with mean θ and variance σ^2. Assume that σ^2 is known and need not be estimated. If the sample average $\delta(X_1, \ldots, X_n) = \sum_i X_i/n$ is used to estimate θ, what is the efficiency of this procedure?

6.1F Consider two procedures δ_1 and δ_2. Assume that their efficiencies are given by $\eta(\delta_k) = g(\theta)/(kn)$, where $k = 1, 2$ and $g(\theta)$ is some function of θ. (a) Does it make any sense to say "δ_1 is twice as efficient as δ_2"? (b) What is the implication in terms of the sample sizes needed for δ_1 and δ_2 to obtain the same precision?

6.1G Consider a set of observations $X_1, \ldots, X_n$ of a random variable X. The procedure $\delta(X) = \sum_i c_i X_i$ is to be an estimator of the mean EX. (a) What constraint must be imposed on the weights c_i for δ to be unbiased? (b) What values for the c_i will result in an unbiased procedure for EX with minimal variance? (*Hint*: The method of Lagrange multipliers might be convenient.) (c) Suppose the X_i are independent variables with identical (but unknown) mean and different but known variance σ_i^2. What values for the c_i will result in an unbiased estimator with minimum variance?

6.1H Let $X_1, \ldots, X_n$ be a set of observations from a normal density with known mean μ and unknown variance θ:

$$f(x|\theta) = \exp[-(x - \mu)^2/2\theta]/\sqrt{2\pi\theta}$$

The procedure $\delta(X) = \sum_{i=1,n} (X_i - \mu)^2/n$ is to estimate θ. (a) Determine the mean and variance of δ. (*Hint:* The quantity $(X - \mu)^2/\theta$ follows a χ^2-distribution with one degree of freedom. Consequently, $n\delta/\theta$ is a χ^2 with n degrees of freedom, whose mean is n and variance is $2n$.) (b) Show that δ is an efficient, unbiased estimator of θ.

6.2A Let $X_1, \ldots, X_n$ be a set of observations from $f(x|\theta) = \theta e^{-\theta x}$. Determine the maximum likelihood estimator for θ.

6.2B Consider the geometric density $f(x|\theta) = \theta(1 - \theta)^x$ for $x = 0, 1, 2, \ldots$. Find the maximum likelihood estimator for θ based on a sample size of n.

6.2C Consider the density $f(x|\theta) = \exp(-x^2/2\theta^2)/\theta\sqrt{2\pi}$. (a) Determine the maximum likelihood estimator for θ based on n observations. (b) Substitute τ for θ^2 in the normal density to yield $f(x|\tau)$. Now derive the maximum likelihood estimator for τ. (c) How do the results compare?

6.2D Let $X_1, \ldots, X_n$ be a random sample from the uniform density $f(x|\theta) = 1/\theta$ for $0 \le x \le \theta$. (a) Verify that the likelihood function is $\mathscr{L}(X|\theta) = 1/\theta^n$. (b) Show that the procedure which maximizes the likelihood function is $\hat{\theta} = \max\{X_1, \ldots, X_n\}$. (c) What characteristic of f renders this problem inappropriate for the usual calculus techniques for maximization?

6.3A Let $X_1, \ldots, X_n$ be a random sample from $f(x|\theta) = \exp[-(x - \mu)^2/2\theta]/\sqrt{2\pi\theta}$, where neither μ nor θ are known. The procedure $\delta(X) = \sum_i (X_i - \bar{X})^2/(n - 1)$ is to estimate θ, where $\bar{X} = \sum_i X_i/n$. (a) Determine the mean and variance of δ. (b) Calculate the efficiency of δ. (c) How do the results compare with Problem 6.2C?

6.4A Consider the normal density with a variance of 9. The mean is estimated by the sample average which happens to be $\bar{X} = 34$, based on $n = 16$ observations. (a) Determine a 90% confidence interval for the mean μ. (b) What value of n will reduce the length of the confidence interval to 1/3 of its current size? (c) Suppose there are k observations for which the corresponding confidence interval is I. What value of the sample size n will halve the length of I?

6.4B Take a coin, toss it 30 times, and note the number of heads. Let $X_i = 1$ if the ith toss is a head, and $X_i = 0$ otherwise. Let $\bar{X} = \sum_i X_i/n$ be the estimator for the probability p of heads. Assume that $\bar{X}$ has a normal density whose standard deviation can be approximated from that of the procedure $\bar{X}$. Construct a 95% confidence interval for p.

6.4C A 90% confidence interval is to be constructed for the mean of a normal density function. The standard deviation σ is assumed to be known. If the confidence interval is to be less than $\sigma/5$, what should be the sample size?

6.4D A 95% confidence interval is to be constructed for the parameter μ of a Poisson density. A sample of $n = 50$ observations results in a sum of $\sum_i X_i = 300$. Assume that the sample average $\sum_i X_i/n$ approximately follows a normal density and that the sample estimator for σ is appropriate.

6.4E Consider the parameter θ from the density $f(x|\theta) = e^{-x/\theta}/\theta$ for $x \geq 0$. A set of $n = 25$ observations yields a total of $\sum_i X_i = 150$. Assume that a normal density is appropriate for $\sum_i X_i/n$. What is a 90% confidence interval for θ?

FURTHER READING

Cramér, H. 1946. *Mathematical Methods of Statistics*. Princeton, NJ: Princeton University Press.

Gauss, C.F. 1887. *Abhandlungen zur Methode der kleinsten Quadrant*, ed., Börsch-Simon. Berlin: P. Stankiewicz.

Kendall M.G., and A. Stuart. 1961, 1966. *The Advanced Theory of Statistics.*, Vols. II and III. New York: Hafner.

Rao, C.R. 1973. *Linear Statistical Inference and Its Applications*. 2nd ed. New York: Wiley.

7

Hypothesis Testing

Decision making under uncertainty involves the selection of a course of action among alternative candidates. The simplest situation arises when there are only two contenders for selection. The selection of one out of two candidates under uncertainty is called *hypothesis testing*.

The task of hypothesis testing is illustrated by the determination of a coin as being fair or not. It also arises in the evaluation of a medical treatment as being effective or not; or in the assessment of the mean time to failure of a machine as exceeding a prescribed number of hours.

In the context of hypothesis testing, one course of action is called the *null hypothesis* and the other the *alternate hypothesis*. For the coin example, suppose that p is the probability of obtaining a head on a toss. Then $p = .5$ might be selected as the null hypothesis, and $p \neq .5$ as the alternate.

The decision procedure for hypothesis testing involves some mapping of the space of outcomes into the binary action space, corresponding to the acceptance or rejection of the null hypothesis. Since the action space contains only two elements, acceptance of the null hypothesis H_0 is tantamount to rejection of the alternate hypothesis H_1, and vice versa.

The selection of one hypothesis over the other is based on a set of observations. The observations are viewed as outcomes of a random variable relating to a random trial or set of such. In other words, a decision procedure δ maps the outcome space into the action space $\mathbf{A} = \{\text{Accept } H_0, \text{Reject } H_0\}$.

The portion of the outcome space which corresponds to the rejection of the null hypothesis is called the *critical region* or *rejection region*. The *acceptance region* is the complement of the critical region. In symbolic form, let $\mathcal{K}$ denote the critical region; then the acceptance region is the rest of the sample space, given by $\Omega - \mathcal{K}$.

7.1   OUTCOME SPACES AND CRITICAL REGIONS

Consider the problem of deciding whether or not a new production process affects the taste of a brand of wine. Let θ denote the proportion of consumers who prefer the wine produced by the older process. The null hypothesis H_0 would imply that the taste is unaffected by the two production processes. Consequently, the preference for the traditional process would be one-half. In other words, the null hypothesis is $H_0: \theta = 0.5$. The alternate possibility then refers to unequal preferences for the two brands of wine. As a result, the alternate hypothesis would be $H_1: \theta \neq 0.5$.

Suppose that the evaluation of the two brands is to be based on a taste test involving three subjects. The random variables are $x_1, x_2,$ and x_3 corresponding to the subjects. The outcome space for each X_i is 0 if the wine produced by the traditional method is preferred, and 1 if the newer process is tastier.

The outcome space for the overall test is given by $\Omega = \{(x_1, x_2, x_3): x_i = 0, 1\}$. A pictorial representation of the outcome space is given in Figure 7-1. The outcome space must be partitioned into a critical region and

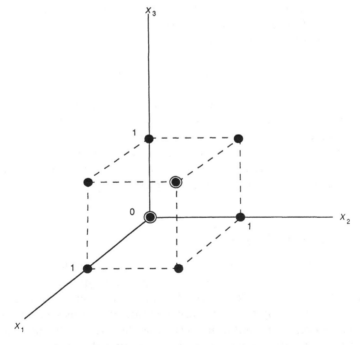

FIGURE 7-1. Example of an outcome space for hypothesis testing. The eight points correspond to $\Omega = \{(X_1, X_2, X_3)\}$ where each X_i takes a binary value of 0 or 1. The critical region might be $\mathscr{K} = \{(0,0,0),(1,1,1)\}$; the acceptance region is then the complement given by $\Omega - \mathscr{K}$.

acceptance region. A plausible policy is to reject the null hypothesis only under overwhelming evidence against it—namely, under unanimous preference for or against the older brand. In this case, the critical region is $\mathcal{K} = \{(0,0,0),(1,1,1)\}$. The acceptance region is defined by the complement:

$$\Omega - \mathcal{K} = \{(0,0,1),(0,1,0),(1,0,0),(0,1,1),(1,0,1),(1,1,0)\}$$

The null hypothesis is accepted when at least one of the tasters disagrees with the others. The decision procedure in this example can be stated in terms of the sample average. Let δ denote the sample average of the subjects' responses:

$$\delta(X_1, X_2, X_3) \equiv \overline{X} = \sum_{i=1}^{3} X_i/3$$

The null hypothesis is rejected for extreme values of δ. The value of δ is 0 under the unanimous preference for the older brand, and 1 for unanimous rejection. Under other conditions, the votes are mixed and lies within 1/3 and 2/3; that is, $1/3 \leq \delta \leq 2/3$. Under the null hypothesis, the proportion of consumers preferring the old brand is $\theta = 0.5$. The null hypothesis is rejected if $|\delta - 0.5| > 1/6$, and is accepted otherwise. We can summarize the decision rule for the hypothesis test as follows:

If $|\delta - \frac{1}{2}| > \frac{1}{6}$, then reject H_0; otherwise accept H_0

In essence, the hypothesis testing problem involves the partitioning of the sample space Ω into two subsets. The critical region $\mathcal{K}$ is the portion of the sample space corresponding to the rejection of the null hypothesis. On the other hand, the acceptance region given by the complement, $\Omega - \mathcal{K}$, corresponds to a positive result. If the observation X falls within $\mathcal{K}$, the null hypothesis is rejected; otherwise it is accepted.

7.2   ERRORS AND RISKS

The basic problem in hypothesis testing involves two simple alternatives. The null hypothesis is H_0: $\theta = \theta_0$, whereas the alternate hypothesis is H_1: $\theta = \theta_1$. The basic problem is a special case of more involved situations, such as the task of selecting among three or more alternatives. In this section, we introduce some concepts and notation to serve as the basis for further discussion in this chapter, as well as in Chapter 10 on multiple decisions.

In the basic testing situation, the comparison of one hypothesis versus another is based on a set of independent observations $X_1, \ldots, X_n$ of some random variable X. The random variable follows a probability function $f(x|\theta)$ whose value of the parameter is unknown.

Suppose A_0 and A_1 are the acceptance regions corresponding respectively to hypotheses H_0 and H_1. In the notation of the previous section, A_1 is tantamount to the critical region $\mathscr{K}$ for rejecting the null hypothesis.

Let $X = (X_1, \ldots, X_n)$ denote the vector of observations. The decision rule is to accept H_0 if $X \in A_0$, and to accept H_1 if $X \in A_1$. The loss function for hypothesis testing was defined in Chapter 5 in terms of a binary function:

$$L(\theta, \delta) = \begin{cases} 0 & \text{decision is correct} \\ 1 & \text{decision is incorrect} \end{cases}$$

The procedure δ is a function of the outcome X; that is, $\delta = \delta(X)$.

The risk for any procedure δ is the mean loss as a function of the parameter θ. In the basic hypothesis testing problem, the parameter space Θ consists of two points, namely, θ_0 and θ_1. When $\theta = \theta_0$, the risk is given by

$$R(\theta_0, \delta) = EL(\theta_0, \delta(X)) = 0 \cdot P\{X \in A_0|\theta_0\} + 1 \cdot P\{X \in A_1|\theta_0\}$$
$$= P\{X \in A_1|\theta_0\} \tag{7-1}$$

In other words, the risk for δ when $\theta = \theta_0$ is the probability of X lying in the acceptance region for H_1. Mistaken acceptance of H_1 is also called *Type I error*. The chance of making a Type I error is denoted by the symbol α.

When the parameter θ has the value θ_1, the risk is

$$R(\theta_1, \delta) = EL(\theta_1, \delta(X)) = 1 \cdot P\{X \in A_0|\theta_1\} + 0 \cdot P\{X \in A_1|\theta_1\}$$
$$= P\{X \in A_0|\theta_1\} \tag{7-2}$$

The risk for δ when the parameter is θ_1 is the chance of entering the acceptance region for H_0. Mistaken acceptance of H_0 is called *Type II error*. The probability of committing a Type II error is indicated by the symbol β.

The fundamental issue in hypothesis testing is to construct a procedure δ to map the outcome space wisely into the acceptance regions for H_0 or H_1. A prudent mapping is tantamount to partitioning the outcome space into acceptance regions A_0 and A_1 in such a way as to minimize the risks under both θ_0 and θ_1. These are opposing goals. Since A_1 is the complement of A_0, we can infer from Eqs. (7–1) and (7–2) that decreased $R(\theta_0, \delta)$ can be achieved only at the expense of increasing $R(\theta_1, \delta)$; and vice versa. For instance, the risk of false acceptance of H_1 can be reduced to 0 by taking $A_1 = \varnothing$ and $A_0 = \Omega$. But in that case, the chance of false acceptance of H_0, namely, a Type II error, is unity.   .

Under these circumstances, a further constraint is needed to relate the risks under θ_0, and θ_1. We have seen in Chapter 5 that one method is to minimize the maximum risk. In other words, the acceptance regions A_0 and A_1 are chosen so that the worst value between $R(\theta_0, \delta)$ and $R(\theta_1, \delta)$ is as small as possible. A second criterion for constructing the acceptance regions is to minimize the average value of the risks.

A third method comes from the traditional approach to hypothesis testing. This approach imposes an arbitrary constraint on the chance of mistaken acceptance of H_1. In other words, the probability α of a Type I error is fixed. Some common constraints on α are .10, .05, or .01.

From Section 7.1, we recall that A_1 is also called the *critical region*. As a result, α is the size of the critical region. The fundamental problem of hypothesis testing is to determine the *least critical region of size α*; namely, the subset of Ω for which the magnitude of β is minimized. A test based on such an approach is called a *best test of size α*.

To summarize, the classical approach to hypothesis testing manages the risks in decision making according to the strategy of partial constraint. The chance α of falsely accepting the alternate hypothesis H_1 is fixed at an arbitrary level. The key is to determine a critical region A_1 to minimize the size β of falsely accepting the null hypothesis H_0. When such a region has been found, it is called a *best critical region of size α*. A hypothesis test based on such an optimal critical region is called a *best test*.

A methodology exists for the determination of best critical regions. This technique is encapsulated in the Neyman–Pearson Lemma, to which we now turn.

7.3 BEST REGIONS FOR BINARY TESTS

The binary test is the basic scenario in hypothesis testing. Here one of two possibilities is to be selected based on a set of observations.

Let $X = (X_1, \ldots, X_n)$ be the vector of n independent observations, each of which emanates from a probability function $f(x|\theta_i)$. We assume that the parameter space consists of two possibilities: $\Theta = \{\theta_0, \theta_1\}$. The task is to determine whether the actual value of parameter θ is θ_0 or θ_1.

The likelihood function for the sample data $X = (X_1, \ldots, X_n)$ is given by $\mathscr{L}_i(x)$, where i corresponds to the parameter value θ_i. More specifically,

$$\mathscr{L}_i(x) \equiv f(x|\theta_i) = f(x_1, \ldots, x_n|\theta_i) = \prod_{j=1}^{n} f(x_j|\theta_i)$$

The last equality depends on the independence of the observations x_j.

Let ρ_{10} denote the ratio of the likelihood functions under θ_1 and θ_0; in algebraic form, $\rho_{10}(x) \equiv \mathscr{L}_1(x)/\mathscr{L}_0(x)$. The best critical region at some level α can be determined by comparing $\rho_{10}(x)$ against some constant k. This idea is embodied in the Neyman–Pearson Lemma, named for two statisticians who proved the result.

Lemma (Neyman–Pearson). Suppose there exists a critical region $\mathscr{K}$ of size α and a non-negative constant k, such that

$$\rho_{10}(x) \geq k \quad \text{if } x \in \mathscr{K}$$

$$\rho_{10}(x) \leq k \quad \text{if } x \notin \mathscr{K}$$

$$(7\text{--}3)$$

Then $\mathscr{K}$ is a best critical region of size α.

Proof. Let $\mathscr{K}$ denote the critical region of size α defined by Eq. (7–3), and $\mathscr{K}_*$ any other critical region of size $\leq \alpha$. We let $\mathscr{R}$ denote the region of $\mathscr{K}$ without points in $\mathscr{K}_*$, $\mathscr{T}$ that part of $\mathscr{K}_*$ excepting $\mathscr{K}$, and $\mathscr{S}$ the intersection of the two critical regions. These regions are depicted in Figure 7–2.

Now $\mathscr{K}$ is a critical region of size α. As a result, we may write

$$P\{x \in \mathscr{K} | \theta_0\} = \int_{\mathscr{K}} f(x|\theta_0)\, dx = \alpha$$

Since $x = (x_1, \ldots, x_n)$ is a vector, dx is actually shorthand for $dx_1\, dx_2 \ldots dx_n$, whereas the integral is actually on n-fold operation over all points $(x_1, \ldots, x_n)$

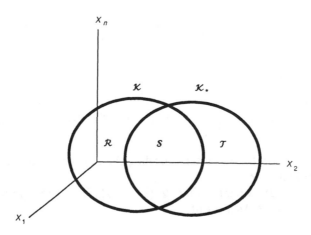

FIGURE 7–2. Illustration of critical regions for Neyman–Pearson lemma. The region $\mathscr{R}$ is $\mathscr{K}$ without $\mathscr{K}_*$; $\mathscr{T}$ is $\mathscr{K}_*$ without $\mathscr{K}$; $\mathscr{S}$ is the intersection of $\mathscr{K}$ and $\mathscr{K}_*$.

lying in $\mathscr{K}$. Similarly, $\mathscr{K}_*$ is a critical region of size $\leq \alpha$:

$$P\{x \in \mathscr{K}_* | \theta_0\} = \int_{\mathscr{K}_*} f(x|\theta_0)\, dx \leq \alpha$$

Combining these two formulas leads to

$$\int_{\mathscr{K}} \mathscr{L}_0(x) \geq \int_{\mathscr{K}_*} \mathscr{L}_0(x)\, dx$$

where $\mathscr{L}_0(x)$ has been substituted for $f(x|\theta_0)$.

Since $\mathscr{K} = \mathscr{R} \cup \mathscr{S}$ and $\mathscr{K}_* = \mathscr{S} \cup \mathscr{T}$, the portion of the two integrals over the common region $\mathscr{S}$ can be cancelled. The result is

$$\int_{\mathscr{R}} \mathscr{L}_0(x)\, dx \geq \int_{\mathscr{T}} \mathscr{L}_0(x)\, dx \tag{7-4}$$

The chances of false acceptance under $\mathscr{K}$ and $\mathscr{K}_*$ can be determined as follows. Suppose β and β_* are the respective probabilities of falsely accepting H_0. The value of β is the chance of falling inside the noncritical region $(\Omega - \mathscr{K})$ when the alternative H_1 holds. The probability is the complement of that for lying in $\mathscr{K}$ under H_1:

$$\beta = 1 - \int_{\mathscr{K}} \mathscr{L}_1(x)\, dx$$

In a similar way,

$$\beta_* = 1 - \int_{\mathscr{K}_*} \mathscr{L}_1(x)\, dx$$

Their difference is

$$\beta_* - \beta = \int_{\mathscr{K}} \mathscr{L}_1(x)\, dx - \int_{\mathscr{K}_*} \mathscr{L}_1(x)\, dx = \int_{\mathscr{R}} \mathscr{L}_1(x)\, dx - \int_{\mathscr{T}} \mathscr{L}_1(x)\, dx \tag{7-5}$$

The second equality follows after canceling the portions of the integral over the common region $\mathscr{S}$.

From Eq. (7–3), we infer that $\mathscr{L}_1(x) \geq k\mathscr{L}_0(x)$ for any point x in $\mathscr{K}$, including those in $\mathscr{R}$. As a result,

$$\int_{\mathscr{R}} \mathscr{L}_1(x)\, dx \geq \int_{\mathscr{R}} k\mathscr{L}_0(x)\, dx$$

Conversely, the region $\mathscr{T}$ lies outside $\mathscr{K}$. From the second part of Eq. (7–3), we know that $\mathscr{L}_1(x) \leq k\mathscr{L}_0(x)$. We can, thus, write

$$\int_{\mathscr{T}} \mathscr{L}_1(x)\,dx \leq \int_{\mathscr{T}} k\mathscr{L}_0(x)\,dx$$

Injecting these results into Eq. (7–5) yields

$$\beta_* - \beta \geq k\int_{\mathscr{R}} \mathscr{L}_0(x)\,dx - k\int_{\mathscr{T}} \mathscr{L}_0(x)\,dx \geq 0$$

The second inequality springs from Eq. (7–4).

Finally, we obtain $\beta_* \geq \beta$. We recall that β_* is the chance of falsely accepting H_0 for any critical region of size $\leq \alpha$. Consequently, $\mathscr{K}$ is a best critical region of size α. ∎

The Neyman–Pearson lemma states that the ratio function $\rho_{10}(X)$ is a sufficient criterion for establishing the critical region for a best test. A converse theorem also exists: the critical region for a best test must satisfy the inequalities specified in the lemma.

To clarify these ideas, we consider the case for a normal distribution. Let μ be the mean for a normal density $f(x|\mu)$ with known variance σ^2. The problem is to test the null hypothesis $H_0: \mu = \mu_0$ against the alternative $H_1: \mu = \mu_1$, where μ_1 exceeds μ_0. The likelihood function for n independent observations is

$$\mathscr{L}(x) \equiv f(x_1,\ldots,x_n|\mu) = \prod_{i=1}^{n} \frac{\exp\{-\frac{1}{2}[(x_i - \mu)/\sigma]^2\}}{\sigma\sqrt{2\pi}}$$

$$= \frac{\exp\{-\frac{1}{2}\sum_i[(x_i - \mu)/\sigma]^2\}}{(2\pi\sigma^2)^{n/2}}$$

The ratio of likelihood functions under H_1 and H_0 is given by

$$\Lambda_{10} \equiv \frac{\mathscr{L}_1}{\mathscr{L}_0} = \frac{\exp[-(1/2\sigma^2)\sum_i(x_i - \mu_1)^2]}{\exp[-(1/2\sigma^2)\sum_i(x_i - \mu_0)^2]}$$

$$= \exp\{-(1/2\sigma^2)[\sum_i(x_i - \mu_1)^2 - \sum_i(x_i - \mu_0)^2]\}$$

$$= \exp\{-(1/2\sigma^2)[2(\mu_0 - \mu_1)\sum_i x_i + n(\mu_1^2 - \mu_0^2)]\}$$

$$= \exp\left(\frac{\mu_1 - \mu_0}{\sigma^2}\sum_i x_i + \frac{n(\mu_0^2 - \mu_1^2)}{2\sigma^2}\right)$$

According to Eq. (7–3), the best critical region $\mathcal{K}$ will satisfy the relation

$$\Lambda_{10} = \exp\left(\frac{\mu_1 - \mu_0}{\sigma^2} \sum_i x_i + \frac{n(\mu_0^2 - \mu_1^2)}{2\sigma^2}\right) \geq k$$

for some non-negative constant k. After taking logarithms and moving a term, we obtain

$$\frac{\mu_1 - \mu_0}{\sigma^2} \sum_i x_i \geq \ln k + \frac{n(\mu_1^2 - \mu_0^2)}{2\sigma^2}$$

The alternate hypothesis H_1 assumed that $\mu_1 > \mu_0$. Consequently, the term $(\mu_1 - \mu_0)/\sigma^2$ is positive, and we can write

$$\sum_i x_i \geq \frac{\sigma^2 \ln k}{\mu_1 - \mu_0} + \frac{n(\mu_1 + \mu_0)}{2}$$

According to the lemma, the value of k can be any non-negative number. When k ranges from 0 to $+\infty$, the value of the right-hand side varies from $-\infty$ to $+\infty$. Consequently the inequality is equivalent to the condition $\sum_i x_i \geq c$, for some real value c.

The geometric interpretation of $\sum_i x_i = c$ is a plane in n-dimensional space, as indicated in Figure 7–3. The inequality $\sum_i x_i \geq c$ encompasses all the points lying on or above this plane. This portion of the n-dimensional space represents the critical region $\mathcal{K}$.

The value of c defines the location of the separating plane. As c increases, the size of the critical region decreases. In the limit as $c \to +\infty$, the acceptance region is the entire sample space Ω.

The selection of a value c for the location of the separating plane is tantamount to the selection of the threshold k for the likelihood ratio ρ_{10}. By properly choosing either threshold, we can construct a critical region $\mathcal{K}$ with any probability of occurrence. In other words, the size α of the acceptance region can be selected to be any value between 0 and 1.

Since the sample average is $\bar{x} \equiv \sum_i x_i/n$, the critical region defined by $\sum_i x_i \geq c$ can also be written as $\bar{x} > c/n$. In other words, the decision procedure is the average $\bar{X}$ of the random variables. The critical region is the portion of n-space given by $\bar{X} \geq b$, for some value of b satisfying the relation $P\{\bar{X} \geq b | H_0\} = \alpha$. In particular, the threshold b beyond which the chance that $\bar{X}$ occurs, is α.

This procedure is remarkable for its simplicity. The information in the random sample data is coalesced into the sample average $\bar{X}$. The test depends only on this scalar function rather than the entire vector of observations given by $(X_1, \ldots, X_n)$.

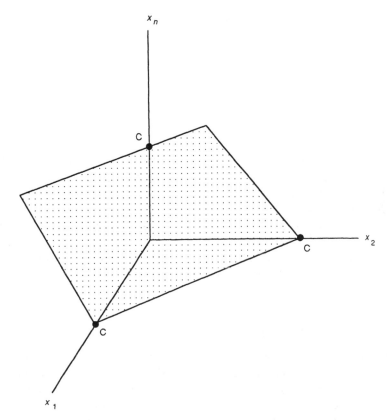

FIGURE 7–3. The equation $\sum_i x_i = c$ defines a plane in n-space. The inequality $\sum_i x_i \geq c$ incorporates all points lying on or above the plane.

Example

An inventory check through a warehouse has resulted in the discovery of a box of microprocessors produced months earlier. Unfortunately, it is not known whether the processors originated from a production line whose mean clock rate is 20 megahertz, or another whose average is 24. However, the clock speed for both production lines is known to be distributed normally with a standard deviation of 3 megahertz. When 36 processors are picked at random and measured, the mean speed is $\bar{x} = 22$ megahertz. Determine a best critical region of size $\alpha = .05$.

The null hypothesis is $H_0: \mu = 20$, whereas the alternate is $H_1: \mu = 24$. The best test is defined by the threshold $\bar{X} > b$, for the value of b corresponding to $P\{\bar{X} > b \,|\, \mu = 20\} = .05$. Under the null hypothesis, $\bar{X}$ possesses a normal density with mean $\mu = 20$ and standard deviation equal to

$\sigma/\sqrt{n} = 3/\sqrt{36} = .5$. We can write

$$.05 = P\{\overline{X} \ge b | \mu = 20\} = P\left\{\frac{\overline{X} - 20}{.5} \ge \frac{b - 20}{.5}\right\} = P\left\{Z \ge \frac{b - 20}{.5}\right\}$$

$$= 1 - P\left\{Z \le \frac{b - 20}{.5}\right\} = 1 - \Phi(z)$$

The quantity Z is the standardized version of $\overline{X}$; that is, Z is a normal variable with a mean of 0 and a standard deviation of 1. Our next task is to find the value z such that $\Phi(z) = 1 - .05 = .95$. According to Appendix B, $\Phi(1.65) = .95$. As a result, we can write $(b - 20)/.5 = z = 1.65$. The desired threshold is $b = 20 + 1.65/2 = 20.82$.

In other words, the critical region of size .05 is the set of all points for which $\overline{X} \ge 20.82$. Since the actual sample average of $\bar{x} = 22$ falls in this region, the null hypothesis is rejected. This is equivalent to accepting the alternate hypothesis, namely, $\mu = 24$ megahertz.   ◆

7.4   COMPOSITE HYPOTHESES

Up to this point, we have focused on *simple hypotheses*; namely, those that specify the value of a parameter θ at a particular point θ_0. For instance, in testing for the mean μ of a normal density, the standard deviation σ is assumed to be known and only the mean is under scrutiny.

A more complex situation arises when a proposition involves more than a single point; this is called a *composite hypothesis*. One class of composite hypotheses relates to a range of values for the unknown parameter. In testing for the parameter p of a binomial distribution, the hypothesis might be $H: p < .5$. In this case p could take any value on the interval from 0 to .5.

A second category of composite hypothesis springs from a vector parameter. In testing for the mean μ of a normal density, suppose the standard deviation σ were unknown. Then the proposition $H: \mu = \mu_0$ would constitute a composite hypothesis for which σ can assume any positive value.

A composite hypothesis can often be reduced to the case of a simple hypothesis. Consider a manufacturer that has developed a new polymer material to enhance the longevity of skis. The developer wishes to determine whether the polymer retains its shape better than the material currently used in the high-performance model. Suppose μ_0 is the average time to loss of shape for a ski made from the existing material, in terms of the number of cycles of bending and flexing. If μ is the average life span for skis made from the new material, the hypothesis test is

$$H_0: \quad \mu = \mu_0$$
$$H_1: \quad \mu > \mu_0 \tag{7-6}$$

The composite alternative H_1 does not require a prior assessment of the precise value of μ, only the assumption that it exceeds μ_0.

Section 7.3 illustrated the Neyman–Pearson lemma for testing simple hypotheses about the mean of a random variable. The testing problem involved $H_0: \mu = \mu_0$ versus $H_1: \mu = \mu_1$. The derivation of the result depended only on the fact that μ_1 exceeds μ_0, but not on the precise value for μ_1. Consequently, the test has the same structure as that specified by Eq. (7–6). The resulting criterion is therefore identical. In other words, the criterion for accepting the alternate hypothesis in Eq. (7–6) is as follows. The sample average $\overline{X}$ must exceed a threshold value:

$$\overline{X} > b$$

for some threshold b.

Another instance of composite tests occurs when both the null and alternate hypotheses involve intervals. More precisely, we have

$$H_0: \quad \mu \leq \mu_0$$
$$H_1: \quad \mu > \mu_0 \tag{7-7}$$

Fortunately, the best tests specified by the Neyman–Pearson lemma can also be applied to this situation. In fact, the criteria defined by the lemma yield best tests for many probability distributions, including the binomial, Poisson, normal, and χ^2-distributions.

The problems specified by Eqs. (7–6) and (7–7) involve values for the alternate hypotheses which lie to the right of that for the null hypotheses. In other words, μ_0 under H_0 is smaller than μ_1 under H_1. The criterion for decision involved a critical region lying to the right-hand side of the real line—that is, a one-sided test to the right.

By the symmetry of the situation, if μ_0 under H_0 exceeds μ_1 under H_1, then the criterion involves a threshold to the left-hand side of the real line. Here, the critical region involves a one-tailed test to the left. In the case of testing for the mean μ_1 against the null value of $\mu_0 > \mu_1$, the criterion requires a sample average lying to the left of some threshold; the test is given by the condition $\overline{X} \leq b$ for some real number b.

7.5   POWER FUNCTION

The probability of accepting the alternate hypothesis—or equivalently, rejecting the null hypothesis—depends on the value of the underlying parameter θ. This probability is called the *power function* of a test, and denoted $\mathscr{P}(\theta)$. In other words, $\mathscr{P}(\theta)$ is the probability that a decision procedure will fall in the critical region when θ is the true value of the parameter.

The size of the critical region depends on the value α for false acceptance of the alternate hypothesis H_1. Strictly speaking, the power function is a function of both α and θ. By assuming a fixed value for α, we can regard the power function $\mathscr{P}(\theta)$ as solely a function of θ.

The power function is useful as a measure of the sensitivity of a decision procedure. Ideally, $\mathscr{P}(\theta)$ should equal 0 when θ lies in the region defined by the null hypothesis H_0, and should equal 1 when θ lies in the jurisdiction of the alternative H_1.

Consider the composite test defined by the relations

$$H_0: \quad \theta < \theta'$$

$$H_1: \quad \theta \geq \theta'$$

for some value of θ'. Ideally, the power function $\mathscr{P}(\theta)$ should be 0 when $\theta < \theta'$ and equal 1 otherwise. The ideal function is shown as a thick dotted path in Figure 7–4. In reality, however, the chance of falling in the critical region is a probabilistic rather than a deterministic phenomenon. The power function will, therefore, tend to be smooth rather than crisp. This is exemplified by the solid curves in the figure.

A good test should have a high chance of accepting the alternate hypothesis in the critical region. Consequently, the thick curve in Figure 7–4 indicates a better decision procedure than does the thin curve.

The power function of a test is directly related to the complement of β, the probability of falsely accepting the null hypothesis. Let Θ_0 denote the subset of the parameter space Θ relating to the null hypothesis H_0; and Θ_1 denote the subspace relating to the alternative H_1. The null hypothesis is falsely accepted when the test procedure $\delta(X)$ falls in the subspace for H_0:

$$\beta(\theta) = P\{\delta \in \Theta_0 | \theta \in \Theta_1\}$$

$$= 1 - P\{\delta \in \Theta_1 | \theta \in \Theta_1\}$$

$$= 1 - \mathscr{P}(\theta) \quad \theta \in \Theta_1$$

Hence, the chance of false acceptance, $\beta(\theta)$, is the complement of the power

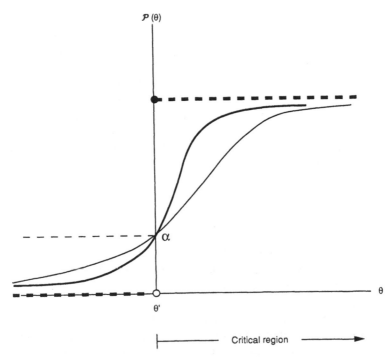

FIGURE 7–4. Examples of power functions $\mathscr{P}(\theta)$. The null hypothesis is $\theta < \theta'$; the alternate is $\theta \geq \theta'$. The thick dotted path is ideal; the solid lines are more realistic. The thin dotted line is the upper limit α in comparing decision procedures.

function within the critical region. If the power function $\mathscr{P}(\theta)$ is high in the critical region, then the probability of false acceptance is low.

A good test procedure should exhibit a low chance of false acceptance $\beta(\theta)$—or equivalently, a high value of power $\mathscr{P}(\theta)$ in the critical region. As indicated previously, we assume that the chance α of false rejection is fixed when comparing test procedures. The thin dotted line in the Figure 7–4 represents the ceiling α for test procedures within the null region specified by H_0.

Uniformly Powerful Test

Consider two test procedures δ and δ' subject to the same probability α of falsely rejecting H_0. In other words, $\mathscr{P}(\theta) \leq \alpha$ when θ lies in the null region Θ_0.

We say that δ is *uniformly more powerful* than δ' if the power function $\mathscr{P}(\theta)$ for δ exceeds that of δ' in the critical region. In Figure 7–4, the thick solid line represents a uniformly more powerful test than does the thin solid line. In addition, we say that a test procedure is *uniformly most powerful* among all tests of significance α if its power function $\mathscr{P}(\theta)$ is maximal for θ in Θ_1, and $\mathscr{P}(\theta) \leq \alpha$ for θ in Θ_0.

A test procedure defined by the Neyman–Pearson lemma is uniformly most powerful. To illustrate, we recall the hypothesis test for the mean μ of a normal density having a known variance. Given the null hypothesis $H_0: \mu = \mu_0$ and the alternate $H_1: \mu > \mu_0$, the test procedure was $\delta(X) = \overline{X}$, while the criterion was $\delta(X) \geq b$. Since the procedure springs from the conditions of the Neyman–Pearson lemma, it is a uniformly most powerful procedure.

7.6 LIKELIHOOD RATIO TESTS

The discussion up to this point has emphasized tests of simple hypotheses. Some of the results are applicable to certain special cases of composite hypotheses. We saw one example in the form of testing $H_0: \mu = \mu_0$ against $H_1: \mu > \mu_0$.

In this section we develop a set of general tools to deal with composite hypotheses. The general testing situation involves composite parameter intervals for both the null and alternate hypotheses. From this perspective, any simple hypothesis can be viewed as a special case of a composite hypothesis in which the endpoints of the interval are identical.

As before, let Θ_0 denote the subset of the parameter space Θ pertaining to the null hypothesis. Conversely, Θ_1 is the subspace for the alternate hypothesis. The general test for two competing hypotheses can then be written as

$$
\begin{aligned}
H_0&: \quad \theta \in \Theta_0 \\
H_1&: \quad \theta \in \Theta_1
\end{aligned}
\tag{7–8}
$$

where the subspaces Θ_0 and Θ_1 are disjoint.

As an illustration, consider the task of assessing the parameter p of a binomial distribution. Suppose the hypotheses are $H_0: p = .5$ and $H_1: p \neq .5$. Then the parameter space is $\Theta = [0, 1]$. The subset of the parameter space relating to the null hypothesis is $\Theta_0 = [.5, .5]$, whereas that for the alternate hypothesis is $\Theta_1 = [0, .5) \cup (.5, 1]$.

From Section 7.3, we recall that the likelihood ratio test leads to optimal procedures for simple hypotheses. It seems plausible that a test for composite hypotheses, as a generalization of that for simple hypotheses, should also incorporate likelihood ratios.

Let $\mathscr{L}_0 \equiv \mathscr{L}(x | \Theta_0)$ represent the likelihood function under the null hypothesis, and $\mathscr{L}_1 \equiv \mathscr{L}(x|\Theta_1)$ the likelihood under the alternative. The likelihood ratio can be constructed as $\rho_{10}(x) = \mathscr{L}(x|\Theta_1)/\mathscr{L}(x|\Theta_0)$. The ratio $\rho_{10}(x)$ happens to be a function of three quantities: the value of the random observation x, the parameter θ_0 in the subspace Θ_0, and the parameter θ_1 in the subspace Θ_1. (In previous sections, when H_0 and H_1 were both simple hypotheses, the values of θ_0 and θ_1 were assumed to be fixed.)

A reasonable way to constrain the likelihood ratio $\rho_{10}(x)$ is to replace the values of θ_0 and θ_1 by their respective estimates. Given the optimal properties of maximum likelihood procedures, the parameters θ_0 and θ_1 could be replaced by their maximum likelihood estimators—namely, $\hat{\theta}_0$ and $\hat{\theta}_1$. If these values lie within their respective domains Θ_0 and Θ_1, then the likelihood can be formed as

$$\rho_{10}(x) = \mathscr{L}(x|\hat{\theta}_1)/\mathscr{L}(x|\hat{\theta}_0)$$

In parallel with the test for simple hypotheses, the critical region can be selected by the criterion $\rho_{10}(x) \geq c$ for some appropriate constant c.

This approach is feasible if the maximum values of the estimators, $\hat{\theta}_0$ and $\hat{\theta}_1$, occur within their respective domains Θ_0 and Θ_1. On the other hand, the maximum value could lie on the boundary; and the boundary might be excluded from consideration if Θ_0 or Θ_1 is an open interval. In the case of the binomial parameter p, for instance, the allowable regions might be $\Theta_0 = (0, .3)$ and $\Theta_1 = [.3, .8]$. If the value of the esimator $\hat{\theta}_0$ under the null hypothesis is maximal at .3, it is not a feasible point in the parameter space for H_0. Under such circumstances, the decision maker can explore other procedures for estimating the value of θ_0 within Θ_0.

High-Order Alternate Subspace

In practical testing situations, the dimension of the parameter subspace under the alternate hypothesis exceeds that for the null hypothesis. Moreover, the subspace under H_1 is given by the complement of the null subspace. In other words, $\Theta_1 = \Theta - \Theta_0$, where Θ_0 is of lower dimension than Θ_1.

To illustrate, consider a test for the mean μ of a normal density with unknown standard deviation σ. The parameter is a vector quantity $\theta = (\mu, \sigma)$. Since μ can be any real number and σ any positive number, the parameter space is given by $\Theta = (-\infty, \infty) \times (0, \infty)$. Suppose the test is for $\mu = \mu_0$ versus $\mu \neq \mu_0$. Since the standard deviation is unknown, the parameter subspace under H_0 is $\Theta_0 = \mu_0 \times (0, \infty)$. Conversely, the parameter subspace for the alternative is $\Theta_1 = [(-\infty, \mu_0) \cup (\mu_0, \infty)] \times (0, \infty)$. The possibilities for μ within Θ_1 is the entire real line, except for one point; hence its dimension equals 1. In contrast, μ can assume only a single point in the subspace Θ_0, from which we conclude that the dimension equals 0. Consequently, the dimension of Θ_1 exceeds that of Θ_0 in this example.

Suppose maximum likelihood procedures are to be used as estimators for the parameters. These are $\hat{\mu} = \overline{X}$ and $\hat{\sigma} = [\sum_i (X_i - \overline{X})^2/n]^{1/2}$. Under Θ_1, the estimator $\hat{\sigma}$ can assume any value; and $\hat{\mu}$ can be any value except μ_0. On the other hand, the chance of observing $\overline{x} = \mu_0$ is zero. (Why?) Consequently, with probability 1, the choice of parameters within Θ_1 will equal that under the entire subspace Θ. In other words, the value of $\hat{\theta}_1$ within Θ_1 is probabilistically equivalent to the value of $\hat{\theta}$ selected from the entire parameter space Θ.

We will focus on problems for which the maximum likelihood estimators $\hat{\theta}_0$, $\hat{\theta}_1$, and $\hat{\theta}$ all exist; and further, $\hat{\theta}_1$ equals $\hat{\theta}$ with unit probability for reasons discussed in the previous paragraph. We may, therefore, replace $\hat{\theta}_1$ by $\hat{\theta}$ in the likelihood ratio to obtain $\rho_{10}(x) = \mathscr{L}(x|\hat{\theta})/\mathscr{L}(x|\hat{\theta}_0)$.

The estimator $\hat{\theta}$ is selected from Θ, a superset of the parameter subspace Θ_0 which contains $\hat{\theta}_0$. The greater range for $\hat{\theta}$ implies that $\mathscr{L}(x|\hat{\theta})$ must be at least as large as $\mathscr{L}(x|\hat{\theta}_0)$. Hence, the ratio $\rho_{10}(x) = \mathscr{L}(x|\hat{\theta})/\mathscr{L}(x|\hat{\theta}_0)$ may range from 1 to ∞.

In general, it is more convenient to work with a function that ranges from 0 to 1 rather than from 1 to ∞. For this reason, it is customary to use the inverse ratio $\lambda_{01} \equiv 1/\rho_{10} = \mathscr{L}(x|\hat{\theta}_0)/\mathscr{L}(x|\hat{\theta})$ for testing composite hypotheses. For notational simplicity, we will refer to λ_{10} as simply λ. The criterion for rejecting the null hypothesis is then $\lambda < \lambda_c$ for some value of λ_c lying between 0 and 1.

Once the values of $\hat{\theta}_0$ and $\hat{\theta}$ are fixed, the quantity λ is a function only of the random sample X. On the other hand, the value of λ might well vary with the actual parameter. In the previous example of testing for the mean μ of a normal density, the likelihood function $\mathscr{L}(x|(\mu, \sigma))$ will depend on the unknown parameter σ even when μ is fixed at μ_0 under the null hypothesis.

For the normal density, for instance, the size of the critical region will depend on the value of σ even when the mean μ is fixed at μ_0 under the null hypothesis. As a result, the threshold λ_c will also depend on the value of σ.

Some problems may involve no unknown parameters other than the one under evaluation. This happens, for instance, in testing for the parameter p of a binomial distribution. The hypotheses might be $H_0: p = p_0$ versus $H_1: p \neq p_0$; there are no free parameters under H_0.

When a distribution is free of unknown parameters under H_0, it is possible to determine the size α for mistakenly rejecting H_0. In other words, we can find the value of λ_c such that $P(\lambda \leq \lambda_c) = \alpha$ under the null hypothesis. The testing problem would then be completely specified.

For many applications, the quantity λ can be shown to follow a familiar distribution function. Consequently, the threshold λ_c can be determined uniquely and the testing problem completely defined.

We summarize these results in the form of a proposition.

Proposition (Likelihood Ratio Test). Let Θ be the parameter space for a parameter θ, and Θ_0 the subset of Θ under the null hypothesis. Assume that the subspace Θ_0 is of lower dimension than Θ. Further, suppose that $\hat{\theta}$ and $\hat{\theta}_0$ are the maximum likelihood estimators for θ within the sets Θ and Θ_0, respectively. The ratio of likelihoods given by $\lambda = \mathcal{L}(x|\hat{\theta}_0)/\mathcal{L}(x|\hat{\theta})$ can be used as a test for the composite hypotheses $H_0: \theta \in \Theta_0$ and $H_1: \theta \in \Theta - \Theta_0$. Let λ_c be the value satisfying the condition $P\{\lambda \le \lambda_c|H_0\} \le \alpha$ for every θ in Θ_0. Then the alternate hypothesis is accepted if and only if $\lambda \le \lambda_c$.

As usual, α is the level of significance, or the chance of falsely accepting the alternate hypothesis.

Asymptotic Distribution of Likelihood Ratio

The precise distribution for λ can often be determined, but the effort usually entails a great deal of work. On the other hand, an approximation to the distribution of λ is available when the sample size n is large and some other conditions are met. In the limit as the sample size increases indefinitely, the likelihood ratio follows a χ^2-distribution.

Proposition (Asymptotic Distribution for λ). Let X be a random sample from the probability function $f(x|\theta)$, whereas $\hat{\theta}$ and $\hat{\theta}_0$ denote respectively the maximum likelihood estimators within the parameter spaces Θ and Θ_0. Consider the likelihood ratio $\lambda = \mathcal{L}(X|\hat{\theta}_0)/\mathcal{L}(X|\hat{\theta})$. Under certain regularity assumptions for f, the random variable given by $-2\ln\lambda$ has an asymptotic χ^2-distribution as the sample size increases. The degrees of freedom for the χ^2-distribution equals the difference between the numbers of unspecified parameters in Θ and Θ_0.

To illustrate the use of this result, we consider once more the previous example of estimating the mean μ of a normal density with unknown standard deviation σ. For the hypothesis $H_0: \mu = \mu_0$ and $H_1: \mu \ne \mu_0$, the null parameter space Θ_0 has only one independent parameter, namely, σ. In contrast, the overall parameter space Θ has two independent parameters. Since their difference equals 1, the quantity $-2\ln\lambda$ follows a χ^2-distribution with one degree of freedom.

To clarify these ideas further, we consider a numerical example.

Example

Let X be a random sample of size $n = 16$ from a normal density with mean μ and variance 25. The hypotheses are $H_0: \mu = 7$ versus $H_1: \mu \ne 7$. If the significance level is $\alpha = .05$, what is the critical region for the quantity $-2\ln\lambda$? The likelihood functions are given by

$$\mathscr{L}(x|\theta) = \frac{1}{(2\pi\sigma^2)^{n/2}} \exp\left(-\frac{1}{2\sigma^2} \sum_i (x_i - \mu)^2\right)$$

$$\mathscr{L}(x|\theta_0) = \frac{1}{(2\pi\sigma^2)^{n/2}} \exp\left(-\frac{1}{2\sigma^2} \sum_i (x_i - \mu_0)^2\right) \ .$$

where $\mu_0 \equiv 7$. We can replace μ by its maximum likelihood estimator $\bar{x}$. Forming the likelihood ratio and simplifying, we obtain $\lambda = \exp[-n(\bar{x} - \mu_0)^2/2\sigma^2]$. A simple transformation yields

$$-2\ln\lambda = n(\bar{x} - \mu_0)^2/\sigma^2 \tag{7-9}$$

The parameter space Θ contains one independent parameter, namely, μ, whereas the subspace Θ contains no free agents. Hence, $-2\ln\lambda$ possesses a χ^2-distribution with 1 degree of freedom. In this example, the asymptotic χ^2-distribution happens to be exact for all values of the sample size. We can show this as follows.

We know from Chapter 3 that $Z \equiv (\bar{X} - \mu_0)/(\sigma/\sqrt{n})$ is a standard normal variable, and that $Z^2 = (\bar{X} - \mu_0)^2/(\sigma^2/n)$ is a χ^2-variable with one degree of freedom. Consequently, the quantity $-2\ln\lambda$ in Eq. (7-9) has a χ^2-distribution for any value of the sampe size n.

The critical region for the hypothesis test is given by $\lambda \le \lambda_c$ for some appropriate value of λ_c. Since we know that $\chi^2 \equiv -2\ln\lambda$ is as χ^2-variable, the equivalent constraint is $-2\ln\lambda \ge \chi_c^2$, where the threshold χ_c^2 is determined by $P\{\chi^2 \ge \chi_c^2\} = .05$. The appropriate value of the threshold can be obtained from Table B-3 in Appendix B as $\chi^2 = 3.84$.

The quantity χ^2 is equal to $-2\ln\lambda = n(\bar{x} - \mu_0)^2/\sigma^2$. As a result, the critical region defined by $\chi^2 \ge \chi_c^2$ is tantamount to

$$|\bar{x} - \mu_0| \ge \sqrt{\chi_c^2}\frac{\sigma}{\sqrt{n}} = 1.96\frac{\sigma}{\sqrt{n}}$$

This result looks familiar. It is, in fact, the two-tailed test for assessing $\mu = \mu_0$ against $\mu \ne \mu_0$ for a normal density function with known variance. After substituting numerical values for each known factor, we obtain $|\bar{x} - 7| \ge 1.96(5)/\sqrt{16} = 2.45$. If the sample average deviates from 7 by 2.45 or more, the null hypothesis is rejected.   ◆

In this example, the asymptotic χ^2-distribution for the random variable $-2\ln\lambda$ was also its exact distribution. The exercises illustrate the more usual case in which the actual distribution for $-2\ln\lambda$ is only approximated by the χ^2-function.

Another application of the decision procedure $-2\ln\lambda$ lies in testing for the equality of parameter values. Let X be the random variable from a density $f(x|\theta_1,\ldots,\theta_k)$; we assume that the density satisfies a number of regularity conditions which are, in fact, often met by common distribution functions. Suppose that the hypotheses are

$$H_0:\quad \theta_1 = \theta_2 = \cdots = \theta_k$$

$$H_1:\quad \theta_i \neq \theta_j \text{ for at least one } i$$

The null hypothesis has only one independent parameter θ_0 because all the other θ_i must match this value. On the other hand, the alternative H_1 allows for k independent parameters. Consequently, the quantity $-2\ln\lambda$ will possess a χ^2-distribution with $k-1$ degrees of freedom.

PROBLEMS

The problems below are numbered according to the sections to which they correspond. For instance, Problem 7.1A is the first ("A") exercise pertaining to Section 7.1.

7.1A   Two candidates are running for office. Let p be the proportion of the population at large which favors the first candidate. A poll of 3 potential voters will be used to test the hypothesis $H_0: p = 1/2$ against $H_1: p < 1/2$. (a) Draw a chart of the outcome space. (b) Determine an appropriate critical region, and explain your choice.

7.2A   Let θ be the parameter from a binomial density function: $p(x) = \binom{n}{x}\theta^x(1-\theta)^{n-x}$. The hypotheses are $H_0: \theta = 1/4$ and $H_1: \theta = 3/4$. The test procedure is $\overline{X} = \sum_i X_i/n$ for a sample size of $n = 4$ and significance level $\alpha = 0.10$. (a) Determine an appropriate critical region. (b) What is β, the size of Type II error?

7.3A   Consider the hypotheses $H_0: \theta = 1$ and $H_1: \theta = 5$ for an exponential density with parameter θ. Use the conditions of the Neyman–Pearson lemma to determine a best test based on a sample size of n.

7.4A   Consider the exponential density with parameter λ. Determine whether each hypothesis below is simple or complex, and explain. (a) $H_0: \lambda = 1$ and $H_1: \lambda = 3$. (b) $H_0: \lambda \leq 1$ and $H_1: \lambda > 1$.

7.5A   Consider a test for the mean μ of a normal density with variance 9. The hypotheses are $H_0: \mu = 20$ and $H_1: \mu > 20$. The test procedure is $\overline{X} = $

$\sum_i X_i/n$ based on a sample size of $n = 36$. The significance level is $\alpha = 0.05$. (a) Determine the critical region. (b) Graph the power function $\mathscr{P}(\theta)$. (c) What is $\beta(\theta)$? Is it identical to $\mathscr{P}(\theta)$?

7.5B   Consider the same situation as Problem 7.5A except for the hypotheses. Here we have $H_0: \mu > 10$ and $H_1: \mu \leq 10$. The questions remain the same.

7.6A   Consider a test at significance level $\alpha = 0.05$ for the mean of a Poisson distribution. The hypotheses are $H_0: \mu = 7$ and $H_1: \mu \neq 7$. A random sample of size 10 yields these values: 13, 5, 6, 8, 16, 11, 4, 5, 9, 12. (a) What are Θ and Θ_0? (b) What is $-2\ln \lambda$? (c) Based on an asymptotic χ^2-distribution, what is the critical region?

7.6B   Consider a test for an exponential density with parameter θ based on a random sample of size n. The hypotheses are $H_0: \theta = \theta_0$ and $H_1: \theta \neq \theta_0$. Answer the same questions as in Problem 7.6A.

FURTHER READING

Fisher, R.A. 1958. *Statistical Methods for Research Workers*. 13th ed. New York: Hafner.

Lehmann, E.L. 1959. *Testing Statistical Hypotheses*. New York: Wiley.

Rao, C.R. 1973. *Linear Statistical Inference and Its Applications*. 2nd ed. New York: Wiley.

III

Decisions

8

Decision Theory

Decision theory deals with the evaluation and selection of alternate courses of action in the face of uncertainty. The conceptual and mathematical foundations of decision theory lie in probability and statistics. The theory focuses on the problem of a single agent attempting to maximize returns—or equivalently, minimize losses—despite incomplete knowledge of the world. The incomplete awareness of the world may be due to many factors, including cognitive limitations, poor understanding of the properties of the environment, or uncertainty about the future.

Decision theory is closely related to *game theory*, which deals with decision making among multiple agents. These agents may interact cooperatively, competitively, or combatively. In cooperative interaction, the agents may join together to maximize the gain to each participant; an example is found in two explorers helping each other through a treacherous terrain. In the competitive scenario, each agent attempts primarily to enhance his own position, while the welfare of the other players is of secondary consequence; this is a common scenario among firms in an industrial environment. In combative interaction, the players actively seek to hinder or harm each other; this is found in the case of two warring nations. The interactions among the agents are usually complicated by imperfect knowledge. Each has only partial information about the state of the world and the intentions of other agents. The various participants may pursue deliberate goals or exhibit inadvertent behavior.

In the decision-theoretic context, an agent must select a course of action without full knowledge of the ultimate consequences. For this scenario, the agent may be regarded as playing a game against nature. Unlike an animate agent, however, nature plays an indifferent game, intending neither to assist nor hinder the decision maker.

On the other hand, the predilections of the decision maker might be such that he, in effect, assigns teleological ends to nature. For instance, an optimistic agent might choose an action with the largest possible payoff, even if the potential penalty is severe; this situation corresponds to a benign view of nature. In contrast, a pessimistic player might choose a strategy entailing only a small penalty, in conjunction with the possibility of only a modest gain; this corresponds to a conservative strategy against nature. A third option is to select an action having the highest expected return; in this case, the agent imputes to nature a natural stance.

Game theory as it is known today is largely the brainchild of the mathematician John von Neumann. His work with Oskar Morgenstern led to the publication in 1944 of *Theory of Games and Economic Behavior*, a book which is still the standard reference in the field.

The field of games gave birth to a theory of decision making. Decision theory owes heavily to the statistician Abraham Wald (1950). Another leading figure is Thomas Ferguson, whose work is a cornerstone in the field (Ferguson, 1967).

8.1   ILLUSTRATIVE PROBLEMS

As explained previously, decision theory deals with the evaluation and selection of alternatives under uncertainty. Examples of problems are the following:

- Diagnosis of equipment malfunction in a flexible manufacturing cell.
- Identification of a moving object based on imperfect information.
- Selection of a mobility system for a sentry robot based on test data, and its impact on system capability.
- Introduction of one or more products based on data from test markets.
- Location of a new warehouse to improve customer service.
- Acquisition of a new corporate division.

Decision theory is based on statistical tools, founded on probabilistic methods. In a recursive fashion, however, decision theory has shaped the development of statistics, a field largely devoted to the evaluation of data and their interpretation under conditions of imperfect knowledge.

One way in which decision theory has influenced statistics lies in clarifying the nature of implicit loss functions used in estimation or hypothesis testing. Another contribution is found in revealing the relationships between the classical and Bayesian approaches to statistical procedures. The *classical* approach relies on the objective interpretation of data, devoid of subjective views. In contrast, the *Bayesian* approach allows for the incorporation of prior information and subjective knowledge. For instance, suppose you were to remove a dime from your pocket and toss it 5 times; it turns up heads on 4 occasions, and yields a tail only once. What is the probability of heads for this coin? The classical view would require an estimate of 4/5; but the Bayesian inference would be closer to 1/2, a number which may well be your prior estimate of obtaining heads.

The decision-theoretic perspective better reconciles the classical and Bayesian views. In particular, the classical approaches can be shown to be optimal according to certain Bayesian criteria. In other words, classical procedures are in some sense implied by the Bayesian approach. These notions will be discussed in greater detail in subsequent sections.

8.2   SOME TERMINOLOGY

The generic decision problem is as follows. An agent is to observe some *variable X*, then select a course of *action A*. The distribution of the random variable depends on a *parameter* θ. The parameter θ can be regarded as the "true state" of nature.

Let Θ denote the *state space*, the set of values that θ can assume. In a similar way, let Ω be the sample space denoting the collection of possible values for X; we may also refer to the sample space as the *data space* or *observation space*. The observation is then transformed into an action A from an action space **A**. The transformation is effected through a procedure δ from a set Δ of decision functions.

These concepts can be clarified through some examples.

- Consider the selection of a mobility system for the robot, where the choice is between wheels or legs. The "true" performance for a sentry robot might take a value θ from the parameter space $\Theta = \{$good, fair, poor$\}$. The observed performance in field tests might be a value X from the variable space $\Omega = \{$excellent, satisfactory, inadequate, miserable$\}$. The decision maker must ultimately select an action A from the action space **A** = $\{$wheels, legs$\}$.

- In identifying moving objects in a factory, consider a vision system which relies solely on the area of the profile. Then the observation X may take on any value in the data space $\Omega = (0, \infty)$, the set of positive real numbers. The sample space may consist perhaps of the set **A** = $\{$shuttle, robot, human$\}$. The parameter space Θ could be any set of characteristics relating to all the moving objects in the factory, including perhaps Fido, the night watchman's dog.

- Consider the task of estimating the mean height of Americans based on a partial sample of the population. In this case, the parameter space Θ, the sample space Ω, and the action space are all equal to the set of positive real numbers.

- How about the location of the center of the Milky Way in relation to a particular corner of my living room at the beginning of the new millenium? In rectangular coordinates, let $X = (X_1, X_2, X_3)$ be the observed position. Then $\Omega = R^3$, the three-dimensional Euclidean space. This is also the parameter space and the action space.

- The general hypothesis-testing task involves the observation of some variable X. Here the null hypothesis H_0 must be accepted or rejected. The action space is $\mathbf{A} = \{accept, reject\}$. The sample space Ω and the parameter space Θ will depend on the particular task, whether testing for defective lightbulbs in a laboratory, innocent defendants in a court of law, or a significant rise in global temperature.

Gains and Losses

The appropriateness of an action A depends on the objective of the decision maker and the true state θ of nature. Consider the problem faced by George: whether or not to hold a garage sale the following Saturday. To announce the sale, he would have to pay \$30 to place a notice in the local newspaper. If the day is sunny, he expects to receive about \$200 from customers; the net return after deducting the cost of the ad is \$170. If it rains, however, no sale is likely to occur, and George suffers a loss due to the newspaper ad. His default action is to hold no sale, for which he neither gains nor loses anything, come rain or shine. The structure of payoffs for this problem is given in Table 8–1.

Often it is convenient to think of a problem in terms of the converse of gains—namely, losses. The *loss function* $L(A, \theta)$ defines the penalty resulting from action A when the state of nature is θ. The loss function is simply the negative of the reward function. The losses for the garage sale example are shown in Table 8–2.

The literature on decision theory usually assumes that losses are nonnegative. In geometric terms, this is equivalent to positioning the origin at or below the minimal value of loss. In other words, the absolute value of the most negative number is added to the loss function.

TABLE 8–1. Structure of payoffs for holding a garage sale.

Action	State	
	θ_1: sun	θ_2: rain
a_1: no sale	0	0
a_2: sale	\$170	$-\$30$

TABLE 8–2. Structure of losses for the garage sale example. The entries are negatives of values in Table 8–1.

Action	State	
	θ_1: sun	θ_2: rain
a_1: no sale	0	0
a_2: sale	$-\$170$	$\$30$

TABLE 8–3. Structure of losses, defined as non-negative values, for the garage sale example. The worst loss in Table 8–2, namely $170, has been added to each entry.

Action	State	
	θ_1: sun	θ_2: rain
a_1: no sale	$\$170$	$\$170$
a_2: sale	0	$\$200$

TABLE 8–4. Structure of regrets for the garage sale example. The minimal loss for each state θ_i in Table 8–2 is subtracted from each entry for that state.

Action	State	
	θ_1: sun	θ_2: rain
a_1: no sale	$\$170$	0
a_2: sale	0	$\$30$

For the garage sale example, the result of adding $170 to each loss is given in Table 8–3. The table indicates that the opportunity cost of not holding a sale is $170, whereas the total penalty for peddling wares on a rainy day is $200.

Another concept of penalty is that of regret. A *regret* function indicates the penalty for choosing one procedure over the others for each possible state of nature. This function can be obtained from the loss function by subtracting the minimal loss for each state θ_i from all the entries for that state.

The structure of regrets for the garage sale example is shown in Table 8–4. The top row, for instance, relates to sunny conditions; in this case, the regret for inaction is $170 in comparison to holding a sale.

As is custom in the field of decision theory, we will focus on non-negative losses as the penalty function of choice. The structure of the loss function depends, of course, on the nature of the task and the predilections of the decision maker.

Perhaps the simplest structure for loss is the *binary function*:

$$L(\theta, A) = \begin{cases} 0 & \text{if } A \text{ is correct} \\ 1 & \text{if } A \text{ is incorrect} \end{cases}$$

This situation arises in hypothesis testing, for instance. Suppose the null hypothesis H_0 refers to the proposition $\theta \in \Theta_0$, whereas the alternate hypothesis H_1 refers to $\theta \in \Theta_1$. Then $L(\theta, A) = 0$ if θ is in the critical region defined by A, namely, Θ_A; otherwise the conclusion is wrong and $L(\theta, A) = 1$.

Another popular structure for the loss function is the *quadratic function* or *squared error*:

$$L(\theta, A) = [f(\theta) - g(A)]^2$$

where f and g are real-valued functions of the state θ and the action a, respectively. The squared error is often used in estimation problems. Suppose that A is the estimate of the mean height of Russians when the true value is θ. Then we may use $f(\theta) \equiv \theta$ and $g(A) \equiv A$ to yield the loss function

$$L(\theta, A) = (\theta - A)^2$$

Decision Rules

After a decision maker observes a random variable X from an outcome space Ω, he must select an action A. This mapping from Ω to $\mathbf{A}$ is called a *decision rule* or *procedure* δ. In other words, when a value $X = x$ is observed, the agent must take some action $a \equiv \delta(x)$.

With the introduction of the decision rule δ, we may write the loss function as $L(\theta, \delta(X))$. This is the loss due to observing variable X and executing procedure δ when the state of nature is θ.

For a particular problem domain and decision maker, one procedure δ may perform better, worse, or equally well in comparison to another procedure δ'. What criteria are we to use in comparing the procedures?

A popular criterion for evaluating the efficacy of a procedure δ is in terms of the average loss it incurs. This metric is called the *risk function*, or simply *risk*:

$$R(\theta, \delta) = E[L(\theta, \delta(X))]$$

The expectation is computed under θ and over X; in other words, θ is considered fixed while X spans its space Ω. The risk $R(\theta, \delta)$ is the average loss due to the use of decision rule δ when θ is the state of nature.

For instance, suppose that θ is to be estimated using a particular procedure $\delta(X)$. Then the risk due to $\delta(X)$ is

$$R(\theta, \delta) = E[L(\theta, \delta(X))] = E[\theta - \delta(X)]^2$$

This is the mean squared error, the same quantity encountered in Section 6.1 as a reasonable criterion for estimating a parameter.

To illustrate further, consider the hypothesis-testing task under binary loss. The null hypothesis is $H_0: \theta \in \Theta_0$, where the alternative is $H_1: \theta \in \Theta_1$. In the language of hypothesis testing, δ is simply the critical function of the test. The procedure δ maps Ω into $\{Accept, Reject\}$, or equivalently $\{0, 1\}$. The risk is given by

$$R(\theta, \delta) = P\{\delta(X) = 0\} \cdot L(\theta, 0) + P\{\delta(X) = 1\} \cdot L(\theta, 1)$$

When $\theta \in \Theta_0$, the penalty is 0 if $\delta(X) = 0$ is chosen, and is 1 otherwise. That is,

$$L(\theta, 0) = 0$$

$$L(\theta, 1) = 1$$

when θ lies in the null region. As a result, the risk is

$$R(\theta, \delta) = P\{\delta(X) = 0\} \cdot 0 + P\{\delta(X) = 1\} \cdot 1$$
$$= P\{\delta(X) = 1\}$$

when $\theta \in \Theta_0$. In a similar way, $R(\theta, \delta) = P\{\delta(X) = 0\}$ if $\theta \in \Theta_1$. In other words, the risk $R(\theta, \delta)$ under binary loss is equivalent to the probabilities of committing Type I and Type II errors.

The following scenarios further clarify the notions of loss, procedures, and risk.

Example

Consider a problem involving three potential actions and two states of nature. We denote the action space as $\mathbf{A} = \{A_1, A_2, A_3\}$ and the state space as $\Theta = \{\theta_1, \theta_2\}$. This situation occurs for a student contemplating her options for the weekend: to go skiing in the local mountains, visit her friend in the next town, or stay home and study. The relevant state space is characterized by whether or not there will be fresh snow during the weekend.

Another situation arises in the contemplation of a new product line for a company. New Age Products, Inc. is considering three options for expansion: home video products, exercise equipment, and sailboats. The performance of each product line will depend on the course of the economy over the next five years. Sailboats will be a highly profitable line if the economy remains strong, but will turn into a fiasco in case of a recession. Exercise equipment will do moderately well regardless of the economic climate, although it will be slightly more profitable in boom times. Home video products, on the other hand, are expected to perform inversely with the economy: in a recession people tend to stay indoors and seek home entertainment rather than venture outside after expensive pursuits.   ◆

Let us consider further this second case relating to corporate strategy. We will assume that the losses are as given in Table 8–5.

In order to make a rational decision, the economic forecasting department is assigned the task of predicting the business climate. The forecasts from this department are better than guesswork, but not entirely reliable. The respective probabilities are given in Table 8–6.

What strategies are available to the company? There are two possible forecasts, after each of which one among three actions may be taken; the number of potential decision rules is therefore 9. The alternative decision rules δ_j are enumerated in Table 8–7.

TABLE 8–5. Losses $L(\theta_i, a_j)$ for the product line example.

	State	
Product Line	θ_1: expansion	θ_2: recession
a_1: sailboat	0	10
a_2: exercisers	6	8
a_3: video	4	3

TABLE 8–6. Probability of accurate forecasts. For instance, when the true state is expansion, the forecast is "growth" with probability .8 and "decline" with probability .2.

	State	
Forecast	θ_1: expansion	θ_2: recession
x_1: growth	.8	.3
x_2: decline	.2	.7

TABLE 8–7. Space of decision rules. For instance, δ_1 involves choosing action a_1 regardless of the forecast; δ_2 specifies a_1 when the forecast is x_1 and a_2 otherwise.

Forecast	Decision Rule								
	δ_1	δ_2	δ_3	δ_4	δ_5	δ_6	δ_7	δ_8	δ_9
x_1	a_1	a_1	a_1	a_2	a_2	a_2	a_3	a_3	a_3
x_2	a_1	a_2	a_3	a_1	a_2	a_3	a_1	a_2	a_3

TABLE 8–8. Values of risk for each state and decision rule. The Bayes' risk is based on a prior density of 0.6 and 0.4 for θ_1 and θ_2, respectively. For instance, $r(\delta_5) = 0.6(6) + 0.4(8) = 6.8$.

State	Decision Rule								
	δ_1	δ_2	δ_3	δ_4	δ_5	δ_6	δ_7	δ_8	δ_9
$R(\theta_1, \delta_j)$	0	1.2	0.8	4.8	6	5.6	3.2	4.4	4
$R(\theta_2, \delta_j)$	10	8.6	5.1	9.4	8	4.5	7.9	6.5	3
$\max_i R(\theta_i, \delta_j)$	10	8.6	5.1	9.4	8	5.6	7.9	6.5	4
$r(\theta, \delta_j)$	4	4.16	2.52	6.64	6.8	5.16	5.08	5.24	3.6

For each value of θ_i, the risk is given by

$$R(\theta_i, \delta) = E[L(\theta_i, \delta(X)]$$
$$= P\{\delta(X) = a_1\}L(\theta_i, a_1) + P\{\delta(X) = a_2\}L(\theta_i, a_2)$$
$$+ P\{\delta(X) = a_3\}L(\theta_i, a_3)$$

To illustrate, the risks for the procedure δ_3 are

$$R(\theta_1, \delta_3) = 0.8(0) + 0.2(4) = 0.8$$
$$R(\theta_2, \delta_3) = 0.3(10) + 0.7(3) = 5.1$$

The complete set of risk values for this example is shown in the first two lines of Table 8–8.

When the state space Θ is discrete and finite, we may represent the values of risk as a point in Euclidean space. In particular, if Θ has n elements, then the risk is a point in R^n.

In the product line example, $n = 2$. Consequently, we can represent the risks on a two-dimensional chart as shown in Figure 8–1.

It is clear from the figure that certain decision rules are superior to a number of others for each value of the state of nature. For instance, δ_3 has lower risk than δ_7 regardless of the state θ_i.

When one course of action δ is better than another procedure δ', we say

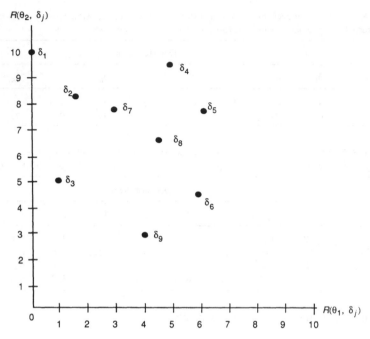

FIGURE 8-1. Values of risk for decision rules, as given in Table 8-8. The state θ_1 denotes economic expansion and θ_2 recession.

that δ *dominates* or *supersedes* δ'. Conversely, we may say that δ' *subsedes* δ. For instance, decision rule δ_3 supersedes all procedures other than δ_1, δ_6, and δ_9.

The set of all actions that are not dominated by any other procedures is called the *efficient frontier* or the *Pareto-optimal* set. In Figure 8-1, the procedures $\{\delta_1, \delta_3, \delta_9\}$ constitute the efficient frontier.

Clearly, a decision maker should choose an action from among those on the efficient frontier. But how is one to select one efficient point over another in the same category? Among these procedures there are no clear-cut victors. For example, δ_3 has lower risk than δ_9 when the state is θ_1, but the converse holds for the state θ_2.

Criteria for selecting one efficient procedure over another is the subject of the next section.

8.3   RANKING DECISION RULES

A decision rule may be better than, worse than, or equivalent to another in terms of performance. As discussed previously, we say that a decision rule δ

dominates or *supersedes* another rule δ' if the relation

$$R(\theta, \delta) \leq R(\theta, \delta')$$

holds for every value of θ, whereas strict inequality holds for at least one θ. In other words, δ supersedes δ' if the risk of the former is smaller for at least one state of nature, and no worse for all other states.

In general, decision rules cannot be ordered so simply; a procedure δ is likely to perform better than another rule δ' under some conditions, and worse under others. To illustrate, δ_3 supersedes δ_9 in Figure 8–1 under conditions of economic growth but subsedes the latter in a recession.

Properties of Estimators

We would like for procedures to exhibit certain properties depending on the domain of application. Examples of desirable properties are unbiasedness and invariance.

- *Unbiasedness* refers to the idea that a decision rule should be accurate. In particular, the expected value of a procedure should equal that of the estimated parameter. For instance, given a set of observations $\{X_1, X_2, \ldots, X_n\}$ from a distribution having a true mean θ, the sample average given by $\overline{X} = \sum_i X_i/n$ is unbiased. In other words, $E(\overline{X}) = \theta$.
- *Invariance* pertains to the notion that the structure of a decision rule should be isotropic; that is, unaffected by the same operations that transform the underlying decision problem. For instance, suppose that the task is to deduce the mean θ of a distribution which generates variables $\{X_1, \ldots, X_n\}$. Suppose that the sample mean is 5, and the procedure accepts this value as the estimate of the state θ. In that case, if the observed value is higher by 3, yielding 8, then the procedure should accept this value instead.

  In other words, suppose that the estimate of a parameter θ given observation X is $\hat{\theta} = \delta(X)$. Let f denote some transformation of the random variable X. Then the estimate resulting from a transformed variable $f(X)$ should be $f(\hat{\theta})$. In the above example, $f(X) = X + 3$ and $f(\hat{\theta}) = \hat{\theta} + 3$.

The importance of properties such as unbiasedness and invariance depend on the application domain. For instance, they are admirably suited for problems of estimation. In contrast, they may be less relevant for sequential decision making in the analysis of stock prices; here, the perceptions of investors could weigh more heavily than accurate estimation of the data.

Preferences among Procedures

Decision rules may be ranked relative to each based on their risk characteristics and the tastes of the decision maker. Some candidate principles for ordering decision rules are the following:

- *Minimin*: choosing the procedure offering the lowest possible risk of average loss. This strategy seeks the rule that offers the lowest possible penalty, or maximum possible reward. It may be appropriate for the inveterate optimist, with little or no concern for negative consequences.
- *Minimax*: selecting the procedure with the most attenuated adverse consequence. In other words, the decision rule should yield the minimum among the maximum losses that may result from the alternative procedures. This is a pessimistic view of life and is recommended for cautious individuals.
- *Bayes'*: favoring the procedure offering the lowest expected risk. To this end, the risks due to each decision rule are consolidated into their expected value, and the rule with lowest expected risk is selected. This strategy is appropriate for an individual with a neutral perspective on life—one for whom large potential losses are not disproportionately painful compared to smaller penalties.

As a group, decision theorists must be more pessimistic than optimistic. The optimistic strategy defined by the minimin principle has received little attention. Rather, the literature focuses on minimax and Bayes' strategies. We will concentrate on these latter principles as well.

Minimax Strategy

According to the minimax principle, a good procedure is one which entails only a small penalty under all circumstances. More specifically, a procedure δ supersedes another rule δ' if its worst risk is smaller:

$$\sup_\theta R(\theta, \delta) < \sup_\theta R(\theta, \delta')$$

A decision rule δ^* is caled *minimax* if its worst risk is the smallest among all candidate rules:

$$\sup_\theta R(\theta, \delta^*) = \inf_\delta \sup_\theta R(\theta, \delta)$$

The minimax principle originates from the theory of games. The decision problem may be regarded as a game between two players whom we will call Nancy and Danny. Nancy, as the first player, picks a state $\theta \in \Theta$. Without any knowledge of the value of θ, Danny selects a procedure δ from the space Δ of candidate procedures. Danny must then pay Nancy the penalty $R(\theta, \delta)$ depending on the choices for θ and δ. Danny can limit his losses by following the minimax strategy. In particular, his loss will not exceed the maximum risk $R(\theta, \delta^*)$ for any value of θ.

The worst loss $\sup_\theta R(\theta, \delta^*)$ that Danny need tolerate is called the *upper pure value* of the game. The "pure" refers to the adoption of a single procedure rather than a combination of such; we will explore hybrid strategies later in this chapter.

In the context of the decision problem, Nancy is viewed as nature and Danny as the decision maker. The minimax procedure represents the most conservative strategy that safeguards against the worst that could happen. Unlike a zealous competitor, nature will not act deliberatively to make life difficult for a decision maker. Even so, given the common desire to limit losses, the minimax rule is a reasonable strategy in many practical contexts.

Example

Let us return to the product line example in the previous section. The first two rows in Table 8–8 show the risk $R(\theta_i, \delta_j)$ corresponding to each state θ_i of the economy and each decision δ_j. The maximum risk $R_*(\delta_j)$ for each decision is listed in the third row.

Among the maximum risks $R_*(\delta_j)$, the minimum value is given by $R_*(\delta_9) = 4$. Consequently, $\delta^* = \delta_9$ is the minimax procedure.

As it happens, δ_9 is the southernmost point on the chart in Figure 8–1. In geometric terms, the minimax procedure lies closest to the origin in each of the northern and eastern directions. ◆

Bayes' Strategy

In the Bayesian view of the world, a state θ of nature is a particular value of a random variable $\tilde{\theta}$ taking values in Θ with some density $\pi(\tilde{\theta})$. Consequently, the observed variable X can be regarded as a random variable having a conditional distribution $f_\theta(X) = f(X | \tilde{\theta} = \theta)$.

According to this view, it is reasonable to take the expectation of the risk as $\tilde{\theta}$ varies. In other words, the risk $R(\theta, \delta)$ can be averaged with respect to the distribution $\pi(\tilde{\theta})$. The result is the Bayes' risk for δ:

$$r(\delta) \equiv E[R(\tilde{\theta}, \delta)]$$

where the expectation is taken over $\tilde{\theta}$. Since $R(\tilde{\theta}, \delta) = E[L(\tilde{\theta}, \delta(X)]$, the Bayes' risk is equivalent to

$$r(\delta) = E[L(\tilde{\theta}, \delta(X)]$$

where the expectation is taken over both X and $\tilde{\theta}$.

The Bayes' principle asserts that a good procedure is one which entails a small average risk under various circumstances. In particular, a procedure δ

supersedes another rule δ' if its expected risk is smaller:

$$r(\delta) = E[R(\theta, \delta)] < r(\delta') = E[R(\theta, \delta')]$$

We say that a decision rule $\delta*$ is *Bayes* if its Bayes' risk is the smallest among all candidate rules:

$$r(\delta*) = \min_\delta r(\delta)$$

The Bayes' strategy is a middling approach, being neither entirely cautious nor thoroughly rash. It is appropriate when a decider can afford to live with the average loss implied by his strategy.

Example

To compute the Bayes' risks in the product line example, we need to determine a probability density for the course of the economy. Assume that, in the considered opinion of the management, the likelihood of economic growth is 0.6 and of decline 0.4. In other words, $\pi(\theta_1) = 0.6$ and $\pi(\theta_2) = 0.4$. Then the Bayes' risk for each decision δ_j is given by

$$r(\delta_j) = 0.6R(\theta_1, \delta_j) + 0.4R(\theta_2, \delta_j)$$

For procedure δ_1, for instance, the Bayes' risk is $r(\delta_1) = 0.6(0) + 0.4(10) = 4$. The complete set of values is given in the last row of Table 8–8. The lowest Bayes' risk is $r(\delta_3) = 2.52$, thereby identifying procedure δ_3 as the Bayes' rule.　◆

Geometric Interpretation of Bayes' Strategy

How is the Bayes' strategy to be interpreted geometrically? To simplify the notation, let ρ_1 and ρ_2 denote the axes of risk—namely, $\rho_1 \equiv R(\theta_1, \delta)$ and $\rho_2 \equiv R(\theta_2, \delta)$. Then the risk for procedure δ is given by

$$r(\delta) = \pi_1 \rho_1 + \pi_2 \rho_2$$

For each value of risk $r(\delta) = c$, the relation

$$\pi_1 \rho_1 + \pi_2 \rho_2 = c$$

defines a line on a chart. This is depicted in Figure 8–2. The slope of the line is given by $-\pi_1/\pi_2$; the horizontal intercept is c/π_1, and the vertical intercept c/π_2. Two procedures δ_a and δ_b lying on the same curve have the same Bayes'

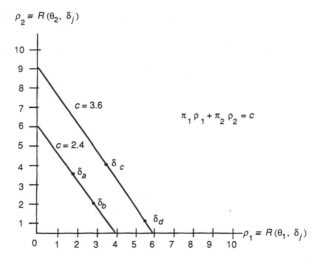

FIGURE 8–2. Isoquants of the Bayes' criterion. Procedures δ_a and δ_b lie on the curve $0.6\rho_1 + 0.4\rho_2 = 2.4$; their Bayes' risks are equal and better than the value of 3.6 for δ_c and δ_d. In general, the slope of a Bayes' curve is $-\pi_1/\pi_2$: if π_1 is large relative to π_2, then the slope is steeper, and procedures closer to the vertical axis are favored.

risk c and are, therefore, equivalent. A decision rule on a lower curve supersedes that on a higher one; for instance, δ_a dominates δ_c.

Issues in Bayesian Analysis

The Bayes' strategy calls for a calculation of the average risk with respect to the density π of the state θ:

$$r(\delta) = \begin{cases} \sum_\theta R(\theta, \delta)\pi(\theta) & \text{if } \theta \text{ is discrete} \\ \int R(\theta, \delta)\pi(\theta)\, d\theta & \text{if } \theta \text{ is continuous} \end{cases}$$

Some individuals, however, find it uncomfortable to think of nature selecting a value θ according to a density function $\pi(\theta)$. In their view, a state of nature simply *is*: Mother Nature has adopted some state of affairs and remains steadfast in this position. Hence, there is no point in thinking about, or working with, density functions for the state θ.

There are two counterarguments to this objection: subjective densities and risk weighting. According to the *subjective density* view, the probability function $\pi(\theta)$ is a prior density defined solely in the mind of the decision maker. Nature may have selected a particular value of state, but this value is unknown to the decider. Under these conditions, the decider may draw on his general knowledge of the world and past experience to regard certain values of θ as being more likely than others. This subjective knowledge is then embodied in the density $\pi(\theta)$.

According to the second argument, the density $\pi(\theta)$ can be regarded simply as a weighting of risks. For instance, the Bayes' risk in the discrete case is given by $r(\delta) = \sum_i R(\theta_i, \delta)\pi(\theta_i)$. If the risk $R(\theta_i, \delta)$ due to state θ_i is considered to be more troublesome for the decision maker than another, then its weight $\pi(\theta_i)$ will be higher. For the two-state case, taking a simple average of the risks

$$\frac{R(\theta_1, \delta) + R(\theta_2, \delta)}{2}$$

corresponds to the situation where the weights are equal at $\pi(\theta_1) = \pi(\theta_2) = 0.5$. In other words, the risk due to state θ_1 is no better or worse than that for θ_2.

Composite Procedures

We have thus far focused on the use of individual strategies in which one particular procedure is employed. This is called a *pure* or *elementary* strategy.

A decider may, however, be able to decrease his greatest risk by selecting randomly among two or more rules. This is called a *randomized* or *composite* strategy. The average risk for the composite technique will then lie somewhere between the risks associated with the pure strategy.

To illustrate, consider rules δ_1 and δ_9 in Table 8–8. Suppose that our composite strategy consists of selecting either of these procedures based on the toss of a fair coin. Then the expected risk associated with state θ_1 is

$$0.5R(\theta_1, \delta_1) + 0.5R(\theta_1, \delta_9) = 0.5(0) + 0.5(4) = 2.0$$

In a similar way, the expected risk due to state θ_2 is

$$0.5R(\theta_2, \delta_1) + 0.5R(\theta_2, \delta_9) = 0.5(10) + 0.5(3) = 6.5$$

The maximum risk of 6.5 is lower than that due to the pure procedure δ_1, although not to δ_9.

In some cases, the composite procedure may fare better than either of its underlying actions. This happens with the use of procedures δ_6 and δ_8 in Table 8–8. If these two rules are employed randomly with equal probability, the expected risk is 5.0 for θ_1, and 5.5 for θ_2. The worst risk of 5.5 is lower than the 5.6 due to δ_6 as well as the 6.5 due to δ_8.

Let Δ denote the class of all elementary decision rules, and Δ^+ the set of composite procedures. The composite class Δ^+ is obviously a superset of the elementary set Δ.

Each composite rule δ in Δ^+ can be viewed as a *convex* combination of the procedures δ_i in Δ. In other words,

$$\delta = \sum_{j=1}^{n} \lambda_j \delta_j$$

where λ_j are non-negative real numbers summing to one: $\lambda_j \geq 0$ and $\sum_j \lambda_j = 1$. The expression $\lambda_j \delta_j$ is to be interpreted as the rule "Choose δ_j with probability λ_j." In particular, if procedure δ is a pure strategy equal to δ_k, then $\lambda_k = 1$ and all other λ_j equal 0.

The risk due to a composite decision rule δ is the convex combination of the elementary risks. That is,

$$R(\theta, \delta) = \sum_{j=1}^{n} \lambda_j R(\theta, \delta_j) \tag{8-1}$$

where the δ_j are the procedures which comprise δ. A *composite minimax* or *randomized minimax* procedure δ^* is one that minimizes $\sup_\theta R(\theta, \delta)$ among all composite procedures. In other words, δ^* minimizes the worst risk under any value of state θ.

The Bayes' risk of a procedure δ is the convex combination of expected risks due to each elementary procedure δ_j. In other words, the Bayes' risk of δ is

$$r(\delta) = \sum_{j=1}^{n} \lambda_j E[R(\tilde{\theta}, \delta_j)]$$

The notation $\tilde{\theta}$ underscores the fact that θ is to be regarded as a variable here. The expectation over $\tilde{\theta}$ is taken with respect to some prior density π on the state space Θ.

A *composite Bayes'* or *randomized Bayes'* procedure δ^* is one that minimizes $r(\delta)$ among all composite procedures. That is, the decision function δ^* exhibits the smallest average risk among all composite procedures.

Elementary versus Composite Procedures

What relationships hold among elementary and composite procedures? We now examine the nature of composite procedures as generalizations of elementary rules and argue for the significance of Bayes' procedures.

For the sake of simplicity, we begin with the product line example presented in earlier sections. In this example, the state of nature takes two values—θ_1 and θ_2. The risk of any rule δ may, therefore, be represented as a pair of values (ρ_1, ρ_2); we recall that $\rho_i \equiv R(\theta_i, \delta)$ is the risk due to procedure δ when the state of nature is θ_i.

The *risk set* S can now be defined as the collection of risks due to all the composite procedures δ in Δ^+:

$$S \equiv \{(r_1, r_2)\} = \{(R(\theta_1, \delta), R(\theta_2, \delta)): \delta \in \Delta^+\}$$

As indicated in Eq. (8–1), the risk of a composite strategy is the convex combination of the elementary risks due to the pure procedures. In other words,

$$r_i \equiv R(\theta_i, \delta) = \sum_{j=1}^{n} \lambda_j R(\theta_i, \delta_j)$$

where $\lambda_j \geq 0$ and $\sum_j \lambda_j = 1$.

For the product line example, the risk set is given by

$$S = \left\{ \sum_{j=1}^{9} \lambda_j R(\theta_1, \delta_j), \sum_{j=1}^{9} \lambda_j R(\theta_2, \delta_j) \right\}$$

The risk set is shown as the shaded region in Figure 8–3: the collection of points δ_j, all line segments joining them, and all line segments joining *these* line segments. The boundary of S consists of the corner points defined by δ_1, $\delta_3, \delta_4, \delta_5, \delta_6,$ and δ_9, as well as "extreme" line segments between pairs of these points. In this context, an "extreme" line segment is one whose associated line partitions the risk space into two regions, one of which contains no risk points. For instance, the segment between δ_1 and δ_3 is part of the boundary, but not that between δ_3 and δ_4.

Finally, each point in the interior of S can be written as a convex combination of two points on the boundary. The collection of all convex combinations of a set of points is also called the *convex hull* of the set.

The minimax strategy calls for the procedure that minimizes the maximum risk. The geometric interpretation is as follows. Consider a family of squares based on the origin and parameterized by a length c, as shown in Figure 8–3. Each box has a diagonal stretching from the origin to the point (c, c). These squares are defined by the relation

$$M(c) = \{(\rho_1, \rho_2): 0 \leq \rho_1 \leq c, 0 \leq \rho_2 \leq c\}$$

Suppose c^* is the minimal value of c for which the square $M(c)$ touches the risk set S. The intersection of these sets may be a point, as illustrated in Figure 8–3; the intersection could also be a horizontal line segment or a vertical line segment, depending on the shape of S. (Why?) Any point lying in this intersection will constitute a minimax rule.

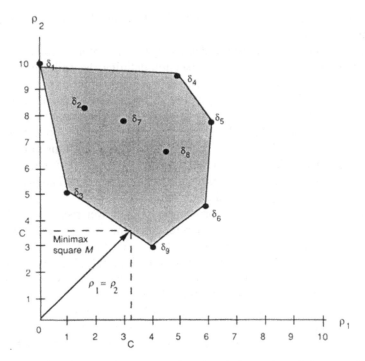

FIGURE 8-3. The risk set S is the convex combination of all procedures δ_i. Any point δ on the boundary of S can be represented as a corner point or the combination of two adjacent corner points; and any point on the interior by combining two boundary points. The set of convex combinations is also called the convex hull of the procedures δ_j. The minimax procedure lies at the first point(s) of intersection of the risk set and the square growing from the origin.

In our example, the minimax rule is a composite procedure based on the elementary rules δ_3 and δ_9. What are the relative contributions of the elementary rules to the composite strategy? This can be determined from the relation $\rho_1 = \rho_2$ where

$$\rho_1 = \lambda R(\theta_1, \delta_3) + (1 - \lambda)R(\theta_1, \delta_9) = \lambda(0.8) + (1 - \lambda)(4)$$
$$\rho_2 = \lambda R(\theta_2, \delta_3) + (1 - \lambda)R(\theta_2, \delta_9) = \lambda(5.1) + (1 - \lambda)(3)$$

The solution is $\lambda \approx 0.19$. The composite rule, therefore, involves selecting procedure δ_3 with about 19% probability, and procedure δ_9 otherwise.

The geometric representation of the Bayes' strategy was discussed in Section 8.2. The criteria involve a family of triangles advancing from the origin, as shown in Figure 8-2. The slope of the hypotenuse depends on the prior density function over the state space Θ. For the two-state case, if the priors for θ_1 and θ_2 are π_1 and π_2, respectively, then the slope of the Bayes' line segments is $-\pi_1/\pi_2$.

The set of Bayes' procedures is defined by the first intersection of the Bayes' triangles with the risk set. The intersection defining the collection of Bayes' rules may be a point or a line segment, just as for the minimax strategy. If the intersection is a line segment, any point on it represents a Bayes' rule.

A Bayes' rule may be elementary or complex. In the former case, it lies on a corner of the risk set S; in the latter, on another boundary point of S.

The minimax and Bayes' criteria imply preference relations among candidate decision rules. A decision rule δ' is called *inadmissible* if there is another rule δ that supersedes δ'. Conversely, a procedure δ is called *admissible* if it is not superseded by any other rule.

In the product line example, procedure δ_7 has lower risk than δ_4 for both states θ_1 and θ_2. Procedure δ_4 is, therefore, inadmissible. In contrast, δ_3 is an admissible procedure since it is superseded by no other rule.

Geometrically, a decision procedure is inadmissible if it lies to the north and east of at least one other procedure. Conversely, the lower left boundary of the risk set S defines the set of admissible rules.

Our discussion has focused on the two-dimensional case for the state θ. Many of the concepts, however, carry over to higher dimensions.

Let $\Theta = \{\theta_1, \theta_2, \ldots, \theta_n\}$ be a finite state space. The risk set may be written as

$$S = \{(R(\theta_1, \delta), \ldots, R(\theta_n, \delta))\}$$

where θ is in Θ and δ is a composite procedure in Δ^+.

In this case, the minimax criterion involves a family of cubes advancing from the origin in n-dimensional space. The intersection of the minimax cube with the risk set S may be a point, line segment, or any bounded convex object having a dimensionality of $n - 1$ or less.

Some characteristics of the Bayes' strategy are as follows:

- All admissible decision rules are Bayes' rules.
- Bayes' procedures lie on the lower boundary of the risk set S.
- For any prior density π, if a composite Bayes' procedure exists, then an elementary Bayes' procedure also exists.
- If Θ is finite and there are minimax procedures, they are Bayes' rules.

For the case where Θ is infinite, there may be admissible decision rules which are not Bayes. But under certain conditions, all admissible rules may be shown to be Bayes' rules or limiting cases of Bayes' rules. In many situations, admissible rules are elementary procedures. The validity of these results including those for elementary procedures relies on the consideration of composite decision rules.

The next section illustrates the use of the Bayes' strategy in the context of parameter estimation.

8.4   BAYESIAN ESTIMATION

In the classical approach to estimation presented in Chapter 6, the parameter θ was regarded as an unknown but fixed quantity. This unknown value was to be determined from a set of observations of a random variable X. In contrast, we now regard the parameter θ as a random variable. The goal of estimation in this context is to determine the characteristics of θ after observing it indirectly in the form of sample data.

More specifically, let $X_1, \ldots, X_n$ be a random sample from a probability function $f(x|\theta)$. The parameter θ is a random variable from a parameter space Θ with a probability function π. From this perspective, $f(x|\theta)$ is the conditional probability function for X given the value of θ. The joint probability for X and θ is defined by $f(x, \theta) = \pi(\theta)f(x|\theta)$. Since x is a vector of random observations, we can write the joint density as

$$f(x_i, \ldots, x_n, \theta) = \pi(\theta)f(x_1, \ldots, x_n|\theta) = \pi(\theta) \prod_{i=1}^{n} f(x_i|\theta) \qquad (8\text{-}2)$$

The purpose of estimation is to assess the attributes of θ based on the indirect observations $x_1, \ldots, x_n$.

Let $\delta(x)$ be the decision procedure for estimating parameter θ. Under squared error loss, the risk for δ at a particular value θ_0 is

$$R(\theta_0, \delta) = \int [\delta(x) - \theta_0]^2 f(x|\theta_0) \, dx$$

The notation dx is a shorthand for $dx_1 \, dx_2 \ldots dx_n$, whereas the integral actually denotes an n-fold integration over the variables x_i.

The Bayes' risk for procedure δ is its expected risk over the entire parameter space Θ:

$$r(\delta) = \int \int [\delta(x) - \theta]^2 f(x|\theta)\pi(\theta) \, dx \, d\theta$$

As usual, each integral should be replaced by a summation if the corresponding probability function for X or θ is discrete rather than continuous. Drawing on Eq. (8-2), the mean risk for procedure δ can be written

$$r(\delta) = \int \int [\delta(x) - \theta]^2 f(x, \theta) \, dx \, d\theta \qquad (8\text{-}3)$$

Our goal is to determine the characteristics of the parameter θ based on the observed data x. This is equivalent to finding some function $h(\theta|x)$ which specifies the value of θ once the random sample $X = x$ is observed. We call $h(\theta|x)$ the *posterior probability distribution* for θ, given the sample data x. In summary, $\pi(\theta)$ is the *prior probability distribution* for parameter θ before observing the random sample, whereas $h(\theta|x)$ is the posterior distribution after the sampling.

Let $g(x)$ denote the unconditional probability function for X. Then the joint density between X and θ can be expressed as

$$f(x, \theta) = g(x)h(\theta|x)$$

Injecting this relationship into Eq. (8–3) and rearranging some factors, we obtain

$$r(\delta) = \int g(x)\left\{ \int [\theta - \delta(x)]^2 h(\theta|x)\, d\theta \right\} dx \qquad (8\text{–}4)$$

Each factor on the right-hand side of the equation is non-negative. Consequently, the Bayes' risk $r(\delta)$ will be minimal if the quantity in braces is minimal. In turn, the inner integral will be minimal if the procedure δ is selected in such a way as to minimize the squared error.

Since $h(\theta|x)$ is the conditional distribution of θ given x, the inner integral represents the second moment of θ about the point $\delta(x)$. The second moment is minimized when $\delta(x)$ equals the mean m of parameter θ. To show this, we note the following:

$$\int [\theta - \delta(x)]^2 h(\theta|x)\, d\theta = E[\theta - \delta(x)]^2 = E\{(\theta - m) + [m - \delta(x)]\}^2$$

$$= E[(\theta - m)^2 + 2(m - \delta x)]E(\theta - m) + [m - \delta(x)]^2$$

$$= \operatorname{Var}\theta + 0 + [m - \delta(x)]^2 \qquad (8\text{–}5)$$

The second equality follows from adding and subracting the mean m of variable θ. The next equality obtains after completing the square. We note from the last expression that Eq. (8–5) is minimized when $\delta(x)$ equals the mean m of θ. Finally, we conclude that the Bayes' risk given by Eq. (8–4) is minimized when the estimator for θ is the mean value of the posterior probability distribution. We have just proven the following result.

Theorem (Bayes' Estimator). Let X be a random sample from a probability function $f(x|\theta)$. The parameter θ originates from a parameter space Θ with probability function π. Suppose $\delta(x)$ is the procedure which estimates θ by the mean of the posterior probability function $h(\theta|x)$; that is, $\delta(x) = E(\theta|x)$. Then δ is a Bayes' estimator.

We illustrate the use of this theorem in the context of the binomial density function.

Example

Let x be a random sample of size n from a binomial distribution with parameter p. The prior density function for p is a β-distribution given by

$$\pi(p) = \frac{\Gamma(\alpha + \beta)}{\Gamma(\alpha)\Gamma(\beta)} p^{\alpha-1}(1 - p)^{\beta-1} \quad 0 < p < 1 \tag{8-6}$$

where Γ denotes the Γ-function. What is a Bayes' estimator for p?
The joint density function for X and p is

$$f(x, p) = \pi(p)f(x|p) = \left(\frac{\Gamma(\alpha + \beta)}{\Gamma(\alpha)\Gamma(\beta)} p^{\alpha-1}(1 - p)^{\beta-1}\right)\left[\binom{n}{k} p^k(1 - p)^{n-k}\right]$$

$$= \binom{n}{k}\frac{\Gamma(\alpha + \beta)}{\Gamma(\alpha)\Gamma(\beta)} p^{\alpha+k-1}(1 - p)^{\beta+n-k-1} \tag{8-7}$$

Here $k = \sum_i x_i$ denotes the number of successes in the n observations. The posterior density function is then

$$h(p|x) = \frac{f(x, p)}{g(x)} = \frac{f(x, p)}{\int_0^1 f(x, p)\,dp}$$

where $g(x)$ denotes the unconditional or marginal density function for x. The Bayes' solution is the mean of the posterior distribution:

$$E(p|x) = \int_0^1 ph(p|x)\,dp = \frac{\int_0^1 pf(x, p)\,dp}{\int_0^1 f(x, p)\,dp}$$

Use of Eq. (8–7) leads to

$$E(p|x) = \frac{\int_0^1 p^{\alpha+k}(1 - p)^{\beta+n-k-1}\,dp}{\int_0^1 p^{\alpha+k-1}(1 - p)^{\beta+n-k-1}}$$

Since π is a density function, its complete integral must equal unity. We, therefore, infer from Eq. (8–6) that

$$\int_0^1 p^{a-1}(1-p)^{b-1} = \frac{\Gamma(a)\Gamma(b)}{\Gamma(a+b)}$$

We can then express the mean of the posterior distribution as

$$E(p|x) = \frac{\Gamma(\alpha+k+1)\Gamma(\beta+n-k)/\Gamma(\alpha+\beta+n+1)}{\Gamma(\alpha+k)\Gamma(\beta+n-k)/\Gamma(\alpha+\beta+n)}$$

$$= \frac{\alpha+k}{\alpha+\beta+n} \qquad\qquad (8\text{–}8)$$

The second equality depends on the recursive property of the Γ-function, namely, $\Gamma(a+1) = a\Gamma(a)$. Equation (8–8) is the Bayes' solution to the estimation problem.

To clarify the ideas further, consider the simple case for which $\alpha = \beta = 1$. Then from Eq. (8–6), the prior density $\pi(p)$ is a constant function over the interval $[0, 1]$; that is, $\pi(p) = 1$ is the uniform density. The mean of the posterior distribution is given by Eq. (8–8) as $E(p|x) = (k+1)/(n+2)$.

This expansion can be contrasted with the result from Chapter 6. The best unbiased estimator from the classical (non-Bayesian) perspective was $\delta(x) = k/n$.

When the sample size n is large, both these procedures yield similar estimates. But they can differ markedly for small sample sizes. Suppose only a single observation is made. According to the classical view, the estimate of p is extreme: the result is 0 or 1, depending on whether a failure or success is encountered. By the Bayesian procedure, however, the estimate is 1/3 or 2/3. The Bayesian approach takes into account prior perceptions of the distribution of p, thereby yielding a more moderate estimate than the classical procedure. For instance, suppose p represents the proportion of a population in favor of a particular candidate running for office. Then the Bayesian estimate clings more closely to 1/2, which is the mean of the prior distribution for p.

To select another special case, let $\alpha = \beta = 10$. Then the probability density in Eq. (8–6) has a high value near the center at $p = 1/2$ and vanishes toward either end at 0 or 1. By Eq. (8–8), the Bayes' solution for a single observation is 10/21 or 11/21. Once again, the Bayes' estimate deviates only slightly from the mean of the prior distribution.   ◆

The next example relates to the estimation of the mean of a normal density function. In this case, the prior density for the mean is itself a normal function.

Example

Let X be a vector of n observations from a normal density $f(x|\mu)$ with known variance σ^2. The mean μ has a normal density $\pi(\mu)$ with mean v and standard deviation τ. What is a Bayesian estimator for μ?

The prior density function for μ is

$$\pi(\mu) = \frac{1}{\tau\sqrt{2\pi}} \exp\left[-\frac{1}{2}\left(\frac{\mu - v}{\tau}\right)^2 \right]$$

The joint density for a vector x of observations and mean μ is

$$
\begin{aligned}
f(x, \mu) &= f(\mu)f(x|\mu) \\
&= \left(\frac{\exp\{-\frac{1}{2}[(\mu - v)/\tau]^2\}}{\tau\sqrt{2\pi}}\right)\left(\frac{\exp\{-\frac{1}{2}\sum_{i=1}^{n}[(x_i - \mu)/\sigma]^2\}}{\sigma\sqrt{2\pi}}\right) \\
&= \frac{1}{2\pi\sigma\tau} \exp\left\{-\frac{1}{2}\left[\left(\frac{\mu - v}{\tau}\right)^2 + \sum_{i=1}^{n}\left(\frac{x_i - \mu}{\sigma}\right)^2\right]\right\}
\end{aligned}
\tag{8-9}
$$

Before calculating $E(\mu|x)$, let us first determine the posterior density $h(\mu|x)$ based on the observation x. Since $h(\mu|x) = f(x, \mu)/g(x)$, we must obtain the marginal density $g(x)$ for x.

The density $g(x)$ results from integrating $f(x, \mu)$ with respect to μ. To this end, we isolate μ in the last expression in Eq. (8–9). We expand the squares, collect common terms for μ^2 and μ, respectively, then complete the square for μ. The expression in braces in Eq. (8–9) becomes $s(\mu) + t$ where

$$s(\mu) = \frac{\sigma^2 + n\tau^2}{\sigma^2\tau^2}\left(\mu - \frac{v\sigma^2 + n\bar{x}\tau^2}{\sigma^2 + n\tau^2}\right)^2$$

$$t = \frac{1}{\sigma^2\tau^2}\left[\frac{\sigma^2 v^2 + \tau^2\sum_i x_i^2}{\sigma^2\tau^2} - \frac{(v\sigma^2 + n\bar{x}\tau^2)^2}{\sigma^2 + n\tau^2}\right]$$

The joint density for x and μ takes the form

$$f(x, \mu) = c\exp\left(-\frac{s(\mu) + t}{2}\right)$$

for some constant c. The marginal density for μ is then

$$h(\mu|x) = \frac{f(x, \mu)}{g(x)} = \frac{f(x, \mu)}{\int_{-\infty}^{\infty} f(x, \mu)\,d\mu} = \frac{e^{-s(\mu)/2}}{\int_{-\infty}^{\infty} e^{-s(\mu)/2}\,d\mu}\tag{8-10}$$

The third equality follows from canceling the c and t terms in the expression. The conditional density $h(\mu|x)$ has the same structure as the normal density. The mean and variance of this normal density can be determined respectively from the subtractive term and the multiplicative factor in the expression for $s(\mu)$. The denominator in Eq. (8–10) is simply $\sqrt{2\pi}$ times the standard deviation of the function. The result is

$$h(\mu|x) = \exp\left[-\frac{1}{2}\left(\mu - \frac{v\sigma^2 + n\bar{x}\tau^2}{\sigma^2 + n\tau^2} \right)^2 \middle/ \frac{\sigma^2\tau^2}{\sigma^2 + n\tau^2} \right] \middle/ \left(\frac{2\pi\sigma^2\tau^2}{\sigma^2 + n\tau^2} \right)^{1/2} \quad (8\text{–}11)$$

The mean of this density function defines the Bayesian estimator:

$$\delta(\mu) = Eh(\mu|x) = \frac{v\sigma^2 + n\bar{x}\tau^2}{\sigma^2 + n\tau^2}$$

We can also write this equation as

$$\delta(\mu) = \frac{\rho}{1 + \rho}v + \frac{1}{1 + \rho}\bar{x}$$

where $\rho \equiv \sigma^2/n\tau^2$ represents the ratio of variances: σ^2/n is the variance of $\overline{X}$ and τ^2 the variance of the prior density π.

Suppose that the variance τ^2 of the prior density $\pi(\mu)$ is large. Then there is little information concerning the pattern of values for μ. The ratio ρ is small, and the decision procedure relies primarily on the sample data: $\delta(x) \approx \bar{x}$. This is the classical, non-Bayesian procedure for estimating the mean μ in $f(x|\mu)$.

On the other hand, assume that the variance σ^2 of each observation is large. Since each observation carries little information regarding the location of μ, the ratio ρ is large and the estimator relies primarily on the mean of the prior density: $\delta(x) \approx v$.

Finally, when the sample size n increases indefinitely, the variance σ^2/n of the sample $\overline{X}$ shrinks. Since ρ vanishes, the estimator again relies primarily on the observed data: $\delta(x) \approx \bar{x}$.

How does the mean risk for the Bayesian estimator compare with that of the classical solution? Since the mean of the classical estimator is $E(\overline{X}) = \mu$, the risk under squared error is simply the variance of $\overline{X}$. In other words, $R(\theta, \overline{X}) = \sigma^2/n$, where $\theta \equiv \mu$ is the parameter being estimated. Since the risk is independent of the parameter, the Bayes' risk is identical: $r(\overline{X}) = \sigma^2/n$.

The mean risk for the Bayes' estimator is as follows. In Eq. (8–4), the inner integral within braces is the variance of the conditional density $h(\theta|x)$ when the procedure δ is used. From Eq. (8–11), the variance of h is $\sigma^2\tau^2/(\sigma^2 + n\tau^2)$.

We can integrate this constant variance with respect to the marginal density $g(x)$, as specified in Eq. (8–4). The result is $r(\delta) = \sigma^2 \tau^2/(\sigma^2 + n\tau^2)$.

The ratio of risks for the Bayesian and classical procedures is

$$\frac{r(\delta)}{r(\overline{X})} = \frac{1}{1 + \rho}$$

When the prior density π is diffuse, for instance, then τ^2 is large and ρ small. In that case, the ratio of mean risks approaches unity, and the Bayes' estimator offers no special advantage. On the other hand, when μ is tightly clustered, τ^2 is small, ρ is large, and the mean risk for the Bayesian solution is relatively small in comparison to the classical procedure.   ◆

A numerical example for the prior density having a normal function is given in the problems.

The Bayesian approach offers the advantages of integrating information from successive observations. More specifically, the prior distribution for a parameter θ can be updated into the posterior distribution based on a set of observations. If further refinement of the distribution is desired, this posterior distribution can be regarded as the prior function for a second round of experiments. The subsequent posterior function can again be considered as a prior for additional observations, thereby refining the original distribution through a succession of experiments.

PROBLEMS

The problems below are numbered according to the sections to which they correspond. For instance, Problem 8.4A is the first ("A") exercise pertaining to Section 8.4.

8.4A   Based on extensive experience with a particular supplier, the net weight of a box of apples is known to possess a normal density with a mean of $v = 8$ kg and a standard deviation of $\tau = 10$ g. The net weight of an individual box is also a normal density function with a standard deviation of $\sigma = 100$ g. A sample of 20 boxes from a new shipment indicates a mean of 7.9 kg. (a) What is the classical estimator γ and the Bayes' estimator δ for the mean μ of the shipment? (b) What is the ratio of mean risks for γ and δ?

8.4B   The parameter μ of a Poisson distribution is to be estimated from a sample of size n. Suppose the prior density for μ is $\pi(\mu) = e^{-\mu}$ for $\mu \geq 0$, whereas the loss function is $L(\mu, a) = (a - \mu)^2$. (a) What is the posterior density $h(\mu|x)$? (b) What is the Bayes' estimator?

8.4C   Consider the parameter p from a binomial distribution. The prior density for p is $\pi(p) = 1$ for $0 \le p \le 1$. (a) Suppose that a sample of m observations yields r successes. What is the posterior density $h_1(p|x)$? (b) Suppose that an additional n trials results in s successes. If the density $h_1(p|x)$ is treated as the prior for this experiment, what is the new posterior density $h_2(p|x)$? (c) What is the Bayes' estimator for p based on h_2? (d) Suppose that a single experiment of $m + n$ trials results in $r + s$ successes. If $\pi(p) = 1$ is the prior density, what is the posterior density $h_3(p|x)$? How does this compare with the result of part (b)?

FURTHER READING

Blackwell, D., and M.A. Girshick. 1954. *Theory of Games and Statistical Decisions.* New York: Wiley.

Chernoff, H., and L. Moses. 1959. *Elementary Decision Theory.* New York: Wiley.

Ferguson, T.S. 1967. *Mathematical Statistics.* New York: Academic Press.

Lehmann, E.L. 1959. *Testing Statistical Hypotheses.* New York: Wiley.

McKinsey, L.C.C. 1952. *Introduction to the Theory of Games.* New York: McGraw-Hill.

von Neumann, J., and O. Morgenstern. 1944. *Theory of Games and Economic Behavior.* 3rd ed. Princeton, NJ: Princeton University Press.

Savage, L.J. 1954. *The Foundations of Statistics.* New York: Wiley.

Wald, A. 1947. *Sequential Analysis.* New York: Wiley.

Wald, A. 1950. *Statistical Decision Functions.* New York: Wiley.

9

Utility Theory

We have partial control of our destiny. Sometimes we decide on one possibility rather than another with full awareness of the consequences, such as choosing tea over coffee. More often, however, our choices involve uncertain outcomes. In a restaurant, we might order swordfish over steak without knowledge of the chef's culinary skills. In weightier circumstances, we may choose to invest in one stock over all the others on the stock exchange, or choose a job in one profession rather than another.

In considering the potential outcomes, our goal is to maximize the anticipated level of satisfaction or happiness. An underlying assumption in this endeavor is the proposition that the outcomes can be ranked relative to each other. More specifically, one outcome can be viewed as being better than, worse than, or equal to another. The relative ranking is also known as a *preference ordering* among the outcomes.

In considering the relative worth of one prospect against another, it is useful to assign a numerical value to indicate the degree of desirability. The mapping of outcomes against a numerical measure of satisfaction is called a *utility function.*

A utility function is a measure of a decision maker's desire for a particular outcome. Since there is no objective measure of human happiness, the numbers assigned to a utility function are arbitrary.

The only significant property of a utility function is that it increases monotonically with the desirability of outcomes. If one outcome is more desired than another, then the value of its utility must be higher. Further, if two outcomes are equally valuable, then their utilities must be identical.

Since the numbers assigned to a utility function are arbitrary, so are the units of measure. In particular, a unit of utility is called a *util* or *utile*. Suppose a pizza has a utility of 10 utils to a hungry diner, and the same valuation applies to a combination of one cheeseburger and french fries; then the two packages are equally desirable.

Advantages of Utility Functions

The notion of a utility function is useful for several reasons: granularity, nonlinearity, and risk aversion. *Granularity* refers to the discrete nature of potential outcomes. A teenager might prefer a stereo to a tennis racket, but by *how much*? Is the stereo worth twice as much? Or 132% as much? Since a utility function maps outcomes into some subset of the real numbers, it is possible to speak of the extent of preference for one outcome over another, rather than being confined to purely categorical rankings.

The second advantage of utility functions relates to the nonlinearity of satisfaction as a function of the amount of a desirable outcome. Two tennis rackets may be better than a single racket, but 20 rackets could be a downright nuisance. In this case, the utility function might increase with the number of rackets, then level off, and eventually decline for large numbers. The nonlinearity of satisfaction with the quantity of a desirable item is evident even when money is concerned. A person may value $2 twice as much as $1; but he is unlikely to value $2 billion twice as much as $1 billion. In a similar way, the incremental worth of $1010 over $1000 is not likely to be identical to that of $10 as compared to $0.

The third advantage of utility functions pertains to risk aversion. Decision makers are often risk averse in that they prefer certain outcomes to probabilistic ones. For instance, most of us would prefer $50,000 for certain rather than a situation which results in $100,000 and $0 with equal probability.

The financial worth of a probabilistic trial is called the *expected monetary value*, or simply EMV. To illustrate, the EMV of a trial which offers $10 and $30 with equal probability is $20.

When a decision maker is risk averse, the utility of a probabilistic outcome with an EMV of $20 is less than that of receiving $20 for certain. Depending on the decider, it may be $17, or $13.27, or some other value.

An experiment or a trial is also called a *lottery*, especially when the outcomes are monetary rewards or penalties. The amount of money which a person would consider to be equally desirable to a particular lottery is called the *certainty monetary equivalent* (CME) of the lottery.

In a risk averse situation, the CME is less than the EMV of the corresponding lottery. For the previous example, the EMV of the lottery is $20, whereas the CME would be $17.00, $13.27, or some other value depending on the predilections of the decision maker.

At times, however, a decider may be *risk prone*: he prefers a lottery over a certain reward that is equal to the expected worth of the lottery. For instance, a person may need to pay off a loan of $2000 the following week. In that case he may prefer a lottery yielding $2000 and $0 with equal probability, over $1000 for certain. He might even be willing to pay $1030, or $1078, or some other amount. In the case of risk affinity, the CME of a lottery exceeds the value of the EMV.

9.1 PREFERENCE ORDERING

Consider a decision maker who can compare and evaluate alternative outcomes or consequences. In this section we will examine a number of plausible principles which a rational decider ought to follow. We also explore a number of consequences implied by the governing principles.

We assume that a decider can rank any two consequences, thereby determining that one is better than, worse than, or equally desirable to another. Given two consequences C and D, we write $C \sim D$ and read "C matches D" if C is equally desirable in relation to D. If C is preferred to D, we write "$C \succ D$" and read "C supercedes D." Conversely, if C is less desirable than D, we write "$C \prec D$" and read "C subcedes D." We call this three-way determination the principle of *trichotomy*.

As an illustration, consider a decision maker facing two possibilities for Saturday evening; these are attending a party or going out to a cinema. We assume that the decider can determine one of the following relationships:

$$\text{Party} \prec \text{Cinema}$$

$$\text{Party} \succ \text{Cinema}$$

$$\text{Party} \sim \text{Cinema}$$

Properties of Relations

The matching of two consequences, indicated by the symbol $\sim$, is called an *equivalence* relation. This relation is symmetric by definition. In particular, to say that "C matches D" is tantamount to declaring that "D matches C." More formally, the principle of *symmetry* refers to the fact that $C \sim D$ is tantamount to $D \sim C$.

Since a consequence C is as desirable as itself, we can write $C \sim C$. A relation that is symmetric for identical objects is said to be *reflexive*. The matching relation $\sim$ is therefore both symmetric and reflexive. The equivalence relation should carry across multiple comparisons. If apples are as desirable as bananas, which in turn are as desirable as cherries, then apples should be as desirable as cherries. This transmission across multiple comparisons is called *transitivity*:

- If $A \sim B$ and $C \sim D$,
  then $A \sim C$

Consequently, the matching relation, symbolized by $\sim$, is symmetric, reflexive, and transitive.

The ordering of two consequences is called a *strict preference* relation, or more simply a *preference* relation. The preference relation is indicated by the symbols $\succ$ or $\prec$, respectively, when the preference is descending or ascending.

A preference relation is not symmetric or reflexive. For instance, if an apple supercedes a banana, then it is not the case that a banana supercedes an apple.

On the other hand, the preference relation is transitive:

- If $A \succ B$ and $B \succ C$,
  then $A \succ C$

For instance, if apples are preferred to bananas, which in turn are preferred to cherries, then apples should be preferred to cherries.

The relation $\prec$ is simply the mirror image of the supercedence relation $\succ$. By a similar argument, it too should abide by the property of transitivity.

The matching and supercedence relations can be combined to yield other properties. Examples of such compound properties are the following:

- If $A \sim B$ and $B \succ C$, then $A \succ C$.
- If $A \prec B$ and $B \sim C$ and $D \sim C$, then $A \prec D$.

A number of such properties are illustrated in the exercises.

Weak Preference

When a consequence C is at least as desirable as D, we write "$C \succsim D$" and read "C weakly supercedes D." This is equivalent to the compound statement

$$C \sim D \quad \text{or} \quad C \succ D$$

In other words, C matches or supercedes D. By the principle of trichotomy, to state that $C \succsim D$ is equivalent to saying that $C \prec D$ is false.

In a similar way, when C is no better than D, we write "$C \precsim D$" and read "C weakly subcedes D." This is equivalent to the compound statement

$$C \sim D \quad \text{or} \quad C \precsim D$$

Another way to define $C \precsim D$ is to claim that it holds precisely when $C \succ D$ is false, and vice versa.

The weak preference relation should hold across multiple comparisons. In other words, the relation is transitive for supercedence:

- If $A \succsim B$ and $B \succsim C$,
  then $A \succsim C$,

as well as subcedence:

- If $A \precsim B$ and $B \precsim C$, then $A \precsim C$.

By combining weak preference in conjunction with the matching and strict preference relations, we can obtain any number of other properties. Among these are the following:

- If $A \precsim B$ and $B \sim C$, then $A \precsim C$.
- If $A \succsim B$ and $B \succ C$, then $A \succ C$.
- If $A \prec B$ and $B \precsim C$ and $C \sim D$, then $A \prec D$.

We now turn to hybrid consequences composed of two or more elementary consequences.

Compound Consequences

A decider does not always have the luxury of selecting a consequence solely according to personal preferences. Often he faces a hybrid consequence in the form of a random trial whose outcomes are elementary consequences.

In complex situations, hybrid consequences may even be nested. In other words, the outcome of a random trial may itself be another trial.

When viewed as a probabilistic consequence, a trial is also called a *lottery*, a *mixture*, or a *compound consequence*. The simplest case of a mixture consists of two consequences C and D. If consequence C occurs with probability p and D with probability $1 - p$, the compound consequence is indicated by the notation $[C, p; D, 1 - p]$.

More generally, let $C_1, \ldots, C_n$ be a set of consequences with respective probabilities $p_1, \ldots, p_n$. Then the compound consequence is written as $[C_1, p_1; \ldots; C_n, p_n]$.

The specification of the probability for each consequence is not entirely necessary, of course. Since each consequence is viewed as an outcome of a trial, the probabilities must sum to unity: $p_1 + \cdots + p_n = 1$. For simplicity of notation we will often eliminate the explicit specification of the last probability p_n. As an example, the compound consequence $[C, p; D, 1 - p]$ will be abbreviated to $[C, p; D]$.

To allow for the consideration of compound consequences in making decisions, we assume that preferences over elementary consequences tend to carry over to probabilistic outcomes. As an illustration, suppose that a car C is preferred over a motocycle M, which, in turn, is preferred to a sailboard S. Consider the lottery L consisting of C and M with equal probability. In other words, $L \equiv [C, 0.5; M]$. A second lottery L' consists of S and M with equal probability; that is, $L' = [S, 0.5; M]$. Since a car supercedes a sailboard to begin with, the first lottery L should supercede the second mixture L'.

The actual value of the probability p for C or S is immaterial because it is specified to be the same for both lotteries. But this is also true of the level of preference for the common element M. Since M appears with equal chance in both lotteries, it is immaterial whether it supercedes, subcedes, or matches either C or S.

This line of reasoning allows us to rank compound consequences. If a consequence B is preferred to A, then for any other consequence C and value of probability p, the lottery $[B, p; C]$ is preferred to $[A, p; C]$. This is the principle of *ranking lotteries*.

A generalization of this binomial lottery involves multiple consequences. Let $A_1, \ldots, A_n$ be a set of consequences and $B_1, \ldots, B_n$ another set. Suppose $A_i \sim B_i$ for all i except for exactly one case k where B_k is preferred to A_k. Then the lottery involving the B_i supercedes that for the A_i.

$$[B_1, p_1; \ldots; B_k, p_k; \ldots; B_n] \succ [A_1, p_1; \ldots; A_k, p_k; \ldots; A_n]$$

Parametric Lotteries

Given the two consequences A and C, the desirability of the lottery $[A, p; C]$ will depend on the value of parameter p. Suppose A is preferred to C. Then a high value of the probability p will make the lottery more appealing than a lower value. In the limit as p approaches unity, the lottery $[A, p; C]$ will, in fact, be identical to the elementary consequence A. Conversely, as p goes to 0, the lottery will be equivalent to the simple consequence C.

If B is another consequence ranked between A and C, then it is plausible that there should exist values for p such that the lottery $[A, p; C]$ supercedes or subcedes B. This is the principle of a parametric lottery:

• Let A, B, and C be three consequences in increasing order of preference: $A \prec B \prec C$. Then there exist values of probability p and q such that the following relations hold:

$$[C, p; A] \prec B$$
$$[C, q; A] \succ B$$

The first formula will hold for a relatively low value of p, and the second for a larger number q.

To illustrate, suppose that Laura is considering the destination for her next vacation. She can go by herself, or with a group of alumni from her high school. If she goes alone, she will settle for Bermuda. If she joins the group, she will, of course, go with the majority. The group plans a trip to the Arctic if they can form a group large enough to charter a ship; otherwise they will go to the lower Carribean. Laura's preferences are the Arctic, followed by Bermuda, and the Carribean in third place. To which course of action should Laura commit herself?

That depends on her expectation concerning the probability p of successfully organizing a trip to the Arctic. If this value p is high, then the lottery [Arctic, p; Caribbean] will supercede the choice of Bermuda. For a small value of p, the group trip will subcede the individual trip. For some intermediate value of p, she will be indifferent between the group and individual course of action.

The precise value of the probability for which the lottery matches the Bahamas will depend on the strengths of her preferences for the three destinations. The matching of a lottery with an elementary consequence will be the focus of the utility analysis in Section 9.2.

Summary

A rational decision maker is assumed to follow a set of principles in evaluating consequences and selecting a course of action. Among these are the ability to compare alternatives and rank them according to a preference ordering.

Further, many courses of action in real life represent composite consequences rather than elementary ones. These hybrid consequences, called mixtures or lotteries, are random trials which may yield one of several outcomes according to a probability function. Once the outcomes are identified, the desirability of a lottery depends on the probabilities corresponding to the outcomes. The lottery represents a composite consequence which may be compared with other consequences, whether elementary or compound.

Assumptions such as the preceding propositions are called *postulates* or *axioms*. Since the postulates deal with the nature of preferences, they are known as the axioms of preference. More formally, the *Preference Axioms* are as follows.

1. *Trichotomy.* Let A and B be consequences. Then either $A \prec B$, or $A \succ B$, or $A \sim B$.
2. *Transitivity.* If $A \prec B$ and $B \prec C$, then $A \prec C$.

3. *Ranking Lotteries.* If $A \prec B$, then $[A, p; C] \prec [B, p; C]$ for any probability p and consequence C.
4. *Parametric Lotteries.* If $A \prec B \prec C$, then there exist values of probability p and q in the interval (0, 1) such that these relations hold:

$$[C, p; A] \prec B$$

$$[C, q; A] \succ B$$

In this section we discussed the direction of preferences but not their magnitude. This feature is addressed by the concept of a utility function. We will see in the next section that the existence of a preference ordering among consequences implies a utility function over the outcomes.

9.2   MEDIATE CONSEQUENCES

We have seen that a mixture of two prospects lies somewhere between them in terms of a preference ordering. More specifically, let A and B denote two consequences for which $A \prec B$. Further, let $M(p)$ denote the mixture of A and B, with p denoting the probability of obtaining B; in other words, $M(p) \equiv [B, p; A]$. Then the mixture $M(p)$ lies between the elementary consequences A and B for any value of p in the open interval (0, 1):

$$A \prec M(p) \prec B$$

Further, the desirability of the mixture $M(p)$ increases with increasing p.

It is possible to show that any intermediate consequence whose desirability lies between A and B can be matched by some lottery $M(p)$. In particular, we can show that there is some probability p such that $M(p)$ is equivalent to the intermediate consequence.

Theorem (Equivalent Lottery). Let A and B be two consequences such that $A \prec B$. Suppose that K is a consequence lying between the first two: $A \prec K \prec B$. Then there is a unique probability p lying between 0 and 1 such that the mixture $[B, p; A]$ is equivalent to K.

The key to the proof lies in contradiction. That is, the consideration of each alternative possibility leads to an inconsistent state of affairs.

Proof. Suppose K is a consequence whose desirability lies between A and B; that is, $A \prec K \prec B$. Let $M(p) \equiv [B, p; A]$ be a nontrivial mixture of consequences A and B. Here, nontriviality refers to the fact that p lies in the open interval (0, 1). The Axiom of Parametric Lotteries from the previous section implies that there is at least one mixture $M(p)$ which subcedes K, and another mixture $M(q)$ which supercedes K:

$$M(p) \prec K \prec M(q)$$

Let L be the set of mixtures which are less desirable than K, and H the set of lotteries which are more desirable. Given any mixture $M(p)$ in the lower set L, the value of p is less than that of q for any $M(q)$ in the higher set H.

Suppose that r is the *supremum* of the probabilities associated with mixtures in the lower set L; that is, r is the smallest upper bound on the collection of probabilities p corresponding to the lotteries $M(p)$ in L. The supremum r will always exist.

We next claim that the supremum r defines the lottery $M(r)$ which is equivalent to the mediate consequence K; in symbolic form, $M(r) \sim K$. This claim can be verified by the following contradiction.

Suppose that $M(r) \sim K$ does not hold. By the Axiom of Trichotomy, either we have $M(r) \prec K$ or $M(r) \succ K$. We consider each possibility in turn.

a. Assume that $M(r) \prec K$. By the Axiom of Parametric Lotteries, there exists a nontrivial mixture of $M(r)$ and B lying below K:

$$M(r) \prec [M(r), s; B] \prec K$$

Since $M(r)$ was defined as $[B, r; A]$, the middle lottery can be written as

$$[B, r + s - rs; A] = M(r + s - rs)$$

However,

$$r + s - rs = r + s(1 - r) > r$$

The inequality above depends on the fact that s and $(1 - r)$ are both positive. Now $M(r + s - rs)$ subcedes K, while $r + s - rs$ exceeds r. This contradicts the assertion that r is the supremum of all the p values in $M(p) \in L$.

b. Assume that $M(r) \succ K$. By the Parametric Lotteries Axiom, we can identify a nontrivial mixture of A and $M(r)$ lying between K and $M(r)$:

$$K \prec [M(r), t; A] \prec M(r)$$

Since $M(r) = [B, r; A]$, the middle lottery is equivalent to $[B, rt; A] = M(rt)$. Since $t < 1$, we infer that $rt < r$. From the first preference inequality above, $M(rt)$ supercedes K; so rt is an upper bound for any p in the set $L = \{M(p)\}$. On the other hand, $rt < r$, so r is not the smallest upper bound, contradicting its title as the supremum.

We have shown that $M(r)$ is equivalent to K. We can conclude r has a unique value, since any other value for r will result in a lottery which subcedes or supercedes K.   ∎

The Equivalent Lottery Theorem implies that any consequence lying between two specified consequences can be associated with a number p between 0 and 1. Of two such intermediate consequences, the one with the higher number is more desirable; if they exhibit the same value, they are equivalent.

In particular, suppose $\{C_1, \ldots, C_n\}$ is a set of consequences with respective probabilities $\{p_1, \ldots, p_n\}$. Then the consequences C_i can be partially ordered according to their indices p_i. The desirability of the outcomes C_i increases with increasing probabilities p_i.

The probabilities p_i can be associated with a measure of desirability known as the utility. We now turn to this topic.

9.3   PROBABILITY AS UTILITY

We have seen that any consequence whose desirability falls between two given alternatives can be ordered by matching it with an equivalent lottery. More specifically, if $\{K_1, K_2, \ldots\}$ is a set of consequences lying between A and B, each K_i can be matched with an equivalent lottery $M(p_i) \equiv [B, p_i; A]$ where p_i is the probability of obtaining the more desirable prospect B. The larger the value of p_i, the greater is the desirability of consequence K_i.

The probability p_i can be used as a measure of desirability called the utility of the associated consequence. That is, the utility u is a function which maps a set C of consequences onto the real interval $[0, 1]$.

- Let K be a consequence lying between two alternatives A and B, such that $A \precsim K \precsim B$. Suppose that the lottery $M(p) \equiv [B, p; A]$ is equivalent to K. Then the *utility* u of K is the probability p of the lottery $M(p)$; in other words, $u(K) = p$.

The utility of a consequence K equals 0 if K is equivalent to A, equals 1 if K matches B, and lies strictly between 0 and 1 for all intermediate prospects.

Example

A teenager is contemplating the acquisition of a set of wheels. The options are a convertible sports car, a coupe, and a dune buggy. Suppose her preferences are

$$\text{buggy} \prec \text{coupe} \prec \text{convertible}$$

Further, she decides that a certain prospect of acquiring a coupe is precisely as desirable as an opportunity to obtain a convertible with probability 2/3 and a buggy with probability 1/3:

$$[\text{convertible}, 2/3; \text{buggy}] \sim \text{coupe}$$

If the value 0 is assigned to the utility of a buggy and a value 1 to a convertible, then the utility of a coupe is 2/3.   ◆

Some Properties of Utility Functions

Utility functions exhibit a number of properties. Among these are monotonicity and linearity. The utility function rises monotonically with the desirability of a prospect. That is, if a consequence is more desirable than another, the value of its utility is higher.

Proposition (Monotonicity). Suppose P and Q are two consequences for which $P \prec Q$. Then the utility of P is less than that of Q: $u(P) < u(Q)$.

To verify this proposition, we consider a prospect A which weakly subcedes P; that is, $A \precsim P$. Conversely, let B be a consequence which weakly supercedes Q; we can write $B \succsim Q$. Moreover, we know that $P \prec Q$. By the Equivalent Lottery Theorem, consequence P can be matched by a mixture $[B, p; A]$, whereas prospect Q can be matched by another mixture $[B, q; A]$. Since the consequence P subcedes Q, the probability p must be less than the value q. So its utility is less than that of q.

A second property of the utility function lies in linearity with respect to lotteries. In particular, the utility of a lottery matches the lottery of the utilities.

Proposition (Linearity). Let P and Q be two consequences. Then

$$u([P, p; Q]) = pu(P) + (1 - p)u(Q)$$

To validate this proposition, we again appeal to two bounding consequences A and B. Suppose A weakly subcedes each of P and Q, whereas B weakly supercedes them. Then there exists a lottery $[B, r; A]$ which matches P; the utility of P, therefore, equals r. In a similar way, there exists a lottery $[B, s; A]$ equivalent to Q and thereby yielding a utility of s for Q.

The compound lottery $[P, p; Q]$ is equivalent to an elementary lottery involving A and B:

$$[P, p; Q] = [[B, r; A], p; [B, s; A]]$$
$$= [B, pr + (1 - p)s; A]$$

We note that the utility of the compound lottery is $pr + (1 - p)s$. We can write

$$pr + (1 - p)s = pu(P) + (1 - p)u(Q)$$

which represents the desired result.

The linearity property of the utility function can be viewed in terms of the expectation operator. The utility of the lottery $[P, p; Q]$ is a random variable which yields $u(P)$ with probability p and $u(Q)$ with probability $(1 - p)$. The expected value of the utility is given by $pu(p) + (1 - p)u(Q)$. In short, the utility of a random consequence is the mean value of the utility.

A third property of the utility function is indifference to linear transformations. Suppose u is a utility function defined by the bounding consequences A and B, for which $u(A) = 0$ and $u(B) = 1$. Then any intermediate prospect P between A and B has a unique utility value p in the open interval $(0, 1)$.

On the other hand, the utility of P can be defined in terms of any other pair of bounding consequences A' and B'. Let v denote this utility function. Then the utility of P is also determined uniquely under v as some number p'.

Any linear transformation between u and v will preserve the ordering of consequences under each utility function. In other words, the conversion

$$v(P) = au(P) + b$$

will lead to the same ordering of the consequences. As a result, the set of prospects $\{P\}$ which supercede both A and A' and subcede both B and B' will be ranked in the same way under utility functions u and v. The term b translates the values of u to the right or left depending on its sign, whereas the scaling factor a magnifies the value of u.

The values of a and b will depend on the choice of the bounds A' and B' in comparison to A and B. Since $u(A) = 0$ and $u(B) = 1$, the two constants can be determined from the relations $v(A) = b$ and $v(B) = a + b$.

The transformed utility function v satisfies the properties of monotonicity and linearity. The applicability of monotonicity for $v(P)$ can be readily verified.

The linearity property can be validated as follows:

$$
\begin{aligned}
v([P, p; Q]) &= au([P, p; Q]) + b \\
&= a\{pu(P) + (1 - p)u(Q)\} + b \\
&= p\{au(P) + b\} + (1 - p)\{au(Q) + b\} \\
&= pv(P) + (1 - p)v(Q)
\end{aligned}
$$

The second equality relies on the linearity property of the utility function u.

Example

Another youth is contemplating the choice of a vehicle, ranging from a buggy to a coupe and a convertible. He determines that his utility for a buggy is 2, and that for a convertible is 5. Suppose he is also indifferent between a coupe and a lottery offering a buggy with probability 2/3 and a convertible with probability 1/3. What is his utility for a coupe?

The bounding utilities are $v(\text{buggy}) = 2$ and $v(\text{convertible}) = 5$. The utility of the lottery can be determined from

$$v([\text{buggy}, 2/3; \text{convertible}]) = \tfrac{2}{3}v(\text{buggy}) + (1 - \tfrac{2}{3})v(\text{convertible})$$

$$= \tfrac{2}{3}(2) + (1 - \tfrac{2}{3})(5) = 3 \qquad \blacklozenge$$

The utility function v in the above example is equivalent to the function u in the preceding example. In fact, they are related by the linear transformation

$$v(P) = 3u(P) + 2$$

Up to this point, we have focused on consequences lying between two boundary prospects A and B. But how can we deal with consequences which lie outside these boundary points?

In the next section, we address the general situation where consequences may lie within or without two points of reference. We will see that the selection of the endpoints A and B have no significant effect on the utility function. The use of different endpoints will lead to equivalent utility functions which differ only by linear transformations.

9.4   UTILITY FUNCTIONS IN GENERAL

In the previous section, we explored the nature of utility functions for consequences which lie between two given prospects. Since the worst consequence is assigned a utility of 0 and the best a value of 1, the utility function in effect has been normalized to unity.

What happens when a new consequence comes along which is worse or better than any of the original prospects? One way to deal with the situation is to renormalize the utility function. This can be achieved by assigning a value of 0 to the new prospect if it is worse than all the original prospects, or a value of 1 if it is better. Next, each prospect in the original set can be reevaluated and assigned a revised utility value.

The problem with the renormalization approach lies in computational inefficiency. Each time a new consequence appears which lies outside the bounds of the existing set of prospects, every utility value has to be recomputed.

A more convenient method is to extend the range of the original utility function. In other words, if a new consequence C is worse than any of the existing prospects, its utility value is negative; or if C is better than all the others, its utility exceeds 1. The precise value of the utility will depend on the level of desirability of the new consequence. The appropriate value of utility can be determined, as usual, by relying on comparisons of certain prospects against lotteries. The approach is clarified below.

Extended Utility Functions

Suppose A and B are two consequences, with respective utilities 0 and 1. Assume that a prospect S is encountered which is more desirable than prospect B. Then its utility value A should exceed 1, as depicted in Figure 9–1. The value of A can be determined by comparing B against a lottery involving A and S.

We can involve the Equivalent Lottery Theorem at this point. Since $A \prec B \prec S$, there exists a lottery involving S and A which is equivalent to the certain prospect B; that is, $[S, p; A] \sim B$ for some probability p. By the Linearity Property of the utility function, we can write

$$pu(S) + (1 - p)u(A) = u(B)$$

Since $u(A) = 0$ and $u(B) = 1$, we immediately obtain the utility for S as $u(S) \equiv s = 1/p$. We see that the utility of S is inversely proportional to its associated probability in the lottery. For instance, if prospect B is equivalent to the lottery between S and A with equal odds, then $p = 1/2$ and the utility of S is $s = 1/(1/2) = 2$.

In an analogous way, suppose that some consequence R is less desirable than prospect A. Then there is some lottery $[B, p; R]$ which matches A. The relation among utilities is

$$pu(B) + (1 - p)u(R) = u(A)$$

Since $u(B) = 1$ and $u(A) = 0$, we obtain $u(R) \equiv r = -p/(1 - p)$.

As an illustration, if A matches the lottery between R and B with equal

FIGURE 9–1. Consequences A and B have utilities 0 and 1, respectively. Prospect S has a utility s which exceeds 1, whereas prospect R has a value r which is negative.

odds, then $p = 1/2$ and $r = -1$. In geometric terms, prospect A lies midway between R and B in utility space.

We can demonstrate that the extended utility function satisfies the properties of monotonicity and linearity. This result is stated in the form of a theorem and its proof.

Theorem (Monotonicity and Linearity). Suppose that Preference Axioms 1 through 4 at the end of Section 9.1 hold. Then there is a finite number u corresponding to any consequence C, which satisfies the following properties:

- *Monotonicity*. If $A \prec B$, then $u(A) < u(B)$.
- *Linearity*. $u([B, p; A]) = pu(B) + (1 - p)u(A)$.

Proof. Let A and B be two consequences in order of preference, and u the utility function for which $u(A) = 0$ and $u(B) = 1$. Suppose R and S are two other consequences for which $R \prec S$. Let C_* denote the worst consequence among $\{A, B, R, S\}$, and C^* the best.

The function u was defined on the pair of consequences A and B. But it could just as easily have been defined on C_* and C^*; we let u^* denote the associated utility. We will first show that a linear relationship exists between u and u^* for any consequence between C_* and C^*.

Consider a prospect T lying between C_* and C^*. We first suppose that $T \succ B$. By the Equivalent Lottery Theorem, there is some probability t such that B matches the lottery between T and A:

$$B \sim [T, t; A] \tag{9-1}$$

We saw earlier in this section that t is the reciprocal of the utility of T; that is, $t = 1/u(T)$. From the linearity property of u^*, we can expand Eq. (9-1) as follows:

$$u^*(B) = tu^*(T) + (1 - t)u^*(A)$$

By solving for $u^*(T)$ we obtain

$$u^*(T) = \frac{u^*(B)}{t} - \left(\frac{1}{t} - 1\right)u^*(A) = [u^*(B) - u^*(A)]u(T) + u^*(A)$$

The second equality relies on the fact that $t = 1/u(T)$. By defining the constants $\alpha^* \equiv u^*(B) - u^*(A)$ and $\beta^* \equiv u^*(A)$, the above equation can be written

$$u^*(T) = \alpha^* u(T) + \beta^* \tag{9-2}$$

The last equation highlights the linear relationship between u^* and u.

For the second case, we assume that $T \prec A$. Then there is some probability b for which A matches the lottery between T and B; that is,

$$A \sim [B, b; T]$$

Using the linearity property, we infer that $b = u(T)/[u(T) - 1]$. We can express the preceding equivalence in terms of the utility function u^*:

$$u^*(A) = bu^*(B) + (1 - b)u^*(T) = \frac{u(T)}{u(T) - 1} u^*(B) + \frac{-1}{u(T) - 1} u^*(T)$$

The second equality depends on the substitution of $u(T)/[u(T) - 1]$ for the probability b. Solving for $u^*(T)$ again leads to the relation

$$u^*(T) = \alpha^* u(T) + \beta^*$$

where $\alpha^* = u^*(B) - u^*(A)$ and $\beta^* = u^*(A)$.

For the third case, suppose that T lies between A and B. Then there exists some lottery between A and B which matches T:

$$T \sim [B, b; A]$$

Since A and B are defining prospects for the utility function u, we know that the utility of T under u is b; that is, $u(T) = b$. If we write out the preceding equivalence in terms of the utility function u^*, we obtain

$$u^*(T) = bu^*(B) + (1 - b)u^*(A) = u(T)[u^*(B) - u^*(A)] + u^*(A)$$
$$= \alpha^* u(T) + \beta^*$$

The second equality depends on the substitution of $u(T)$ for b.

We have just seen that no matter whether T lies above, below, or between A and B, the linear relation $u^*(T) = a^* u(T) + \beta^*$ ties together the utility functions u^* and u. The relation can also be verified for the special cases where T matches A or B. For instance, if $T \sim A$, then $u(T) = u(A) = 0$. The result is

$$u^*(T) = [u^*(B) - u^*(A)](0) + u^*(A) = u^*(A)$$

In this way, we can demonstrate that Eq. (9–2) holds for any T between C_* and C^*.

Equation (9–2) can be expressed in terms of u^* by writing

$$u(T) = \alpha u^*(T) + \beta \tag{9–3}$$

Here $\alpha \equiv 1/[u^*(B) - u^*(A)]$, while $\beta = -u^*(A)/[u^*(B) - u^*(A)]$.

The monotonicity property is now simple to verify. Suppose that R and S are any two prospects between C_* and C^* for which $R \prec S$. We can write $u^*(R) < u^*(S)$. Use of Eq. (9–3) yields

$$u(R) = \alpha u^*(R) + \beta < \alpha u^*(S) + \beta = u(S)$$

We conclude that the extended utility function u is monotonic.

The property of linearity can be verified in the following way. Once again, we let R and S be two consequences for which $C_* \precsim R \prec S \precsim C^*$.

Equation (9–3) is valid for any consequence T between C_* and C^*. It is, therefore, valid for any lottery $[S, s; R]$. Use of the equation leads to

$$\begin{aligned}
u([S, s; R]) &= \alpha u^*([S, s; R]) + \beta \\
&= \alpha[su^*(S) + (1 - s)u^*(R)] + \beta + (\beta s - \beta s) \\
&= su(S) + (1 - s)u(R) \tag{9–4}
\end{aligned}$$

The second equality relies on the linearity property of the normalized utility function u^*. We have also added and subtracted the quantity βs in preparation for the third equality. Equation (9–4) demonstrates the linearity property of the extended utility function u.

We have verified the monotonicity and linearity properties of the extended utility function u for arbitrary consequences. The consequences may lie within the pair of prospects A and B, or outside; the extended utility function u will handle all possibilities.

On the other hand, it might seem that the function u depends on the initial pair of prospects A and B. In reality, a change in the reference points will lead to a distinct utility function v which differs from u by only a linear transformation.

To verify this, consider the utility function v defined on prospects G and H: $v(G) = 0$ and $v(H) = 1$. For simplicity's sake, we address the particular case for which $A \prec G \prec B \prec H$.

Now consider any other prospect T. In particular, suppose that T lies between G and B:

$$A \prec G \prec T \prec B \prec H$$

There are probabilities t and x such that

$$T \sim [B, t; A] \quad \text{and} \quad T \sim [H, x; G] \tag{9-5}$$

The utilities for T under u and v are

$$u(T) = t \quad \text{and} \quad v(T) = x$$

Expanding the second equivalence in Eq. (9–5) in terms of the utility function u yields

$$u(T) = xu(H) + (1 - x)u(G)$$

Solving for x and using the relation $v(T) = x$ leads to

$$v(T) = \frac{u(T) - u(G)}{u(H) - u(G)}$$

The new utility function $v(T)$ is a linear transformation of the old function $u(T)$. More specifically, v is constructed in such a way that it takes value 0 at $T = G$ and value 1 at $T = H$.

We have verified the linear relationship for u and v for one ordering among the prospects—namely, $A \prec G \prec T \prec B \prec H$. The other cases can be handled in similar fashion.   ∎

The linearity property of utility functions applies not only to elementary lotteries involving two prospects, but to those composed of a finite number of multiple prospects. This demonstration is left as an exercise.

The linearity property applies even to a countably infinite number of prospects. To this end, we first demonstrate that utility functions are bounded. Then we can show that the utility function is linear for lotteries involving a countably infinite number of prospects.

Theorem (Boundedness). Suppose that Preference Axioms 1 to 4 hold, as described at the end of Section 9.1. Then the utility function is bounded:

$$|u(C)| \leq M$$

for some number M and any consequence C.

We verify the theorem by an argument of contradiction.

Proof. Let A and B be two consequences with respective utilities a and b. Suppose u is not bounded. Then there exists at least one consequence Z for which $u(Z) > M$ for any constant M. Thus we have $A \prec B \prec Z$.

By the Axiom of Parametric Lotteries, there exists a positive value p for which $[Z, p; A] \prec B$. This would imply

$$u([Z, p; A]) < u(B)$$

The left-hand side can be expanded by the linearity property. We obtain

$$pu(Z) + (1 - p)a < b$$

where $u(A)$ and $u(B)$ have been replaced by a and b, respectively. Solving for $u(Z)$ yields

$$u(Z) < \frac{b + (p - 1)a}{p}$$

Since p is positive, the right-hand side is a finite number. This contradicts the assumption that u is unbounded. ∎

We can now verify the linearity property of the utility function for countably infinite prospects.

Theorem (Countable Prospects). Assume Axioms 1 to 4. Let $C_1, C_2, \ldots$ be a sequence of consequences with respective probabilities $c_1, c_2, \ldots$. Then, the utility of the corresponding lottery is the mean of the individual utilities:

$$u([C_1, c_1; C_2, c_2; \ldots]) = \sum_{i=1}^{\infty} c_i u(C_i)$$

Proof. Let the entire lottery be indicated by $C \equiv [C_1, c_1; C_2, c_2; \ldots]$. We denote the tail of the lottery as $D_m \equiv [C_m, c_m; C_{m+1}, c_{m+1}; \ldots]$ and its chance of occurrence as d_m. The value of d_m is

$$d_m = \sum_{i=m}^{\infty} c_i$$

Since $\sum_{i=1}^{\infty} c_i = 1$, we can infer that $d_m \to 0$ as $m \to \infty$.

The entire lottery can be written as $C = [C_1, c_1; \ldots; C_n, c_n; D_{n+1}]$. The corresponding utility is

$$u(C) = u([C_1, c_1; \ldots; C_n, c_n; D_{n+1}])$$

$$= \sum_{i=1}^{n} c_i u(C_i) + d_{n+1} u(D_{n+1})$$

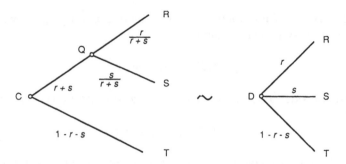

FIGURE 9-2. The composite lottery C is equivalent to the elementary lottery D.

From the Boundedness Theorem, we know that $u(D_{n+1})$ is finite. Since $d_{n+1} \to 0$ as $n \to \infty$, we know that the last term vanishes as n increases. We obtain $u(C) = \sum_1^\infty c_i u(C_i)$, which is the desired result. ∎

A simple consequence of this theorem is the equivalence of composite and elementary lotteries. Consider the composite lottery

$$[[R, r/(r + s); S], r + s; T]$$

which contains a sublottery as one of its consequences. This mixture matches the direct lottery $[R, r; S, s; T]$ consisting of elementary prospects.

Corollary (Lottery Reduction). The composite lottery $[[R, r/(r + s); S], r + s; T]$ is equivalent to the elementary lottery $[R, r; S, s; T]$.

Proof. We first define the following lotteries: $Q \equiv [R, r/(r + s); S]$, $C \equiv [Q, r + s; T]$, and $D \equiv [R, r; S, s; T]$. These lotteries are depicted in Figure 9-2. The utility of lottery C is

$$u(C) = u([Q, r + s; T]) = (r + s)u(Q) + (1 - r - s)u(T)$$

$$= (r + s)\left(\frac{r}{r + s}u(R) + \frac{s}{r + s}u(S)\right) + (1 - r - s)u(T)$$

$$= ru(R) + su(S) + (1 - r - s)u(T)$$

For the third equality, we have replaced $u(Q)$ by $u([R, r/(r + s); S])$ and invoked the linearity property of the utility function.

By the Countable Prospects Theorem, the utility of lottery D is

$$u(D) = u([R, r; S, s; T]) = ru(R) + su(S) + (1 - r - s)u(T)$$

Since $u(C) = u(D)$, we conclude that C matches D. ∎

9.5   MONETARY UTILITY FUNCTIONS

For many practical domains, the consequences of a decision can be denominated in financial terms. This is often the case in business, engineering, and economics, as well as some personal affairs.

Decision theory focuses on the analysis of alternative outcomes to a solution. Underlying the methodology is an implicit assumption that gains and losses are isotropic or symmetric. This implies, for instance, that $2 is twice as good as $1; in a similar way, a lottery which offers even odds of winning $10 or $30 should be worth $20 to a player.

In reality, of course, people are seldom indifferent to absolute levels of payoff or penalty. One reason for the lack of isotropy lies in *risk aversion*: we often prefer certain outcomes to chancy lotteries. This preference for certitude occurs when we choose $50 for certain over a lottery offering even odds of $100 or nothing.

A second reason for the asymmetry is due to the difference in the *marginal* or *incremental utility* in the level of any good. We might enjoy the use of a telephone in each room at home; but having 2 in every room would be of little incremental value, and having 20 would be a downright nuisance.

The law of *decreasing returns to scale* claims that the marginal utility of a good decreases with its quantity. A little iron is good for the body, but too much can be toxic.

On the other hand, the nonlinearity can work in the opposite direction as well: *synergism* results in an outcome that exceeds that of the sum of the parts. A pinch of sugar may have no perceptible effect on the taste of a cup of coffee, but 10 pinches might. In a comparable way, one or even three wheels could be powerless to move an automobile, but four will serve admirably.

Difficulties such as these are the proper domain of utility theory. In other words, the theory deals with decision making in the face of nonlinearities in the worth of consequences, whether due to the lack of certitude or synergism. In this section we examine the relationships between money and utility, for both the linear and nonlinear cases.

In practice, the utility for money is a bounded function. A numerical example is described below.

Example (St. Petersburg Paradox)

A decision maker is offered the following gamble involving tosses of a fair coin. If the first head occurs on the nth trial, he is offered $M_n = \$2^n$. For instance, if the first head arises on the third toss, the decider receives $\$2^3$, or $8; then the game ends. How much should the decider be willing to pay for this opportunity?

Consider the expected payoff for this gamble. The probability of observing the first head on the nth toss is $p_{M_n} = 1/2^n$. The expected value of the gamble is, therefore,

$$E(M) = \sum_{n=1}^{\infty} p_{M_n} M_n$$

$$= \sum_{n=1}^{\infty} \left(\frac{1}{2^n}\right)(2^n) = \sum_{n=1}^{\infty} 1 = \infty$$

The expected payoff is infinite!

In contrast, real people would not be willing to a pay an infinite amount for this gamble. In fact, you might not be willing to exchange a mere million for this grand opportunity.   ◆

Suppose that the value of money were directly proportional to its amount. Then the utility function for M dollars is $u(M) = kM$ for some positive constant k. This relationship is shown as the dashed line in Figure 9–3.

When the utility function rises less than proportionally with the amount of money, it takes a concave shape. An example is found in the function $u(M) = k(1 - e^{-M})$, which starts from $u(0) = 0$ and increases asymptotically toward k.

In certain situations, the utility of money can rise more than proportionately with the amount. For instance, a decider may be unable to purchase anything with 3 cents, and the corresponding utility would be practically zero. On the other hand, 25 cents may be enough to make a phone call, for which the utility could be very high.

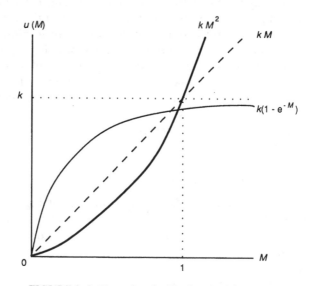

FIGURE 9–3. Examples of utility functions for money.

A utility function which rises more than proportionately with the amount of money is convex as illustrated by the function $u(M) = kM^2$. The constant k in the preceding utility function is largely immaterial. We saw in the previous section that utility functions are uniquely determined only up to a linear transformation. The choice for k is tantamount to selecting the size of the unit of utility, which is called a *util*. Since the value for k is insignificant as long as it is positive, we can just as well choose $k \equiv 1$.

Some people may claim that the possession of money is a curse, as it invites an entourage of headaches. This may be a romantic notion, but most individuals would feel otherwise. Money in excess of one's needs could be stashed away or given to charity. In practice, then, the utility function rises monotonically with the amount of money.

We turn to a numerical example to illustrate the use of utility functions.

Example

Little Lila wants to run a lemonade stand after school. The weather will be the primary determinant of revenues. Her gross revenues will be $0, $1, or $25, depending respectively on rain, cloud, or sun; these conditions are equally likely. Lila has determined that her utility function is $u(M) = \sqrt{M}$ over the entire range from $0 to $25. What is the maximum amount she should be willing to invest for a huge sign and lemonade supplies?

Lila has a nonlinear utility function, as depicted in Figure 9–4. The proper quantity to evaluate is the utility of the consequences, rather than the money itself. In essence Lila faces the lottery $L = [\$0, 1/3; \$1, 1/3; \$25]$. The corre-

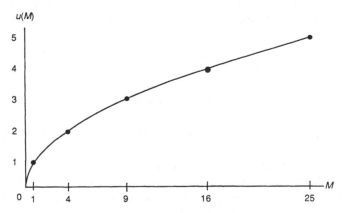

FIGURE 9–4. Utility function given by $u(M) = \sqrt{M}$.

sponding utility is

$$u(L) = \tfrac{1}{3}u(\$0) + \tfrac{1}{3}u(\$1) + \tfrac{1}{3}u(\$25) = \tfrac{1}{3}(0) + \tfrac{1}{3}(1) + \tfrac{1}{3}(5) = 2$$

The elementary consequence which has a utility of 2 is \$4. In other words, \$4 $\sim L$ since $u(\$4) = 2 = u(L)$. If Lila's required investment is less than \$4, she should proceed with the enterprise.   ◆

In the preceding example, the expected value of the revenues is

$$\tfrac{1}{3}(\$0) + \tfrac{1}{3}(\$1) + \tfrac{1}{3}(\$25) \cong \$8.67$$

This amount is the expected monetary value of the lottery.

In contrast, we have seen that the worth of the lottery is far less, valued at a certainty monetary equivalent of only \$4.

- Let $C = [C_1, c_1; \ldots; C_n]$ be a lottery involving monetary amount C_i with probability c_i. The mean worth of the mixture C, namely, $\sum_1^n c_i C_i$, is the *expected monetary value* (EMV) of the lottery.
- Suppose that M is the amount of money which matches the lottery C; that is $M \sim C$. The amount M is called the *certainty monetary equivalent* (CME) of the lottery.

In general, the CME will differ from the EMV due to the nonlinear nature of the utility function. In the lemonade example, the utility function $u(M) = \sqrt{M}$ was concave. In that case, the CME will always fall below the EMV:

$$\text{CME}(C) < \text{EMV}(C)$$

for any lottery C when the utility function is concave. Since the decision maker is willing to pay less than the expected value of the lottery, we say that he is *risk averse*.

In contrast, suppose that the utility function is convex. Then the CME will always exceed the EMV:

$$\text{CME}(C) > \text{EMV}(C)$$

for any lottery C with a convex utility function. Here we say that the decider is *risk prone*.

As discussed previously, the utility function for a decider could be both concave and convex, as well as linear. In other words, one portion of the utility function might be concave, while others are linear or convex. Such a utility function is illustrated in Figure 9–5.

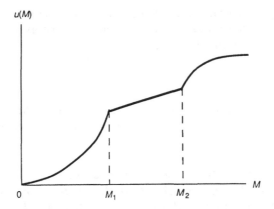

FIGURE 9–5. Example of a utility function which is piecewise convex, linear, and concave.

9.6 SUMMARY

A decision maker's attitudes toward alternative consequences leads to a preference ordering among the prospects. The strength of the preference can be defined in terms of a utility function.

The preference ordering is invariant under linear transformations of the utility function. In other words, the utility function is unique up to a linear transformation.

The field of decision theory often considers the desirability of outcomes in terms of the converse notion of a penalty or loss associated with each consequence. A consideration of the risk or expected loss due to each decision procedure is tantamount to a linear relationship between the loss and lack of desirability.

In many situations, however, the desirability of an outcome is not directly proportional to the amount of loss or gain. This complication arises even in the context of monetary considerations, where rewards and penalties can be manipulated with relative simplicity.

The nonlinear character of preference as a function of a set of outcomes can be encoded explicitly when utility functions are employed. In this way, utility functions are appropriate when the desirability of certain consequences takes a nonlinear form.

The comparison of alternative procedures can then be accomplished by a consideration of the associated utilities. In other words, the minimization of the loss function associated with a procedure corresponds to the maximization of the associated utility function. For instance, the minimax strategy for minimizing the loss among a set of procedures is tantamount to a maximin strategy for maximizing the smallest gain. In a similar way, the Bayes' strategy for minimizing the average risk when the underlying performance measure is a loss function, corresponds to a maximization of the average utility under the new criterion of utility functions.

PROBLEMS

The problems below are numbered according to the sections to which they correspond. For instance, Problem 9.1A is the first ("A") exercise pertaining to Section 9.1.

9.1A   Using the properties of the various relations, prove the following: (a) If $A \sim B$ and $B \succ C$, then $A \succ C$. (b) If $A \precsim B$ and $B \sim C$, then $A \precsim C$. (c) If $A \precsim B$ and $B \prec C$, then $A \prec C$. (d) If $A \prec B$ and $C \succsim B$, then $A \prec C$.

9.1B   Suppose $A \succ B$, while $p \geq q$ and $p \neq 0$. Consider the lottery whose consequences are $[A, p; B]$ with probability q/p and B with probability $1 - q/p$. Show that this lottery is equivalent to the simpler lottery $[A, q; B]$. In other words,

$$[[A, p; B], q/p; B] \sim [A, q; B]$$

(a) Draw the event tree. (b) Demonstrate the equivalence.

9.2A   Consider the Equivalent Lottery Theorem. Its proof asserts that the supremum r of the p values in $L = \{M(p)\}$ leads to $M(r) \sim K$. Prove that the infimum w of the q values in $H = \{M(q)\}$ leads to the same result; that is, $M(\sup\{p\}) \sim M(\inf\{q\}) \sim K$.

9.3A   Wendy prefers ice cream to cheesecake. But she would be indifferent between a fruit cup and a lottery which offers ice cream with probability 3/4 and cheesecake otherwise. If her utility for cheesecake is -2, and that for fruit is $+1$, what is her utility for ice cream?

9.3B   Xenia's utility for cheesecake and ice cream are 0 and 8, respectively. (a) If she is indifferent between a fruit cup and the lottery [ice cream, 3/4; cheesecake], what is her utility for a fruit cup? (b) Is there a simple algebraic relationship between Xenia's utility function v and Wendy's function u in Problem 9.3A?

9.3C   Prove that two consequences are equally desirable if and only if they have the same utility:

$$R \sim S \text{ if and only if } u(R) = u(S)$$

9.4A   Consider the lottery $[A, 1/4; B, 1/2; C]$ where $u(A) = -1, u(B) = 1$, and $u(C) = 2$. What is the utility of the lottery?

9.4B   Show that the linearity property of utility functions applies to lotteries composed of three prospects: $u([R, r; S, s; T]) = ru(R) + su(S) + (1 - r - s)u(T)$.

9.4C   Consider the second part of the proof for the Monotonicity and Linearity Theorem. The proof considered the case where $A \prec G \prec T \prec B \prec H$. Show that u and v are related linearly for some other ordering of the prospects, for instance the case where $A \prec B \prec G \prec T \prec H$.

9.4D   Use induction to show that the linearity property of the utility function applies to any *finite* number of consequences: $u([C_1, c_1; \ldots; C_n, c_n]) = \sum_1^n c_i u(C_i)$. (*Hint*: Define $B_n \equiv [C_1, d_1; \ldots; C_n, d_n]$ and consider the lottery $[B_n, b_n; C_{n+1}]$. Note that $d_i = c_i / \sum_1^n c_i$ in the sublottery B_n.)

9.5A   Consider the St. Petersburg game. Suppose the game is offered by a casino which will pay up to $2^{14} = \$4,096$ for $n \geq 14$. In other words, the payment is $\$2^n$ for $1 \leq n \leq 13$, and $\$4,096$ if $n \geq 14$. What is the EMV of the game?

9.5B   Consider the lemonade example in Section 9.5. Suppose Little Lila's utility function is as follows: $u(M) = M/4$ for $M \leq 16$, and $u(M) = \sqrt{M}$ for $M > 16$. (a) Draw the chart for the utility function. (b) What is the EMV? (c) What is the CME?

9.5C   Draw a utility function having the following characteristics: $u(\$0) = 0$; $u(\$10) = 1$; concave for $M < 0$; convex for $\$0 < M < \$10$; linear for $M > \$10$.

9.5D   Tom and Jerry have conceived of a household device to wash cars automatically. To sell their invention, they need to prepare a demonstration model at an investment of $3000. The idea might sell for up to $13,000 or it might not sell at all. (a) What is the relevant range of net payoffs? (b) Suppose the net outcome of this affair, in thousands of dollars, is: $-3, -1, 0, 2, 7, 10$. If each outcome is equally likely, what is the EMV of the lottery? (c) Tom's utility for the outcomes in (b) are $-.5, -.4, 0, .5, .9, 1$, respectively. Jerry's corresponding utilities are $-.5, -.1, 0, .1, .4, 1$. Plot these points on a chart and connect the data with piecewise linear line segments. (d) What are the CME values for each of Tom and Jerry?

FURTHER READING

Chernoff, Herman, and Lincoln E. Moses. 1959. *Elementary Decision Theory*. New York: Wiley.

Lindgren, B.W. 1971. *Elements of Decision Theory*. New York: Macmillan.

von Neumann, John, and Oskar Morgenstern. 1980. *Theory of Games and Economic Behavior*. Princeton, NJ: Princeton University Press.

10

Multiple Decisions

Up to this point, we have focused on highly constrained problems such as estimation, or tasks involving a choice between two alternatives. But the problems we face are rarely so simple.

In this chapter, we examine the nature of decision making among multiple alternatives. To this end, we begin with the interpretation of binary decisions involving simple hypotheses. This is followed by a discussion of the general multiple decision problem.

10.1 BINARY DECISIONS
AND SIMPLE HYPOTHESES

We consider the hypothesis testing problem based on n observations of a random variable X. The random variable follows a probability function $f(x|\theta)$. The parameter θ takes a value in the parameter space Θ with a probability function π.

For the case of simple hypotheses, the propositions are $H_0: \theta = \theta_0$ versus $H_1: \theta = \theta_1$. The prior probabilities are given by the discrete mass function $\pi_0 \equiv \pi(\theta_0)$ and $\pi_1 \equiv \pi(\theta_1)$. Since the parameter space Θ contains only two points, the probabilities sum to unity: $\pi_0 + \pi_1 = 1$.

A reasonable criterion for selecting one decision procedure over another is the mean risk, also known as the Bayes' criterion. Suppose A_i is the acceptance region corresponding to hypothesis H_i. Then the risk for a procedure δ is given by

$$r(\delta) = ER(\theta, \delta) = \pi_0 R(\theta_0, \delta) + \pi_1 R(\theta_1, \delta)$$

$$= \pi_0 P\{X \in A_1 | \theta_0\} + \pi_1 P\{X \in A_0 | \theta_1\}$$

This is the average chance of incorrectly accepting the null or alternate hypotheses. The Bayes' approach thus seeks a procedure δ which provides minimal chance of making an incorrect decision.

The Neyman–Pearson lemma stipulated the conditions for determining the best critical region, once the chance of falsely accepting the alternate hypothesis is fixed. More precisely, an assumption lies in the prior stipulation of the level of significance—namely, the chance of falsely accepting H_1, given by $\alpha \equiv P\{X \in A_1 | \theta_0\}$. The lemma then specifies the conditions for minimizing the probability $P\{X \in A_0 | \theta_1\}$ of falsely accepting H_0.

In the current approach to the construction of decision procedures, we no longer have any artificial constraint on the chance of falsely accepting the alternate hypothesis. Rather, our task is to find the acceptance regions A_0 and A_1 in such a way as to minimize the average risk of falsely accepting either hypothesis.

Despite the lack of a prior constraint on the chance of falsely accepting H_0, the two situations exhibit many features in common. It seems plausible, then, to expect a theorem for the new, Bayesian context which resembles the Neyman–Pearson lemma for the classical hypothesis–testing formulation. This is, in fact, the case.

Theorem (Binary Decision). Let X be a random sample of size n from a probability function $f(x|\theta)$. The parameter θ can assume values θ_0 and θ_1 with probabilities π_0 and π_1, respectively. Consider the hypotheses $H_0 : \theta = \theta_0$ versus $H_1 : \theta = \theta_1$. The critical region which satisfies the constraint

$$\rho_{10}(x) \geq \frac{\pi_0}{\pi_1} \tag{10-1}$$

defines a Bayes' solution to the testing problem.

As usual, $\rho_{10}(x)$ is the ratio of likelihood functions given by $\mathcal{L}(x|\theta_1)/\mathcal{L}(x|\theta_0)$. The theorem is a special case of a more general result which is stated and proved in the next section.

We demonstrate the use of the theorem by drawing on the same example as that in Section 8.3 to illustrate the Neyman–Pearson lemma. The problem is to determine the mean μ from a normal density with known standard deviation σ. The hypotheses are $H_0 : \mu = \mu_0$ and $H_1 : \mu = \mu_1$, where we assume $\mu_1 > \mu_0$. Suppose that π_i is the prior probability for μ_i. The use of Eq. (10–1) leads to

$$\exp\left(\frac{\mu_1 - \mu_0}{\sigma^2} \sum_i x_i + \frac{n(\mu_0^2 - \mu_1^2)}{2\sigma^2}\right) \geq \frac{\pi_0}{\pi_1}$$

where the left-hand side is the likelihood ratio $\rho_{10}(x)$ obtained in Section 8.3. Replacing $\sum_i x_i$ with $n\bar{x}$ and isolating $\bar{x}$, we obtain

$$\bar{x} \geq \frac{\mu_0 + \mu_1}{2} + \frac{\sigma^2}{n(\mu_1 - \mu_0)} \ln\left(\frac{\pi_0}{\pi_1}\right) \qquad (10\text{–}2)$$

when the prior probabilities are equal, $\pi_0 = \pi_1$, and the threshold for the critical region is $(\mu_0 + \mu_1)/2$. In other words, the lack of preference between μ_0 and μ_1 leads to a threshold which lies midway between the two values.

On the other hand, suppose that μ_0 is more likely to occur than μ_1. Then $\pi_0 > \pi_1$ and the threshold in Eq. (10–1) moves to the right, away from μ_0. As a result, the null hypothesis is less likely to be rejected.

The extent of the shift to the right depends on the ratio π_0/π_1 as well as the variance σ^2, the sample size n, and the difference in means, $\mu_1 - \mu_0$. In particular, as the sample size n increases, the rightmost term in Eq. (10–2) vanishes and the threshold reduces to $(\mu_0 + \mu_1)/2$. In other words, for large sample sizes, the observed data takes precedence over prior conceptions of the likelihood of μ_0 over μ_1.

10.2   MULTIPLE DECISIONS

We can extend the results of the preceding section for problems involving three or more alternatives. Consider a random sample of size n for a variable X from a probability function $f(x|\theta)$. The parameter θ can take any value in the space $\Theta = \{\theta_1, \theta_2, \ldots, \theta_n\}$ with probability π_i for θ_i. The hypotheses are $H_i: \theta = \theta_i$, each of which has a loss binary loss function given by

$$L(\theta_i, a) = \begin{cases} 0 & \text{if decision is correct} \\ 1 & \text{otherwise} \end{cases}$$

Our task is to develop a procedure to minimize the average risk.

The risk is minimized by selecting the acceptance region A_i for the ith hypothesis if its weighted likelihood is high. More specifically, let $\mathscr{L}_i(x) \equiv \mathscr{L}(x|\theta_i)$ denote the likelihood function for x under θ_i. The *weighted likelihood* $\pi_i\mathscr{L}_i(x)$ defines the relative probability of observing a sample point x after it is weighted by the prior probability π_i of having θ_i occur. The point x should be included in the acceptance region for H_i if the corresponding weighted likelihood is high. In other words, the condition

$$\pi_i\mathscr{L}_i \geq \pi_j\mathscr{L}_j \quad \text{for all } j$$

defines the acceptance region for A_i. We encapsulate this result in the form of a theorem and its proof.

Theorem (Multiple Decisions). Let X be a random variable from a density $f(x|\theta)$. The parameter θ can assume value θ_i, for $i = 1, \ldots, k$, with probability π_i. Suppose the hypotheses take the form $H_i\colon \theta = \theta_i$. Further, let A_i be the set of points x for which

$$\pi_i \mathcal{L}_i(x) \geq \pi_j \mathcal{L}_j(x) \quad \text{for all } j \tag{10-3}$$

If a point x satisfies the condition for more than one parameter θ_i in $\mathcal{L}_i(x) \equiv \mathcal{L}(x|\theta_i)$, it can be assigned to any appropriate region A_i. The acceptance regions A_i constitute a Bayes' solution to the multiple decision problem.

As indicated in the theorem, a point x might satisfy Eq. (10–3) for more than one index i. In that case it may be assigned to any of the candidate regions A_i, perhaps by random assignment or by selecting the region with the lowest index i. In this way, the regions A_i partition the sample space Ω into a collection of exhaustive and disjoint subsets. If a point x falls in the region A_i, then hypothesis H_i is accepted. The theorem can be validated as follows.

Proof. Let $A_1, \ldots, A_k$ be a partition of the sample space as indicated in the theorem. Suppose $B_1, \ldots, B_k$ is any other partition of the sample space. We define an indicator function ϕ_i corresponding to region A_i, and ψ_i corresponding to region B_i. More specifically,

$$\phi_i(x) = \begin{cases} 1 & \text{if } x \in A_i \\ 0 & \text{otherwise} \end{cases}$$

$$\psi_i(x) = \begin{cases} 1 & \text{if } x \in B_i \\ 0 & \text{otherwise} \end{cases}$$

Let δ be the decision procedure which maps an observation x into the acceptance regions A_i; and similarly, ε is the procedure mapping x into the regions B_i.

The likelihood function $\mathcal{L}_i(x)$ is the probability function corresponding to hypothesis H_i. In algebraic terms

$$\mathcal{L}_i(x) \equiv f_i(x) \equiv f(x|\theta_i) = f(x_1, \ldots, x_n|\theta_i) = \prod_{q=1}^{n} f(x_q|\theta_i)$$

The third equality relies on the fact that x is a vector of n observations; the last equality depends on the independence of the observations.

Suppose $L(\theta_i, \delta)$ is the loss for procedure δ when the value of the parameter is θ_i. The risk under θ_i is the average value of the loss. For the case when $f(x|\theta_i)$ is a continuous density function, we can express the risk as

$$R(\theta_i, \delta) = EL(\theta_i, \delta) = \int L(\theta_i, \delta(x_i)) f_i(x) \, dx$$

The derivative dx is shorthand for $dx_1 \ldots dx_n$. Similarly, the integral actually represents an n-fold integral over all values of $x_1, \ldots, x_n$. The mean risk for procedure δ under all possible values of θ_i is

$$r(\delta) = ER(\theta_i, \delta) = \sum_{i=1}^{k} \pi_i R(\theta_i, \delta) = \sum_{i=1}^{k} \pi_i \int L(\theta_i, \delta(x)) f_i(x) \, dx$$

The mean risk $r(\varepsilon)$ for procedure ε exhibits the same formal structure. Consequently, the difference in mean risks due to ε and δ can be expressed as

$$\Delta r \equiv r(\varepsilon) - r(\delta) = \sum_{j=1}^{k} \pi_j \int L(\theta_j, \varepsilon) f_j(x) \, dx - \sum_{i=1}^{k} \pi_i \int L(\theta_i, \delta) f_i(x) \, dx$$

We have defined our penalty function as a binary construct. In particular, the loss function L assumes value 0 when θ_i lies in the proper acceptance region, and equals 1 otherwise. Substituting these values for L in the previous equation yields

$$\Delta r = \sum_{j=1}^{k} \pi_j \int_{\Omega - B_j} f_j(x) \, dx - \sum_{i=1}^{k} \pi_i \int_{\Omega - A_i} f_i(x) \, dx \qquad (10\text{–}4)$$

Here Ω denotes the entire sample space. The right-hand side of the equation takes account of the fact that the integral vanishes over the acceptance regions A_i and B_j. By the linearity property of integrals, the integral over $(\Omega - A_i)$ can be written as the difference of integrals:

$$\int_{\Omega - A_i} f_i(x) \, dx = \int_{\Omega} f_i(x) \, dx - \int_{A_i} f_i(x) \, dx$$

In a similar way, the integral over $(\Omega - B_j)$ can be expressed as the difference of integrals over Ω and over B_j. When Eq. (10–4) is expanded in this fashion, the integrals over Ω—in conjunction with their weights π_i—cancel from the expression. The result is

$$\Delta r = \sum_{i=1}^{k} \pi_i \int_{A_i} f_i(x) \, dx - \sum_{j=1}^{k} \pi_j \int_{B_j} f_j(x) \, dx$$

We can use the indicator functions to obtain integrals over the entire sample space Ω:

$$\Delta r = \sum_{i=1}^{k} \pi_i \int_{\Omega} \phi_i(x) f_i(x)\, dx - \sum_{j=1}^{k} \pi_j \int_{\Omega} \psi_j(x) f_j(x)\, dx \qquad (10\text{--}5)$$

Since the acceptance regions represent a collection of exhaustive and mutually exclusive subsets of the sample space, their corresponding indicator functions sum to unity. In algebraic form, $\sum_i \phi_i(x) = 1$ and $\sum_j \psi_j(x) = 1$ for every value of x. We can inject these expressions as unit factors into Eq. (10–5) without affecting the results:

$$\Delta r = \sum_{i=1}^{k} \pi_i \int_{\Omega} \phi_i(x) \left(\sum_{j=1}^{k} \psi_j(x) \right) f_i(x)\, dx - \sum_{j=1}^{k} \pi_j \int_{\Omega} \psi_j(x) \left(\sum_{i=1}^{k} \phi_i(x) \right) f_j(x)\, dx$$

$$= \sum_{i=1}^{k} \sum_{j=1}^{k} \int_{\Omega} \phi_i(x) \psi_j(x) [\pi_i f_i(x) - \pi_j f_j(x)]\, dx$$

At this point, we can replace the integral over Ω by the sum of integrals over $A_1, A_2, \ldots, A_k$. For any point $x \in A_i$, we know from Eq. (10–3) that the last expression in brackets is non-negative. Since ϕ_i and ψ_j are also non-negative, each integral is non-negative, and we infer that $\Delta r \geq 0$. Finally, we obtain the result $r(\varepsilon) \geq r(\delta)$. In other words, procedure δ is the Bayes' solution to the problem of multiple decisions. ∎

When the parameter space Θ contains only two points, $k = 2$ and the multiple decision problem simplifies to the binary problem discussed in the previous section. For consistency with the notation in Section 9.1, we can define a constant m as the upper limit of the index for the case of two alternatives. Since $k = 2$ here, the new upper limit is $m = k - 1 = 1$. Then the two hypotheses are $H_0: \theta = \theta_0$ and $H_1: \theta = \theta_1$. According to Eq. (10–3), the acceptance region for H_1 is

$$\pi_1 \mathscr{L}_1(x) \geq \pi_0 \mathscr{L}_0(x)$$

We can rewrite this inequality as

$$\rho_{10} = \mathscr{L}_1(x)/\mathscr{L}_0(x) \geq \frac{\pi_0}{\pi_1}$$

which is Eq. (10–1).

We now turn to a numerical example to illustrate the use of the Multiple Decisions Theorem.

Example

Let X be a normal variable with a variance of 1 and unknown mean θ. Suppose the three possibilities for θ, and their respective probabilities, are as follows:

$$H_1: \theta_1 = 0 \quad \text{with } \pi_1 = 1/4$$

$$H_2: \theta_2 = 1 \quad \text{with } \pi_2 = 1/2$$

$$H_3: \theta_3 = 2 \quad \text{with } \pi_3 = 1/4$$

Further, assume that our decision will depend on a single trial.

The acceptance region A_i for θ_i is defined by relations of the form $\pi_i \mathcal{L}_i(x) \geq \pi_j \mathcal{L}_j(x)$. In our case, the inequality is

$$\pi_i \frac{1}{\sigma\sqrt{2\pi}} \exp\left(-\frac{(x-\theta_i)^2}{2\sigma^2}\right) \geq \pi_j \frac{1}{\sigma\sqrt{2\pi}} \exp\left(-\frac{(x-\theta_j)^2}{2\sigma^2}\right)$$

for each value of j. After taking logarithms, setting $\sigma^2 = 1$, and simplifying, we obtain

$$x(\theta_i - \theta_j) \geq \frac{\theta_i^2 - \theta_j^2}{2} + \ln\left(\frac{\pi_j}{\pi_i}\right)$$

For the first acceptance region A_1, the two relations are

$$x(0 - 1) \geq \frac{0^2 - 1^2}{2} + \ln\left(\frac{1/2}{1/4}\right) \quad \text{for } j = 2$$

$$x(0 - 2) \geq \frac{0^2 - 2^2}{2} + \ln\left(\frac{1/4}{1/4}\right) \quad \text{for } j = 3$$

The resulting inequalities are $x \leq 1/2 - \ln 2$ and $x \leq 1$. Combining these results, the acceptance region for H_1 is $A_1 = (-\infty, 0.5 - \ln 2]$.

In a similar way, the resulting inequalities for A_2 are $x \geq 1/2 - \ln 2$ and $x \leq 3/2 + \ln 2$. The lower boundary point $x = 1/2 - \ln 2$ has already been assigned to A_1. So the second acceptance region is $A_2 = (0.5 - \ln 2, 1.5 + \ln 2]$.

In this example, the assignment of boundary points to one or another acceptance region is actually inconsequential. Since X is a continuous random variable, the probability that its value will be precisely $0.5 - \ln 2$ is zero.

The remainder of the real line must then belong to the third acceptance region. In other words the last region is $A_3 = (1.5 + \ln 2, \infty)$. ◆

PROBLEMS

The problems below are numbered according to the sections to which they correspond. For instance, Problem 10.1A is the first ("A") exercise pertaining to Section 10.1.

10.1A Let X be a random sample of size n from a binomial distribution with parameter p. The parameter can take values $1/4$ and $3/4$ with respective probabilities $2/3$ and $1/3$. Consider the hypotheses $H_0: p = 1/4$ and $H_1: p = 3/4$. (a) What is the likelihood ratio ρ_{10}? (b) What are the acceptance regions A_0 and A_1?

10.2A Consider the three-way test for the normal random variable discussed in Section 10.2. Assume, as before, that the π_i are $1/4$, $1/2$, and $1/4$ for θ_i equal to 0, 1, and 2, respectively. However, suppose that the variance is $\sigma^2 = 2$. What is the new set of values for the acceptance regions? Explain the results.

10.2B Let $x_1, \ldots, x_n$ be a set of observations from a normal density with variance $\sigma^2 = 4$. The mean is unknown, but is assumed to take one of the values $\{-1, 0, 5\}$; the prior probabilities for these values are $\{.4, .4, .2\}$, respectively. What are the acceptance regions?

10.2C Let $x_1, \ldots, x_n$ be a set of observations from an exponential density $\lambda e^{-\lambda x}$ for $x \geq 0$. The anticipated values of λ are $\{1, 4, 8\}$ with respective prior probabilities $\{1/4, 1/4, 1/2\}$. What is the acceptance region A_i corresponding to each λ_i?

FURTHER READING

Ferguson, Thomas S. 1967. *Mathematical Statistics, A Decision Theoretic Approach.* San Diego, CA: Academic Press.

Hoel, Paul G., Sidney C. Port, and Charles J. Stone. 1971. *Introduction to Statistical Theory.* Boston: Houghton Mifflin.

11

Sequential Methods

Sequential procedures are based on the idea that approaching a decision-making task one step at a time may be better than attempting to attack the problem in one fell swoop. A stepwise procedure may be superior to a monolithic one in two senses—in terms of effectiveness as well as efficiency.

When a problem is complex, it may exceed our cognitive capabilities. Some problems may lie beyond the reach of allocated time and financial resources even with the assistance of computer programs. In this scenario, sequential procedures may be *effective* in the sense that complex problems that would otherwise be intractable are cut down to size. In fact, in many arenas such as scientific research, sequential procedures represent the only realistic approach. The results of an experiment must be analyzed to determine the appropriateness of the working assumptions and to motivate the next round of investigations. For innovative research, it would be infeasible to chart out all the possible consequences and potential results beforehand. The world around us often turns out to be simpler than we expected; but it is not so simple that we can imagine all the alternatives beforehand.

Sequential methods can be helpful even for smaller problems. A sequential procedure can be *efficient* in that the amount of data or computation required to resolve a problem may be less than that for a monolithic approach.

11.1 ILLUSTRATION OF MONOLITHIC VERSUS SEQUENTIAL APPROACHES

Consider a manufacturer of computer chips. The products are large-scale integrated circuits that are complex and prone to production defects. The proportion of products that currently fail inspection is 10%. A new production process has just been introduced to cut operating costs. Unfortunately, the defect rate for the new process seems to be about 20%. How many chips should the plant manager test in order to determine whether the failure rate for the new process is, in fact, 20%?

Let p be the proportion of defective chips. Then the hypotheses are as follows:

$$H_0: p_1 = 0.1$$

$$H_1: p_0 = 0.2$$

For a monolithic experiment, the manager must expect to collect numerous data. Suppose he were to decide on a sample size of 15. Is this sample large enough if he wants to keep the chances of false rejection and false acceptance down to 5%?

Assume for the moment that $n = 15$ is chosen. It is conceivable that all 15 could be defective. Can the manager reject H_0 in this case? Under the null hypothesis, the chance of observing 15 defects is

$$p(15|H_0) = \binom{15}{15} p_0^{15}(1 - p_0)^0 = (1)(.1)^{15}(1 - .1)^0$$

$$= 10^{-15}$$

The chance of false rejection is 10^{-15}, much less than the threshold of $\alpha = .05$. Hence, the manager could reject H_0 with impunity.

Now let us consider the other extreme. Suppose 0 defects are observed, can the manager safely conclude that H_0 is true? The chance of false acceptance is

$$p(0|H_1) = \binom{15}{0} p_1^0 (1 - p_1)^{15} = (1)(.2)^0(1 - .2)^{15} = .03520$$

This is less than the threshold for false acceptance, namely, $\beta = 0.05$. Hence, the manager may accept H_0.

Suppose that the acceptance region for H_0 consists of the set $\{0, 1\}$. The chance of observing one defect under the alternate hypothesis is

$$p(1|H_1) = \binom{15}{1} p_1(1 - p_1)^{14} = 15(.2)(.8)^{14} = .1319$$

The chance of false acceptance can be calculated as

$$p(X \le 1|H_1) = p(0|H_1) + p(1|H_1) = .0352 + .1319 = .1671$$

or 16.71%. This exceeds the threshold of $\beta = 0.05$. In other words, the manager can neither accept nor reject the null hypothesis if the number of defects is 1.

The reason for this dilemma is that the sample size is too small for monolithic decision making. By increasing the sample size, we could reduce the standard deviation of the test procedure, and thereby discriminate between the hypotheses for all values of the experimental outcome.

To be more precise, we have assumed that the probability of the ith item being defective is a binomial variable X_i with parameter p. The test statistic is the sample mean $\overline{X} = \sum_i X_i / n$. The variance of $\overline{X}$ is

$$\text{Var}(\overline{X}) = \text{Var}\left(\frac{\sum_i X_i}{n}\right) = \frac{\sum_i \text{Var}(X_i)}{n^2} = \frac{n[p(1-p)]}{n^2} = \frac{p(1-p)}{n}$$

The second equation follows from the linearity of the variance operator for independent random variables.

The standard deviations under the null and alternate hypotheses are, respectively,

$$\sigma_0 = \left(\frac{p_0(1-p_0)}{n}\right)^{1/2} = \left(\frac{.1(1-.1)}{15}\right)^{1/2} = .0775$$

$$\sigma_1 = \left(\frac{p_1(1-p_1)}{n}\right)^{1/2} = \left(\frac{.2(1-.2)}{15}\right)^{1/2} = .1032$$

For $n = 15$, we see that the standard deviations are roughly as large as the difference we are attempting to establish—namely, $p_1 - p_0 = .2 - .1 = .1$. Here lies the source of difficulty for the production line example. From the preceding discussion, however, it is clear that the standard deviation of $\overline{X}$ can be reduced by increasing the sample size n.

Now let us consider a scenario from the sequential approach. Suppose we take the first chip, which we deduce to be defective. Can we reject or accept the null hypothesis? Under H_0, the probability of observing 1 defect out of a batch of 1 is

$$p(1|H_0) = p_0 = .1$$

Since this value exceeds the threshold of $\alpha = .05$ for false rejection, we cannot reject H_0.

The corresponding probability under H_1 is

$$p(1|H_1) = p_1 = .2$$

Since this exceeds the threshold of $\beta = 0.05$ for false acceptance, we cannot accept H_0 either.

We, therefore, continue and test the second item. Suppose it is again defective. Under H_0, the probability of obtaining 2 defects out of 2 trials is

$$p(2|H_0) = p_0^2 = (.1)^2 = .01$$

This value is less than the threshold of $\alpha = .05$ for false rejection. We may, therefore, reject H_0.

In this particular scenario, the sequential procedure led to a decision after only two trials. This is in contrast to the monolithic approach, where even 15 trials are not necessarily adequate to resolve a hypothesis-testing problem.

The main difficulty with the monolithic approach is that *all* potential outcomes have to be accounted for. Each eventuality must be identified and potentially resolved beforehand.

The sequential procedure, on the other hand, allows the decision maker to capitalize on a serendipitous set of outcomes. If the particular string of data happens to be extremely favorable or unfavorable to either hypothesis, the procedure can be terminated immediately, thereby conserving on the amount of data required. On average, sequential methods will lead to sample sizes that are roughly one-half of those required for monolithic procedures. This phenomenon, however, is probabilistic rather than certain. The requisite sample size is, therefore, a random facet with which the sequential decider must contend.

11.2 STOPPING AND TERMINAL RULES

Consider a sequence of observations $X_1, X_2, \ldots$. The sequence may be finite, as in the case of sampling without replacement from a finite population; or it might be infinite, as in the position of an atom over time.

We assume that the distribution is known for each finite segment $(X_1, \ldots, X_k)$ of the sequence. The simplest case arises when the variables $X_1, \ldots, X_k$ represent independent observations from a fixed population. Then the joint probability function is

$$f_k(x_1,\ldots,x_k|\theta) = \prod_{i=1}^{k} f(x_i|\theta)$$

Here, $f(x|\theta)$ denotes the probability function for each member of the population.

A sequential decision procedure must consider explicitly the cost of sampling. If the cost of obtaining additional data were zero, then each procedure would call for the largest possible data set. Unfortunately, such a requirement is unrealistic in most practical applications.

The need to account for the cost of observations is a key differentiator between sequential and monolithic procedures. In the monolithic approach, the sample size is fixed, thereby resulting in the same sampling cost for all the candidate procedures.

The cost of data acquisition in a sequential procedure will depend on the state of nature θ, the sample size k, and perhaps even on the values of the observations. In symbolic form, the cost c_k is

$$c_k = c(\theta; k; X_1, \ldots, X_k)$$

A simplifying assumption is that the cost of sampling depends only on the number of observations. Then the cost is given by $c_k = kc$ for some constant c. This simplification is reasonable for many practical applications.

A sequential methodology can be partitioned into two stages. The first stage relates to the acquisition of a sufficient amount of data, whereas the second pertains to the action to be taken as a result of the observations. The respective procedures are called stopping rules and terminal rules.

Stopping Rule

A *stopping rule* specifies the conditions under which sampling is to halt. A stopping rule can be characterized in terms of indicator functions which denote whether or not the data acquisition phase is to terminate.

One such indicator function specifies whether the procedure has halted at any point up to the current observation k. We call this kind of function a *running indicator*.

A second type of indicator specifies whether the sequential procedure has halted at *precisely* the ith observation. We call this type of function a *point indicator*.

We define both types of indicators more precisely and will determine that they are equivalent.

Running Indicator

A stopping rule can be specified in terms of the observations $X_1, \ldots, X_n$ in the following way. We define the *running indicator* ρ_k to denote whether the procedure is to halt, based on the cumulative data up to, and including, the kth observation:

$$\rho_k \equiv \rho_k(X_1, \ldots, X_k) = \begin{cases} 1 & \text{if stopping rule says halt, given } X_1, \ldots, X_k \\ 0 & \text{otherwise (take another sample)} \end{cases}$$

For instance, ρ_0 denotes whether or not any observation should be made at all: $\rho_0 = 1$ if no data is to be obtained, whereas $\rho_0 = 0$ if an observation is to be made. In a similar way, $\rho_1 = 1$ if the sequential procedure is to halt after a single observation, whereas $\rho_1 = 0$ if a second sample should be obtained.

The set of running indicators $\rho \equiv \{\rho_0, \rho_1, \rho_2, \ldots\}$ characterizes the behavioral aspects of the stopping rule. In other words, the functions ρ_k determine when the sampling process is to terminate, based on the values of observations $X_1, \ldots, X_k$.

Point Indicator

A second way to describe a stopping rule is in terms of the current step. The *point indicator* ψ_k specifies whether or not the sampling should halt for the first time at the kth observation:

$$\psi_k \equiv \psi_k(X_1, \ldots, X_k) = \begin{cases} 1 & \text{if stopping rule says halt for the first time} \\ & \text{at step } k, \text{ given } X_1, \ldots, X_k \\ 0 & \text{otherwise (halting occurred before step } k, \\ & \text{or sampling continues after step } k) \end{cases}$$

In particular, ψ_0 assumes the value 1 if no data at all is to be acquired, or takes value 0 if at least one observation is to be made. Similarly, $\psi_1 = 1$ if the procedure halts after a single observation, whereas $\psi_1 = 0$ if no data at all was collected, or if a second sample is to be obtained.

Equivalence of Running and Point Indicators

The set of point indicators $\psi = \{\psi_0, \psi_1, \ldots\}$ characterize the stopping rule, serving a role similar to the running indicators $\rho = \{\rho_0, \rho_1, \ldots\}$. In fact, these two types of functions are equivalent.

We can show this by first noting that $\psi_0 = \rho_0$. Each of these functions takes the value 1 if no data is collected at all, and 0 if at least one sample is obtained.

We can express the running indicator in terms of the point indicator in the following way:

$$\rho_k = \sum_{i=1}^{k} \psi_i$$

for $k = 1, 2, \ldots$. In other words, the sampling process terminates by the kth observation if it has stopped at any point up to, and including, that step. Consequently, we can write each point indicator ψ_k in terms of the running indicators:

$$\psi_k = \left(\prod_{i=0}^{k-1} (1 - \rho_i) \right) \rho_k$$

If the sampling stops before the kth observation, then $\rho_i = 1$ for at least one value of i; consequently the right-hand side is 0, at the same time when ψ_k is defined to equal 0. In the second scenario, assume that the sampling stops for the first time at step k. Then $\rho_k = 1$, whereas every other $\rho_i = 0$; in this case $\psi_k = 1$ as well. Finally, if sampling continues after the kth step, then both sides of the equality assume the value 0.

Mean Sample Size

The decision to halt with a particular sample depends on the values of the sample. Since each observation is a random variable, so is the number of observations.

Let N be the random variable denoting the sample size. Given observations $X_1, \ldots, X_k$, the conditional probability of stopping at that point is 0 or 1. These values are precisely the outcomes of the point indicator ψ_i. We can, therefore, write

$$P\{N = k | x_1, \ldots, x_k\} = \psi_k(x_1, \ldots, x_k)$$

What about the prior probability of stopping at step k, before the sample is taken? This is equal to the expected value of ψ_k:

$$P\{N = k\} = 0 \cdot P\{\psi_k = 0\} + 1 \cdot P\{\psi_k = 1\} = E(\psi_k)$$

For practical applications, we would require that the sequential procedure terminate at some point. We can state this requirement by the condition

$$\sum_{i=1}^{M} P\{N = i\} = 1$$

for some $M < \infty$.

Terminal Rule

Once a sequential procedure halts, a decision procedure must map the observed data into an action space. This mapping is called a *terminal decision procedure* or simply a *terminal rule*.

The input to a terminal rule is a set of observations, the same type of data as that for a monolithic rule. The terminal procedure for a sequential methodology, therefore, has the same formal structure as that for a monolithic approach for the same sample size.

More precisely, let δ_k be a decision procedure based on observations $X_1, \ldots, X_k$. Then, $\delta_k(X_1, \ldots, X_k)$ maps the observations into some action in the action space **A**.

The terminal rule for a sequential procedure is a sequence

$$\delta \equiv \{\delta_0, \delta_1, \delta_2, \ldots\}$$

where each δ_k is a function of $X_1, \ldots, X_k$. The first rule δ_0 represents a default mapping: if no data are acquired, some action $a_0 \in \mathbf{A}$ is to be undertaken.

We will see in subsequent sections that terminal rules have properties similar to those of decision procedures for monolithic techniques. In the meantime, we clarify the ideas in this section with an example.

Example

An important class of sequential procedures relates to the problem of hypothesis testing. Let hypotheses H_0 and H_1 specify two states of nature, θ_0 and θ_1, respectively. Further, assume that $f_i(x) \equiv f(x|\theta_i)$ is the density function corresponding to state θ_i.

We form the likelihood ratio based on n independent observations:

$$\Lambda_n \equiv \frac{\mathscr{L}_0(x_1, \ldots, x_n)}{\mathscr{L}_1(x_1, \ldots, x_n)} = \frac{\prod_{i=1}^n f_0(x_i)}{\prod_{i=1}^n f_1(x_i)}$$

Let a and b denote two constants for which $a < b$. We define the stopping rule in terms of running indicators. At least one sample should be taken, from which we require that $\rho_0 = 0$. At step $n \geq 1$,

$$\rho_n = \begin{cases} 1 & \text{if } \Lambda_n < a \text{ or } \Lambda_n > b \\ 0 & \text{if } a \leq \Lambda_n \leq b \end{cases}$$

In other words, continue sampling if the likelihood ratio Λ_n falls within the thresholds a and b; otherwise stop the procedure.

The terminal rule might be defined as follows. Suppose k is some constant lying between a and b. Then the decision rule is

$$\delta_n = \begin{cases} \text{Accept } H_0, & \text{if } \Lambda_n \geq h \\ \text{Reject } H_0, & \text{otherwise} \end{cases}$$

According to this rule, the decider should accept the null hypothesis if Λ_n is at least as large as some criterion h; and assume the alternative hypothesis in the converse case.

The sequential procedure is defined by the pair (ρ, δ), where each of ρ and δ represents a family of rules.   ◆

11.3 BAYESIAN PROCEDURES

The overall cost for a sequential procedure depends on both the cost of sampling and the loss due to the terminal decision. Let c_i be the cost of the ith observation and $L(\theta, \delta_N)$ the loss due to the terminal decision. The terminal decision rule $\delta_N = \delta_N(X_1, \ldots, X_N)$ is a function of the observations $X_1, \ldots, X_N$.

The overall cost is given by

$$C(\theta; N; X_1, \ldots, X_N) = L(\theta, \delta_N(X_1, \ldots, X_N)) + \sum_{i=1}^{N} c_i$$

The sample size N is a random variable which depends in part on the stopping rule ψ. It also depends on the observations $X_1, \ldots, X_N$, which, in turn, are influenced by the state of nature θ. When the cost of making an observation is fixed at value c, the overall sampling cost is $\sum_1^N c_i = Nc$.

Consider a sample of size N. The risk for the sequential procedure (ψ, δ_N) is the expected loss due to both terminal error and sampling cost:

$$R(\theta, (\psi, \delta_N)) = EC(\theta; N; X_1, \ldots, X_N) = E\left(L(\theta, \delta_N(X_1, \ldots, X_N)) + \sum_{i=1}^{N} c_i\right)$$

For a given stopping rule ψ, the sample size N might be 0, 17, or any other value. If the probability of an infinite size is positive, the overall risk is defined as infinite. On the other hand, if the chance of a finite sample is unity, the overall risk is the sum of the risks due to each sample size:

$$R(\theta, (\psi, \delta)) = \sum_{j=0}^{\infty} E\{\psi_j[L(\theta, \delta_j) + c_j(\theta; X_1, \ldots, X_j)]\} \qquad (11\text{-}1)$$

Suppose the parameter θ has a prior probability function $\pi(\theta)$. Then averaging the risk over θ with respect to the prior yields the Bayes' risk:

$$r(\psi, \delta) = ER(\theta, (\psi, \delta)) \qquad (11\text{-}2)$$

The heart of the sequential problem is to find the procedure (ψ, δ) which minimizes the Bayes' risk. This can be achieved by determining the optimal terminal rule δ^* for each stopping rule ψ_i, then selecting the stopping rule that yields the minimal total risk.

The choice of an optimal stopping rule is usually a complicated affair. On the other hand, the optimal terminal rule is straightforward. It is, in fact, the familiar Bayes' procedure when the sample size at each step is regarded as fixed.

Theorem (Terminal Procedure). Suppose $\pi(\theta)$ is the prior probability function for a parameter θ, and ψ a particular stopping rule. Let δ_i^* be the Bayes' rule corresponding to the sample $X_1, \ldots, X_i$, where the size i is regarded as fixed. Then the Bayes' risk $r(\psi, \delta^*)$ is minimal for rule ψ when the terminal rule is the collection of monolithic Bayes' procedures; that is,

$$\delta^* = (\delta_0^*, \delta_1^*, \delta_2^*, \ldots)$$

The task of finding an optimal sequential procedure can be partitioned into the determination of a good stopping rule and the terminal rule. More specifically, the optimal terminal rule δ^* is independent of the choice of the stopping rule ψ.

Proof. From Eqs. (11-1) and (11-2), we can write

$$r(\psi, \delta) = E^{(\pi)} \sum_{i=0}^{\infty} E^{(x)}\{\psi_i[L(\theta, \delta_i) + c_i(\theta; X_1, \ldots, X_i)]\} \tag{11-3}$$

Here, $E^{(\pi)}$ represents expectation over the prior probability function π, and $E^{(x)}$ denotes expectation over the sample $X_1, \ldots, X_N$.

The risk $r(\psi, \delta)$ is minimized when each term on the right-hand side of Eq. (11-3) is minimized. To this end, we must minimize

$$E^{(\pi)} \sum_{i=0}^{\infty} E^{(x)}[\psi_i L(\theta, \delta_i)]$$

The stopping rule ψ and the prior probability function π are given beforehand. Hence, the above expression can be optimized by choosing each δ_i to minimize $E^{(\pi)}E^{(x)}[\psi_i L(\theta, \delta_i)]$.

The point indicator ψ_i assumes the value 1 if the data set $X_1, \ldots, X_i$ terminates the sampling process; and ψ_i vanishes for all other sample sizes. Hence, we may write

$$E^{(\pi)}E^{(x)}[\psi_i L(\theta, \delta_i)] = E^{(x)}\{\psi_i E^{(\pi)}[L(\theta, \delta_i)|(X_1, \ldots, X_i)]\}$$

As a result, each δ_i must minimize $E^{(\pi)}[L(\theta, \delta_i)|X_1, \ldots, X_i]$; this corresponds to the Bayes' rule δ_i^* with respect to the prior function π when the sample size is fixed at i. The reasoning holds for each value i, thereby completing the proof. ∎

A sequential procedure may potentially require an infinite number of observations. The theoretical foundation for such infinite situations is involved, but the procedure itself is often easy to describe.

On the other hand, a sequential procedure might be constrained to be finite. This situation arises, for instance, when a finite population is sampled without replacement; or the decision maker is limited by temporal or financial resources. The case of finite procedures has a relatively simple theoretical basis, but may be involved and tedious to implement in practice.

When a sequential procedure is truncated at the finite number of observations, the preceding Terminal Procedure Theorem can be used to determine the optimal strategy. This strategy will represent the Bayes' procedure (ψ^*, δ^*), where the terminal rule δ^* is obtained through a process of backward induction.

More specifically, suppose that the truncated strategy permits a sample size up to n. If the Bayes' procedure (ψ^*, δ^*) requires all n observations, the terminal rule δ^* is simply the Bayes' rule.

On the other hand, if the stopping rule requires at least $n - 1$ samples, the decision to continue or halt depends on a comparison of two conditional expected losses:

a. Conditional expected loss based only on $X_1, X_2, \ldots, X_{n-1}$.
b. Conditional expected loss including sampling cost, if sample X_n is obtained.

In each scenario, the losses to be compared are the minimal values resulting from the use of Bayes' procedures. If the expected loss due to the current sample size is less than or equal to the mean loss due to a larger sample, then the sampling halts; otherwise another observation is taken.

This type of reasoning continues in a backward chain for $n - 2$ observations, then $n - 3$, and so on. Since n is finite, the process of backward induction eventually terminates. The result is the Bayes' sequential strategy for the particular stopping rule.

11.4 LIKELIHOOD RATIO PROCEDURES

As discussed in the previous section, sequential procedures are used to minimize the amount of data acquisition needed to fulfill some objective. The basic approach is to collect data, determine whether enough information is available to make a decision, and continue if a decision would be premature.

At each step we can decide to accept the null hypothesis, reject it, or take another observation. Since we are dealing with probabilistic phenomena, any decision is, of course, subject to a chance of error. It is possible, however, to maintain the probabilities of false acceptance and rejection to predetermined levels.

To describe the sequential procedure, we assume that the random variable X takes values in the outcome space Ω with a density function $f(x|\theta)$. Our task is to discriminate between simple hypotheses:

$$H_0: \theta = \theta_0$$

$$H_1: \theta = \theta_1$$

Both hypotheses are simple values rather than composite intervals. Hence, our test procedure should resemble that used in the Neyman–Pearson lemma for a monolithic strategy where the sample size n is fixed:

$$\Lambda_n \equiv \frac{\mathscr{L}_{1n}}{\mathscr{L}_{0n}} = \frac{\prod_{i=1}^n f(x_i|\theta_1)}{\prod_{i=1}^n f(x_i|\theta_0)}$$

According to the lemma, the acceptance regions are determined in the following way. Each point x in Ω for which the ratio $\Lambda_n(x)$ exceeds some constant c, belongs to the acceptance region for the alternate hypothesis; and each point for which $\Lambda_n(x)$ falls below c belongs to the acception region for the null hypotheses.

The procedure suggests a way to extend the methodology for small data sets. Since our assessment will be tentative when the sample size is small, we could decide to accept H_1 when the ratio satisfying the relation Λ_n is much larger than c, accept H_0 when Λ_n is much smaller than c, and take another sample if Λ_n takes a middling value.

To be more precise, we should decide in advance on a lower threshold c_0 and an upper threshold c_1. If the likelihood ratio satisfies the relation $\Lambda_n \le c_0$, we accept H_0; if $\Lambda_n \ge c_1$, we accept H_1; if Λ falls in the interval (c_0, c_1), then we sample once more. The strategy is depicted in Figure 11–1.

The thresholds c_0 and c_1 can be selected in such a way that they constrain the chances of Type I and Type II errors, namely, α and β. No algorithm is available for calculating c_0 and c_1 exactly from values of α and β. On the other hand, the formulas

$$c_0 = \frac{\beta}{1 - \alpha} \quad \text{and} \quad c_1 = \frac{1 - \beta}{\alpha}$$

yield remarkably good results.

We summarize our discussion in the form of a proposition.

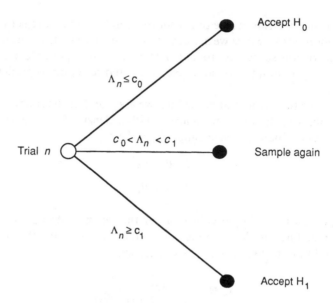

FIGURE 11–1. Decision tree for sequential sampling. The procedure halts at the first trial n for which Λ falls on or below c_0, or on or above c_1.

Proposition (Sequential Ratio Test). Let $x_1, \ldots, x_n$ be a sequence of observations for testing simple hypotheses $H_0: \theta = \theta_0$ and $H_1: \theta = \theta_1$. Suppose the likelihood ratio at the nth step is given by $\Lambda_n = \mathscr{L}_{1n}/\mathscr{L}_{0n}$. Further, let $c_0 \equiv \beta/(1 - \alpha)$ and $c_1 \equiv (1 - \beta)/\alpha$, where α and β are respective probabilities of false rejection and acceptance. Then an appropriate sequential test is the following:

- If $\Lambda_n \leq c_0$, accept H_0.
- If $\Lambda_n \geq c_1$, accept H_1.
- If $c_0 < \Lambda_n < c_1$, take another observation.   (11–4)

The utility of this test is enhanced by the fact that it requires no knowledge of the distribution of any test procedure. In earlier applications of the Neyman–Pearson lemma, the distribution of the test statistic—such as $\overline{X}$ for a normal density—was required in order to determine the critical region for a given level of α. The distribution was also required to compute β. For the sequential test, however, the decision maker chooses the values of α and β beforehand, then blithely follows the procedure.

To illustrate the use of the sequential test, consider the test for an unknown mean from a normal density. The variance is assumed to be known, and the hypotheses are $H_0: \mu = \mu_0$ versus $H_0: \mu = \mu_1$ where $\mu_1 > \mu_0$. The likelihood ratio is

$$\Lambda_n \equiv \frac{\mathscr{L}_{1n}}{\mathscr{L}_{0n}} = \frac{\prod_{i=1}^{n} (1/\sigma\sqrt{2\pi})\exp[-(x_i - \mu_1)^2/2\sigma^2]}{\prod_{i=1}^{n} (1/\sigma\sqrt{2\pi})\exp[-(x_i - \mu_0)^2/2\sigma^2]}$$

After taking logarithms and simplifying, the relation $c_0 < \Lambda_n < c_1$ takes the form

$$\ln c_0 < \frac{\mu_1 - \mu_0}{\sigma^2} \sum_i x_i + \frac{n(\mu_0^2 - \mu_1^2)}{2\sigma^2} < \ln c_1$$

Since we have assumed that $\mu_1 > \mu_0$, we obtain

$$\frac{\sigma^2 \ln c_0}{n(\mu_1 - \mu_0)} + \frac{\mu_0 + \mu_1}{2} < \bar{x} < \frac{\sigma^2 \ln c_1}{n(\mu_1 - \mu_0)} + \frac{\mu_0 + \mu_1}{2} \qquad (11\text{–}5)$$

where $\bar{x}$ is substituted for $\sum_i x_i/n$. If our test had assumed that $\mu_0 > \mu_1$, then the inequalities above would simply be reversed.

The preceding relation is the criterion for continuing the observations for a sequential test of simple means from a normal density. If $\bar{x}$ falls on or above the right-hand expression, then H_0 would be rejected; or if $\bar{x}$ falls on or below the left-hand expression, then H_0 would be accepted. Note that the upper and lower bounds both converge toward the average of the means, namely, $(\mu_0 + \mu_1)/2$, as $n \to \infty$. In other words, the boundaries get tighter as the sample size increases, and converges to a single point in the case of an infinite procedure. A numerical example is given below.

Example

For the test of means described above, suppose we had selected error limits of $\alpha = .10$ and $\beta = .05$. Further, assume that the variance is 1 and the means under evaluation are $\mu_0 = 5$ and $\mu_1 = 6$.

Then the values for the c_i are

$$c_0 = \frac{\beta}{1 - \alpha} = \frac{.05}{1 - .10} = 0.0555$$

$$c_1 = \frac{1 - \beta}{\alpha} = \frac{1 - .05}{.1} = 9.5$$

TABLE 11–1. Illustration of the sequential test. Each x is from a normal density with mean 5 and variance 1. The values of d_{0n} and d_{1n} are the lower and upper thresholds for the test statistic $\bar{x}_n$. The sample average of 4.96 for $n = 6$ falls below the lower threshold of 5.018.

n	x_n	$\bar{x}_n$	d_{0n}	d_{1n}
1	3.34	3.34	2.609	7.751
2	6.80	5.07	4.054	6.626
3	4.32	4.82	4.536	6.250
4	5.94	5.10	4.777	6.063
5	5.43	5.17	4.922	5.950
6	3.95	4.96	5.018	5.875

The lower and upper bounds in Eq. (11–5) are respectively

$$d_{0n} = \frac{1^2 \ln(0.0555)}{n(6-5)} + \frac{5+6}{2} = -\frac{2.891}{n} + 5.5$$

$$d_{1n} = \frac{1^2 \ln(9.5)}{n(6-5)} + \frac{5+6}{2} = \frac{2.251}{n} + 5.5$$

A sequence of trials is represented in Table 11–1. At each step n, a sample x_n is obtained at random from a normal density with mean 5 and standard deviation of 1.[1] The table indicates that at step $n = 6$, the sample average of $\bar{x}_6 = 4.96$ falls below the lower threshold of $d_{06} = 5.018$. The sequential procedure, therefore, terminates at this step.

The data for this exercise are plotted in Figure 11–2. The figure highlights the attenuation of the envelope with the sampling process and the convergence of the bounds toward the common asymptote of 5.5.

How does this procedure compare with the monolithic approach? The variance of the test statistic is

$$\mathrm{Var}(\bar{X}) = \mathrm{Var}\left(\frac{\sum_i X_i}{n}\right) = \frac{\sum_i \mathrm{Var}(X_i)}{n^2} = \frac{\sigma^2}{n}$$

---

[1] The simulation was performed using the Monte Carlo method. The purpose of this technique is to obtain values for a random variable X from a known distribution $F(X)$. First, a number r is generated randomly from a uniform density between 0 and 1. Second, the desired value of X is obtained as the number x that satisfies the relation $F(x) = r$.

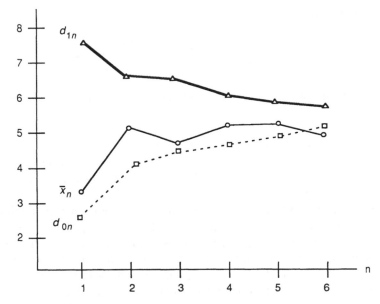

FIGURE 11–2. Chart of the data from Table 11–1.

Since $(\overline{X} - \mu)/(\sigma/\sqrt{n})$ has a normal density, the probability of false rejection is

$$P\left\{\frac{\overline{x} - \mu}{\sigma/\sqrt{n}} > z_{.90}\middle| H_0\right\} = P\left\{\frac{\overline{x} - 5}{1/\sqrt{n}} > z_{.90}\right\} = \alpha = .10$$

Here $z_{.90}$ is given by the relation $\Phi(z) = .90$. Hence, the threshold for the test procedure due to α is $\overline{X}_\alpha \equiv 5 + z_{.90}/\sqrt{n}$. The probability of false acceptance is

$$P\left\{\frac{\overline{X} - \mu}{\sigma/\sqrt{n}} \le z_{.05}\middle| H_1\right\} = P\left\{\frac{\overline{X} - 6}{1/\sqrt{n}} \le z_{.05}\right\} = \beta = .05$$

from which the threshold associated with β is $\overline{X}_\beta = 6 + z_{.05}/\sqrt{n}$. For small values of n, we have $\overline{X}_\alpha > \overline{X}_\beta$. The gap $(\overline{X}_\beta, \overline{X}_\alpha)$ represents an indeterminate region in which the null hypothesis can be neither rejected nor accepted.

For the monolithic procedure to have a clear-cut outcome, no such gap can exist. In other words, n must be large enough that $\overline{X}_\beta = \overline{X}_\alpha$:

$$6 + \frac{z_{.05}}{\sqrt{n}} = 5 + \frac{z_{.90}}{\sqrt{n}}$$

The resulting value of n is

$$n = \left(\frac{z_{.90} - z_{.05}}{6 - 5}\right)^2 = \left(\frac{1.28 - (-1.65)}{6 - 5}\right)^2 = 8.58$$

We see that this value of n is larger than the actual value from the sequential method.   ◆

The sequential task has been described for the case of simple hypotheses—namely, testing $H_0: \theta = \theta_0$ against $H_1: \theta = \theta_1$. The test, as specified, does not allow for composite hypotheses such as discriminating $H_0: \theta = \theta_0$ against $H_1: \theta > \theta_0$.

In practice, however, this limitation is of minor significance. As an illustration suppose that a new fuel additive has been developed to increase gas mileage in automobiles. Assume that the test vehicle normally attains 30 miles per gallon. Then the straightforward test would involve $H_0: \mu = 30$ mpg versus $H_1: \mu > 30$ mpg. In reality, a tiny increment over the default hypothesis will be of no interest to prospective consumers. For instance, when m is actually 30.2 rather than 30 mpg, the difference is unlikely to be of consequence to the potential buyer. As a result, the experimenter can establish a lower bound μ_1 for the alternate hypothesis that *would* be of consequence; perhaps this value is 32 or 35 mpg, depending on the application. The new testing scenario is then $H_0: \mu = 30$ mpg versus $H_1: \mu = 32$ mpg. In this way, a composite test can be converted into a simple one that is amenable to the sequential method.

11.5   DETERMINATION OF THRESHOLDS

The previous section presented approximate formulas for the thresholds c_0 and c_1 in terms of the errors α and β. In this section we examine the rationale behind the formulas. Although the argument used here is not entirely formal, it provides some insight into the relationships behind the thresholds.

For each observation x_n in the sequential procedure, the outcome space Ω is effectively partitioned into three regions. These are R_{0n}, the acceptance region for H_0; R_{1n}, the acceptance region for H_1; and R_{2n}, the continuation region calling for an additional observation. If an observation falls in R_{0n} or R_{1n}, its predecessors must have lain in R_{2m} for all $m < n$; otherwise a decision would have been made earlier.

The continuation region R_{2n} might not vanish completely as $n \to \infty$. Consequently, there is no prior upper bound on n: the observations could possibly continue indefinitely. An accurate model of this process would, therefore, require the consideration of an infinite-dimensional outcome space.

To simplify the analysis, however, we will assume that H_0 or H_1 will ultimately be accepted, and that finite-space arguments are applicable.

Let X denote the n-dimensional sample space $(X_1, \ldots, X_n)$. The probability of correctly accepting H_0 is

$$1 - \alpha = P\{\text{Accept } H_0 | H_0\} = \sum_{n=1}^{\infty} P\{X \in R_{0n} | H_0\}$$

$$= \sum_{n=1}^{\infty} \int_{R_{0n}} \mathcal{L}_{0n}(x)\, dx$$

Since X is of dimension n, dx is simply a shorthand for $dx_1, dx_2 \ldots dx_n$; the single integral shown above stands for n integrals corresponding to the observations $x_1, \ldots, x_n$.

According to the sequential test, the null hypothesis is accepted only if $\Lambda = \mathcal{L}_{1n}/\mathcal{L}_{0n} \le c_0$. In other words, $\mathcal{L}_{0n} \ge \mathcal{L}_{1n}/c_0$. We can, therefore, write

$$1 - \alpha \ge \sum_{n=1}^{\infty} \int_{R_{0n}} \frac{\mathcal{L}_{1n}(x)}{c_0}\, dx$$

$$= \frac{1}{c_0} \sum_{n=1}^{\infty} \int_{R_{0n}} \mathcal{L}_{1n}(x)\, dx = \frac{1}{c_0} P\{X \in R_{0n} | H_1\}$$

$$= \frac{1}{c_0} P\{\text{Accept } H_0 | H_1\} = \frac{1}{c_0} \beta$$

For any $\alpha < 1$, we obtain the result

$$c_0 \ge \frac{\beta}{1 - \alpha} \tag{11-6}$$

A similar analysis for the quantity $1 - \beta$ will yield the result

$$c_1 \le \frac{1 - \beta}{\alpha} \tag{11-7}$$

as long as $\alpha > 0$.

Assume that α and β are both small, each being about .10 or less. Figure 11-3 shows the two lines $c_0(1 - \alpha) = \beta$ and $c_1 \alpha = 1 - \beta$ in (α, β) space. The shaded region represents all positive values of α and β which satisfy Eqs. (11-6) and (11-7).

Suppose that the values of c_0 and c_1 are fixed. In that case, the largest values of α and β will lie on the northeastern border of the shaded region. The

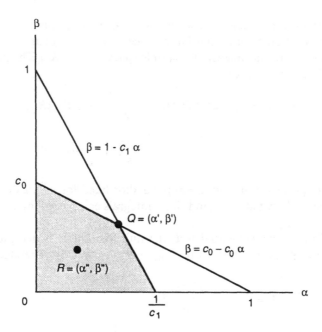

FIGURE 11–3. Thresholds c_0 and c_1 for the sequential test as parameters in (α, β) space.

corner point (α', β') happens to satisfy the relations in Eqs. (11–6) and (11–7) without any slack in the inequalities. The actual values of α and β should lie somewhere near (α', β'); from this point, neither parameter can be increased without entailing a corresponding decrease in the other.

Conversely, if the maximum errors α and β are reduced, then the shaded region can be shrunk. Suppose that α and β are decreased to (α'', β'') at point R in the chart. Then the constraining lines $\beta = 1 - c_1\alpha$ and $\beta = c_0 - c_0\alpha$ can be shifted closer to the origin. This corresponds to decreasing the intercepts c_0 and $1/c_1$.

Decreasing c_0 implies that the lower threshold for the test $\Lambda \leq c_0$ is moved farther to the left. Decreasing $1/c_1$ means increasing c_1. The result is a rightward shift in the threshold for accepting $H_1: \Lambda \geq c_1$. In other words, the acceptance regions R_{0n} and R_{1n} become smaller.

This phenomenon is in harmony with intuition. If the allowable errors α and β are small, the test should be more stringent. This corresponds to shrinking the acceptance regions to raise the required level of discrimination among the observations.

Suppose the specified values of the errors, namely, α' and β', are small, and the mean sample size n needed to reach a decision is sizable. Then the test thresholds indicated in the Sequential Ratio Test,

$$c_0 = \frac{\beta'}{1 - \alpha'} \quad \text{and} \quad c_1 = \frac{1 - \beta'}{\alpha'}$$

lead to actual errors which are close to the specified values of α and β.

Before selecting a sequential method over a monolithic one, the decision maker may wish to assess the prospective sample sizes required. A formula is available for the mean sample size $E(n)$:

$$E_\theta(n) \cong \frac{P(H_1|\theta)\ln c_0 + [1 - P(H_1|\theta)]\ln c_1}{E_\theta[\ln f(X|\theta_1) - \ln f(X|\theta_0)]}$$

The subscript θ in E_θ indicates that the expected value depends on the true value of θ, which, of course, can only be approximated beforehand. The quantity $P(H_1|\theta)$ is the probability of accepting H_1 when θ is the actual value of the parameter.

The preceding formula is approximate rather than exact. Its derivation is involved and is omitted here. For small values of α and β, the formula will indicate that $E(n)$ for the sequential test is about one-half that of the monolithic test.

The sequential procedure as specified in the Sequential Ratio Test can be shown to possess an optimal feature. Among all sequential tests having identical values of α and β, the specified procedure minimizes $E_{\theta_0}(n)$ and $E_{\theta_1}(n)$. In other words, if the value of θ happens to be one of the hypothesized values, then the Sequential Ratio Test minimizes the expected number of observations needed to reach a decision.

11.6 SUMMARY

The sequential likelihood ratio test is essentially similar to the Bayes' sequential method implied by the Terminal Procedure Theorem in Section 11.3. For the ratio test described here, however, no mention is made of considering cost explicitly. Suppose that an explicit cost structure is assumed, and the possibility of taking no samples is considered. Then the likelihood ratio test is comparable to a Bayes' sequential procedure.

In fact, the following relationship can be shown between the two methods. Consider any likelihood ratio test with constants $A = \alpha/(1 - \beta)$ and $B = (1 - \alpha)/\beta$, with $A < 1 < B$. Further, the prior distribution for parameter θ is assumed to be $\pi(\theta)$. Let a and b denote the losses due respectively to false rejection and false acceptance, whereas c is the unit cost of sampling. Then there exist values of a, b, and c for which the Bayes' sequential procedure for $\pi(\theta)$ corresponds to the given sequential likelihood ratio test.

The sequential likelihood ratio test, as described, represents the classical approach. In this approach the prior distribution and losses are not explicitly addressed. The methodology will allow for errors of arbitrarily small size since the cost of such accuracy is ignored.

In contrast, the decision theoretic approach takes explicit account of the cost of data acquisition and calls for a balance between the utility of additional data against its cost. Taking such trade-offs into account, however, can be a nontrivial task in practice.

PROBLEMS

The problems below are numbered according to the sections to which they correspond. For instance, Problem 11.1A is the first ("A") exercise pertaining to Section 11.1.

11.1A　Consider the semiconductor example discussed in Section 8.1. For the given level of Type I and Type II errors, what is the minimum sample size which defines the monolithic (nonsequential) problem? In other words, what is the smallest number n such that each outcome leads to a rejection or acceptance of the null hypothesis within the error levels $\alpha = \beta = .05$?

11.1B　For the semiconductor plant example, assume that the test statistic $\overline{X}$ can be viewed as a random variable from a *normal* density. Suppose that the mean is estimated by p_i and variance by $p_i(1 - p_i)/n$, where $p_0 = .1$ for the null hypothesis and $p_1 = .2$ for the alternate. For the monolithic procedure, what is the minimum sample size n for which the conclusion abides by the error levels $\alpha = \beta = .05$? How does this compare with the exact result from the previous question?

11.5A　An approximation for the upper threshold of the sequential test is $c_1 \leq (1 - \beta)/\alpha$ for any $\alpha > 0$. Derive this result when the observation space X is assumed to be finite and the continuation region R_{2n} vanishes with n.

11.5B　Let $X = 1$ represent a head on the toss of a coin, and $X = 0$ a tail. Let the allowable errors be $\alpha = \beta = .1$. (a) Describe a sequential likelihood ratio test for $H_0: p = .5$ versus $H_1: p = .3$. (b) Use a coin and carry out the test.

11.5C　Consider the same scenario as in Problem 11.5B, except that p is the probability of obtaining a 4, 5 or 6 on the toss of a die. (a) How would you define X? (b) Describe a sequential likelihood ratio test for H_0 and H_1. (c) Use a die and conduct the test.

FURTHER READING

Degroot, Morris H. 1970. *Optimal Statistical Decisions*. New York: McGraw-Hill.

Ferguson, Thomas S. 1967. *Mathematical Statistics: A Decision Theoretic Approach*. New York: Academic.

Lippman, S.A., and J.J. McCall (Eds.). 1979. *Studies in the Economics of Search*. New York: North-Holland.

Wald, Abraham. 1947. *Sequential Analysis*. New York: Wiley.

Appendices

Appendices

Appendix A

Set Theory

Set theory is the kernel of mathematics. It provides the basic concepts and terminology that are used to build up the entire structure of mathematics. This appendix provides an introduction to the cornerstones of this foundation.

A.1 MEMBERSHIP

A *set* is any collection of objects. The objects in a set are called the *elements* or *members* of the set. When an object a is a member of the set $\mathbf{A}$, we write $a \in \mathbf{A}$ and read it as "a is in $\mathbf{A}$." Conversely, when a is not a member of $\mathbf{A}$, we write $a \notin \mathbf{A}$ are read "a is not in $\mathbf{A}$."

A set is enumerated by listing its members within parentheses. Hence, if the set $\mathbf{A}$ consists of members a, b and c, we write $\mathbf{A} = \{a, b, c\}$.

The elements of a set are viewed as abstract rather than concrete objects. Hence, repetitions are immaterial. As an illustration, the set $\{a, b, b\}$ is the same as the set $\{a, b\}$.[1]

The set containing no objects is called the *empty set* or *null set*. It is denoted by the symbol $\varnothing$.

A set is an unordered collection: the relative placement of objects in the enumeration is immaterial. Hence, the set $\{a, b\}$ is equal to the set $\{b, a\}$.[2]

A set may be defined by describing the nature of its members rather than

---

[1] A collection for which repetitions are significant is called a *bag*. Hence, the bag $\{a, b, b\}$ is unequal to the bag $\{a, b\}$. A shopping bag containing an apple and two bananas differs from another containing one instance of each fruit.

[2] An ordered set is called a *list*. The list $\langle a, b \rangle$ differs from the list $\langle b, a \rangle$.

itemizing them explicitly. The notation used is $A = \{x: D\}$, which is read as "A is the set of all x such that D holds."

Suppose $A = \{x: x$ is an integer between 1 and 4$\}$. If B is the set $\{1, 2, 3, 4\}$, then $A = B$.

The number of members in a set is called its *cardinality*. The cardinality of a set A is denoted $|A|$, and read as the "cardinality of A." In the case of $A = \{a, b, c\}$, the cardinality is given by $|A| = 3$.

The descriptive approach to defining a set is especially helpful when the cardinality of the set is large. Consider the set

$$P = \{1, 3, 5, 7, 9, 11, 13, 15, 17, 19, 21, 23, 25, 27, 29, 31\}$$

It is simpler to define P in the following way:

$$P = \{x: x \text{ is an odd integer, and } 1 \leq x \leq 31\}$$

When the cardinality of a set is infinite, there is little choice but to use the descriptive approach. For instance, one would be hardpressed to enumerate all the members of the set Q, defined as the real numbers between 0 and 1: $Q = \{x: 0 < x < 1\}$.

A.2   SUBSETS AND SUPERSETS

A *subset* of a set is another set whose elements are all in the original set. If A is a subset of B, we write $A \subset B$ and read "A contained in B." In other words, if $x \in A$ implies $x \in B$ for each x in A, then we say that A is a *subset* of B. Consider the following sets:

$$R = \{a, c\}$$
$$S = \{b, d\}$$
$$T = \{a, b, c, d\}$$

Then $R \subset T$ and $S \subset T$.

A set is a subset of itself: $A \subset A$ since each element of A is a member of A.

A set which is strictly smaller than another is called a *proper* subset of the latter. More precisely, a set A is a proper subset of B if the following conditions hold: $A \subset B$ and $A \neq B$. As an illustration, the set $\{a\}$ is a proper subset of the set $\{a, b\}$.

The empty set is a subset of every other set: $\emptyset \subset A$ for every set A since the definition of a subset holds by default. The reasoning is as follows: the set $\varnothing$ contains no elements, so the "If" part of "If $x \in A$ implies $x \in B$, then $A \subset B$" is never true. Consequently, the rule is never activated, and it is not possible then to disprove that $\varnothing \subset A$.

The subset operator carries across a chain of pairwise relationships. If a set is contained in another, which is itself contained in a third, then the first is contained in the third. This property is called *transitivity*:

- Let **A**, **B**, and **C** be any sets. If $A \subset B$ and $B \subset C$, then $A \subset C$.

We can write the last statement more compactly by introducing some conventions from mathematical logic:

$$(A \subset B) \mathbin{\&} (B \subset C) \to A \subset C$$

The arrow is called the *implication* symbol. The quantity to the left of the arrow is the *premise* or *antecedent*; the quantity to the right is the *conclusion*. The entire expression is called a *rule*.

The validity of this rule can be verified as follows. The premise is that $A \subset B$ and $B \subset C$. From the definition of the subset operator, the premise is equivalent to the statements "For any element $x \in A$, it is also the case that $x \in B$" and "For any $x \in B$, it is also true that $x \in C$." We can, therefore, infer that "For any $x \in A$, it is the case that $x \in C$." Since this is the conclusion of the above proposition, we have just verified the rule.

The equality of sets may be defined in terms of bidirectional inclusion. More specifically, we say that two sets **A** and **B** are *equal* if and only if $A \subset B$ and $B \subset A$. In other words, two sets are equal if each contains all the elements of the other.

To prove the equality of two sets **A** and **B**, the usual strategy is the following. If some element x is in **A**, then it is in **B**; moreover, if some element y is in **B**, it is in **A**. In this way, we demonstrate that **A** is a subset of **B**, and vice versa. As a result, the two sets are equal.

The converse of the notion of a subset is the *superset*. A set **A** is a superset of the set **B** if and only if **B** is a subset of **A**. When **A** is a superset of **B**, we write $A \supset B$ and read "A contains B." In symbolic form, $A \supset B$ if and only if $B \subset A$.

Since the superset operator is the converse of the subset operator, it also possesses the property of transitivity. More specifically,

$$(A \supset B) \mathbin{\&} (B \supset C) \to A \supset C$$

If a set contains a second set which contains a third, then the first contains the third.

A.3   UNION AND INTERSECTION

The *union* of two or more sets is the set containing elements of each of the original sets. The union of A and B is denoted $A \cup B$ and read as "A union B." To illustrate, let $A = \{a, b, c\}$ and $B = \{b, d\}$; then $A \cup B = \{a, b, c, d\}$.

The union operator can be applied to any number of sets. Let $A_1, A_2, \ldots, A_n$ denote sets. Their union is denoted by

$$\bigcup_{i=1}^{n} A_i$$

The union operator has a number of interesting properties. Among these are commutativity, associativity, and absorption.

- *Commutativity.* For any two sets A and B

$$A \cup B = B \cup A$$

In other words, the order of union is immaterial.
- *Associativity.* For any three sets A, B, and C

$$(A \cup B) \cup C = A \cup (B \cup C)$$

The sequence of union is immaterial. Hence, we may drop the parentheses and simply write $A \cup B \cup C$.
- *Absorption.* The union of a subset with a superset is the superset:

$$A \subset B \rightarrow (A \cup B = B) \tag{A-1}$$

The absorption property has several special cases. One of these occurs when A is the empty set $\varnothing$. The union of the empty set with any set is the latter set: $\varnothing \cup B = B$.

At the other extreme, let Ω be the largest set under discussion. The set Ω is known as the *total set*; it is also called the *universe* or the *complete set*. When $B = \Omega$, then we have $A \cup \Omega = \Omega$ for any set A. In other words, the union of the total set with any set is the total set.

A third special case occurs when B is A itself. Since $A \subset A$, the premise of the rule in Eq. (A–1) is satisfied. The result is $A \cup A = A$: the union of a set with itself is itself. This is called the property of *idempotence*.

The *intersection* of two or more sets is the set containing elements that occur in each of the original sets. The intersection of two sets A and B is denoted $A \cap B$ and read as "A intersection B." The intersection of sets A_1,

$A_2, \ldots, A_n$ is denoted by

$$\bigcap_{i=1}^{n} A_i$$

Suppose $A_1 = \{a, b, c\}$ and $A_2 = \{b, d\}$. Then $A_1 \cap A_2 = \{a, b, c\} \cap \{b, d\} = \{b\}$.

However, suppose that $A_3 = \{e\}$. Then the intersection of A_1 and A_3 is empty: $\{a, b, c\} \cap \{e\} = \varnothing$.

If the intersection of two sets is empty, we say that the sets are *disjoint*. In the example above, A_1 and A_3 are disjoint since $A_1 \cap A_3 = \varnothing$.

The intersection operator has properties similar to the union operator. These are as follows.

- *Commutativity.* For any two sets A and B

$$A \cap B = B \cap A$$

- *Associativity.* For any three sets A, B, and C

$$(A \cap B) \cap C = A \cap (B \cap C)$$

Hence, we may ignore parentheses and simply write $A \cap B \cap C$.
- *Absorption.* The intersection of a subset with a superset is the subset:

$$A \subset B \rightarrow A \cap B = A \qquad \text{(A--2)}$$

A special case of Eq. (A–2) occurs when A is the empty set $\varnothing$. The intersection of the null set with any set is the null set: $\varnothing \cap B = \varnothing$.

At the other extreme, suppose B is the total set Ω. The intersection of any set with the total set is the former set: $A \cap \Omega = A$.

The third subcase arises for $B = A$. Since $A \subset A$, the premise of Eq. (A–2) is satisfied and we have $A \cap A = A$. The intersection operator is idempotent, just like the union operator.

A.4   DIFFERENCE AND COMPLEMENTATION

The *difference* of a set against a second set is a new set containing all the elements of the first set which are not in the second set. The difference of A by B is denoted by $A - B$ and read "A minus B."

If $A = \{5,6,7,8\}$ and $B = \{6,8\}$, then $A - B = \{5,7\}$. On the other hand, $B - A = \varnothing$ since each element in B is eliminated by some element in A.

A special case of the difference between two sets is that of complementation. This operation is defined only between two sets for which one is a subset of the other.

Often the superset under discussion is the total set Ω. Then the *complement* of A with respect to Ω is the set of all elements in Ω which are not in A. The complement of A within Ω is denoted A^c and reads as "A complement." In symbolic notation, the complement of A is given by

$$A^c \equiv \Omega - A$$

where A is a subset of Ω.

To illustrate, consider the set $R = \{a,c\}$ and the total set $\Omega = \{a,b,c,d\}$. Then the complement of R is given by

$$R^c = \Omega - R = \{a,b,c,d\} - \{a,c\} = \{b,d\}$$

One property of the complement is pairwise identity. More specifically, the complement of a complement is the original set: $(A^c)^c = A$. In the previous example,

$$(R^c)^c = (\{a,c\}^c)^c = \{b,d\}^c = \{a,c\} = R$$

A set and its complement are disjoint since they share no elements in common: $A \cap A^c = \varnothing$. In addition, the union of two complementary sets is the total set: $A \cup A^c = \Omega$. We, therefore, say that a set and its complement are *mutually exclusive* and *collectively exhaustive*.

If A is the entire set Ω, we obtain $A^c = \Omega - \Omega = \varnothing$. Hence, the complement of the total set is the empty set.

Conversely, if A is the empty set $\varnothing$, then $\varnothing^c = \Omega - \varnothing = \Omega$. Thus, the complement of the empty set is the total set.

Some interesting properties of the complementation operator arise in conjunction with the union and intersection operators. These properties are called De Morgan's Laws:

- The complement of a union is the intersection of complements:

$$(A \cup B)^c = A^c \cap B^c \tag{A-3}$$

- The complement of an intersection is the union of complements:

$$(A \cap B)^c = A^c \cup B^c \tag{A-4}$$

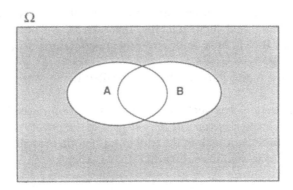

FIGURE A–1. Example of a Venn diagram. The shaded area represents $(A \cup B)^c$ which is also equivalent to $A^c \cap B^c$.

We will explore the first of these formulas, and leave the second one as an exercise.

Suppose **A** and **B** are two sets within the total set Ω. This situation is depicted in Figure A–1, which portrays a chart called a *Venn diagram*. The sets **A** and **B** are represented by ovals within the rectangle, which denotes the total set Ω. The white area in the center of the diagram represents $A \cup B$. Its complement $(A \cup B)^c$ is the shaded region. Now consider the quantity A^c. This is the entire rectangle except the oval **A**. A similar situation holds for B^c. The intersection of these two regions is any area which is not in **A** and not in **B**—namely, the shaded region in the Venn diagram. Hence, we have verified Eq. (A–3) through an informal appeal to intuition.

A rigorous proof of Eq. (A–3), however, should rely on the definitions and more basic arguments. To illustrate the ways of formal derivations, we present Eq. (A–3) in the form of a theorem and its proof.

Theorem. Let **A** and **B** be any two sets contained within the total set Ω. Then

$$(A \cup B)^c = A^c \cap B^c \tag{A–5}$$

Proof. Let **L** and **R** denote respectively the left-hand and right-hand sides of Eq. (A–5). In other words

$$\mathbf{L} \equiv (A \cup B)^c$$

$$\mathbf{R} \equiv A^c \cap B^c$$

Here the symbol $\equiv$ stands for "defined as."

From Section A.2, we know that the equality of two sets can be demonstrated by bidirectional inclusion. In other words, we need to show that $L \subset R$ and that $R \subset L$. Whenever two sets are contained in each other, they are equal: $L = R$.

There are two major parts to the proof, called the forward and backward directions.

Forward Direction. The verification of a proposition in the forward direction is also called the *necessary* or *only-if* part of the proof.

For this theorem, our task in the forward direction is to show that $L \subset R$. This is equivalent to demonstrating that if $x \in L$, then $x \in R$. So we begin with the assumption

$$x \in L \equiv (A \cup B)^c$$

Suppose x were in $A \cup B$. Then x would be in A, B, or both. But x is actually in the *complement* of $A \cup B$. Therefore, x is neither in A nor B.

Since x is not in A, x is in the complement A^c. In addition, since x is not in B, it is in the complement B^c.

To summarize the results so far, x is in A^c and in B^c. Therefore, x is in their intersection, $A^c \cap B^c$. We have just shown that $x \in L$ implies $x \in R$; in other words, $L \subset R$.

Backward Direction. The verification of a proposition in the backward direction is also called the *sufficient* or *if* part of the proof. In this segment of the proof, we wish to show that $R \subset L$. This is equivalent to demonstrating that $x \in R$ implies $x \in L$.

We, therefore, assume that

$$x \in R \equiv A^c \cap B^c$$

In other words, x is in the complement of A *and* x lies in the complement of B. Therefore, x lies in neither A nor B. In that case, x cannot be in $A \cup B$. Consequently, x must be in the complement of $A \cup B$—namely, $(A \cup B)^c$.

We have shown that $x \in R$ implies $x \in L$. This is equivalent to saying that $R \subset L$.

Summary. In the forward direction, we have demonstrated that $L \subset R$. Conversely, we showed that $R \subset L$ in the backward direction. We conclude that $L = R$, thereby verifying Eq. (A–5). ∎

The rationale behind the second of De Morgan's Laws, as given in Eq. (A–4), is similar. The proof is left as an exercise.

A.5 SETS OF SETS

The elements of a set can themselves be sets. If $\mathbf{G} = \{a\}$ and $\mathbf{H} = \{b, c\}$, then the set $\mathbf{I} = \{\mathbf{G}, \mathbf{H}\}$ is a set of sets:

$$\mathbf{I} = \{\{a\}, \{b, c\}\}$$

This is in contrast to the union of $\mathbf{G}$ and $\mathbf{H}$. The union of two sets is simply a set in which the elements are the elements of the original set. For the above example, the union of $\mathbf{G}$ and $\mathbf{H}$ is

$$\mathbf{J} = \mathbf{G} \cup \mathbf{H} = \{a, b, c\}$$

Of course, the union of two sets can itself be a set of sets if the members of the original set are themselves sets. Suppose $\mathbf{F}$ is the set $\{\varnothing, \{d\}\}$. Then the union of $\mathbf{F}$ and $\mathbf{I}$ is

$$\mathbf{K} = \mathbf{F} \cup \mathbf{I} = \{\varnothing, \{d\}, \{a\}, \{b, c\}\}$$

The *power set* of a particular set is the set of all its subsets. Given a set Ω, its power set is denoted by the notation 2^{Ω} and read as "power set of Ω." For any set Ω, the empty set $\varnothing$ and the set itself are subsets; therefore, both these elements are members of the power set.

As an illustration, consider the set $\Omega = \{a, b, c\}$. The power set of Ω is

$$2^{\Omega} = \{\varnothing, \{a\}, \{b\}, \{c\}, \{a, b\}, \{a, c\}, \{b, c\}, \{a, b, c\}\}$$

Note that the set Ω contains 3 members, but its power set contains 8 members, which happens to be 2^3. This exponential relationship is a general characteristic of the power set. If a set Ω has n elements, then its power set has 2^n elements. Hence, the cardinality of the power set of Ω is 2 raised to the cardinality of Ω; in symbolic form, $|2^{\Omega}| = 2^{|\Omega|}$. This is the rationale for the notation 2^{Ω} to denote the power set of Ω.

Where does the number 2^n come from? In considering the subsets of Ω, each element may be included in a particular clustering—or it may not. For each element there are two possibilities. Given n elements in Ω, there is an n-fold number of combinations, each of which has 2 alternatives. The total is $2 \cdot 2 \cdot \ldots \cdot 2$, where the 2s appear n times; in other words, 2^n.

This appendix has provided a brief introduction to set theory. The material covered here will provide sufficient background in set theory for many introductory texts on applied mathematics, including the current volume.

PROBLEMS

The problems below are numbered according to the sections to which they correspond. For instance, Problem A.1 is the first ("A") exercise pertaining to Section A.1.

A.1A Consider the following sets: $A = \{1, 3, 3, 5\}$, $B = \{x: x$ is an integer; x is odd; $0 < x \leq 5\}$; $C = \{x: x = 2y + 1;\ y = 0, 1, 2\}$. Is it true that $A = B = C$? Explain.

A.2A Let $B_i = \{x: 0 < x < 1/i\}$. (a) Does $B_{i+1} \subset B_i$ hold? (b) Does $B_{i+1} \supset B_i$ hold? Explain.

A.2B Suppose $A \subset B$. (a) Prove that $A \cup B = B$. (b) Show that $A \cap B = A$.

A.2C Prove that the superset relation is transitive: if $A \supset B$ and $B \supset C$, then $A \supset C$.

A.3A Let $A = \{1, 2, a, b\}$; $B = \{2, 3, c, d\}$. (a) What is $A \cup B$? (b) What is $A \cap B$?

A.3B Let $A_i = \{n: n$ is an integer; $n \geq 1\}$. (a) What is $\bigcup_{i=1}^{M} A_i$? (b) What is $\bigcap_{i=1}^{M} A_i$?

A.3C Let $B_i = \{x: 0 < x < 1/i\}$. We define the following:

$$C \equiv \lim_{n \to \infty} \bigcup_{i=1}^{n} B_i$$

$$D \equiv \lim_{n \to \infty} \bigcap_{i=1}^{n} B_i$$

What are C and D in descriptive notation?

A.3D Let $B_i = \{x: 0 \leq x < 1/i\}$. Define C and D as in Problem A.3C. What are C and D in descriptive notation?

A.4A Let $G_i = \{x: 0 \leq x \leq 1/i\}$ and $H_i = \{x: 0 \leq x < 1/i\}$. (a) If $J_i = G_i - H_i$, express J_i in descriptive form. (b) Let $K_i = G_i^c$ with respect to $\Omega = [0, 1] \equiv \{x: 0 \leq x \leq 1\}$. What is K_i in descriptive form?

A.4B Consider any two sets A and B. Let $R \equiv A - B$, $S \equiv B - A$, and $T \equiv A \cap B$. (a) Draw a Venn diagram representing these sets. (b) Prove that R, S, and T are all disjoint. (c) Prove that $A \cup B = R \cup S \cup T$.

A.4C For any sets A, B, and C, (a) show that $(A \cup B) \cup C$; (b) show that $(A \cap B) \cap C = A \cap (B \cap C)$.

A.4D Consider De Morgan's Law relating to the complement of an intersection: $(A \cap B)^c = A^c \cup B^c$. (a) Draw the Venn diagram. (b) Prove the formula.

A.4E For any sets A, B, and C, (a) show that $A \cup (B \cap C) = (A \cup B) \cap (A \cup C)$; (b) show that $A \cap (B \cup C) = (A \cap B) \cup (A \cap C)$.

A.5A Let $A = \{a\}$, $B = \{b\}$, $C = \{c, d\}$, $D = \{A, B\}$, and $E = \{B, C\}$. (a) Does $D = A \cup B$? (b) Does $D \cup E = A \cup B \cup C$? (c) Does $C = E - D$? (d) Let $\Omega = D \cup E$, and $F = D^c$ with respect to Ω. Does $F = E - D$?

A.5B (a) Let $\Omega = \{1, 5, 7\}$. What is 2^Ω? (b) Let $\Omega' = \{x: x$ is an integer; $1 \leq x \leq 100\}$. What is the cardinality of $2^{\Omega'}$?

Appendix B

Tables of Cumulative Distributions

TABLE B–1. Cumulative distribution for the standard normal density. For instance, the second entry is $\Phi(z) = 0.5040$ where $z = .01$; thus, 50.40% of the area under the curve lies to the left of $z_c = .01$. If X is any normal random variable having mean μ and standard deviation σ, it can be converted into a standard normal variable by the transformation $Z = (X - \mu)/\sigma$.

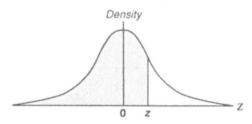

z	0	1	2	3	4	5	6	7	8	9
.0	.5000	.5040	.5080	.5120	.5160	.5199	.5239	.5279	.5319	.5359
.1	.5398	.5438	.5478	.5517	.5557	.5596	.5363	.5675	.5714	.5753
.2	.5793	.5832	.5871	.5910	.5948	.5987	.6026	.6064	.6103	.6141
.3	.6179	.6217	.6255	.6293	.6331	.6368	.6406	.6443	.6480	.6517
.4	.6554	.6591	.6628	.6664	.6700	.6736	.6772	.6808	.6844	.6879
.5	.6915	.6950	.6985	.7019	.7054	.7088	.7123	.7157	.7190	.7224
.6	.7257	.7291	.7324	.7357	.7389	.7422	.7454	.7486	.7517	.7549
.7	.7580	.7611	.7642	.7673	.7703	.7734	.7764	.7974	.7823	.7852
.8	.7881	.7910	.7939	.7967	.7995	.8023	.8051	.8078	.8106	.8133
.9	.8159	.8186	.8212	.8238	.8264	.8289	.8315	.8340	.8365	.8389
1.0	.8413	.8438	.8461	.8485	.8508	.8531	.8554	.8577	.8599	.8621
1.1	.8643	.8665	.8686	.8708	.8729	.8749	.8770	.8790	.8810	.8830
1.2	.8849	.8869	.8888	.8907	.8925	.8944	.8962	.8980	.8997	.9015

(*continued*)

TABLE B–1. (*Continued*)

z	0	1	2	3	4	5	6	7	8	9
1.3	.9032	.9049	.9066	.9082	.9099	.9115	.9131	.9147	.9162	.9177
1.4	.9192	.9207	.9222	.9236	.9251	.9265	.9278	.9292	.9306	.9319
1.5	.9332	.9345	.9357	.9370	.9382	.9394	.9406	.9418	.9430	.9441
1.6	.9452	.9463	.9474	.9484	.9495	.9505	.9515	.9525	.9535	.9545
1.7	.9554	.9564	.9573	.9582	.9591	.9599	.9608	.9616	.9625	.9633
1.8	.9641	.9648	.9656	.9664	.9671	.9678	.9686	.9693	.9700	.9706
1.9	.9713	.9719	.9726	.9732	.9738	.9744	.9750	.9756	.9762	.9767
2.0	.9772	.9778	.9783	.9788	.9793	.9798	.9803	.9808	.9812	.9817
2.1	.9821	.9826	.9830	.9834	.9838	.9842	.9846	.9850	.9854	.9857
2.2	.9861	.9864	.9868	.9871	.9874	.9878	.9881	.9884	.9887	.9890
2.3	.9893	.9896	.9898	.9901	.9904	.9906	.9909	.9911	.9913	.9916
2.4	.9918	.9920	.9922	.9925	.9927	.9929	.9931	.9932	.9934	.9936
2.5	.9938	.9940	.9941	.9943	.9945	.9946	.9948	.9949	.9951	.9952
2.6	.9953	.9955	.9956	.9957	.9959	.9960	.9961	.9962	.9963	.9964
2.7	.9965	.9966	.9967	.9968	.9969	.9970	.9971	.9972	.9973	.9974
2.8	.9974	.9975	.9976	.9977	.9977	.9978	.9979	.9979	.9980	.9981
2.9	.9981	.9982	.9982	.9983	.9984	.9984	.9985	.9985	.9986	.9986
3.0	.9987	.9990	.9993	.9995	.9997	.9998	.9998	.9999	.9999	1.0000

TABLE B–2. Student's t-distribution: thresholds for a two-tailed test. For instance, the first entry indicates that 10% of the area under the density function for $v = 1$ degree of freedom, lies beyond $t_c = \pm 6.3138$.

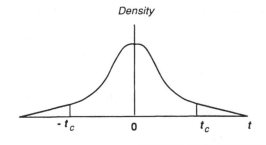

Density

	Fractile c		
Degrees of Freedom v	0.10	0.05	0.01
1	6.3138	12.706	63.657
2	2.9200	4.3027	9.9248
3	2.3534	3.1825	5.8409

(*continued*)

TABLE B–2. (*Continued*)

Degrees of Freedom v	Fractile c		
	0.10	0.05	0.01
4	2.1318	2.7764	4.6041
5	2.0150	2.5706	4.0321
6	1.9432	2.4469	3.7074
7	1.8946	2.3646	3.4995
8	1.8595	2.3060	3.3554
9	1.8331	2.2622	3.2498
10	1.8125	2.2281	3.1693
20	1.7247	2.0860	2.8453
30	1.6973	2.0423	2.7500
40	1.6839	2.0211	2.7045
60	1.6707	2.0003	2.6603
120	1.6577	1.9799	2.6174
∞	1.6449	1.9600	2.5758

TABLE B–3. Cumulative distribution for the χ^2 random variable. For instance, the last entry indicates that $\chi^2_{0.95}(100) = 124.3$; for $v = 100$ degrees of freedom, 95% of the area lies to the left of 124.3.

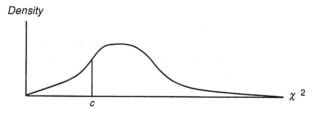

Degrees of Freedom v	Fractile c			
	0.05	0.10	0.90	0.95
1	0.00393	0.0158	2.71	3.84
2	0.103	0.211	4.61	5.99
3	0.352	0.584	6.25	7.81
4	0.711	1.064	7.78	9.49
5	1.145	1.61	9.24	11.07
6	1.64	2.20	10.64	12.59
7	2.17	2.83	12.02	14.07
8	2.73	3.49	13.36	15.51
9	3.33	4.17	14.68	16.92
10	3.94	4.87	15.99	18.31
20	10.85	12.44	28.41	31.41

(*continued*)

TABLE B–3. (*Continued*)

Degrees of Freedom v	Fractile c			
	0.05	0.10	0.90	0.95
30	18.49	20.60	40.26	43.77
40	26.51	29.05	51.81	55.76
50	34.76	37.69	63.17	67.50
60	43.19	46.46	74.40	79.08
70	51.74	55.33	85.53	90.53
80	60.39	64.28	98.58	101.9
90	69.13	73.29	107.6	113.1
100	77.93	82.36	118.5	124.3

Appendix C

Sufficiency of Procedures

Data are often redundant in the sense that the results they imply can be obtained from less information. In fact, a plethora of data too often succeeds in obscuring rather than clarifying a situation. For this reason, data sets are usually condensed into a more manageable form.

We first define the notions of statistics and sufficiency. Suppose X is a random variable defined over the sample space Ω. Then any real-valued function $T(X)$ is called a *statistic*. Here both the observation X and the statistic T may be scalar or vector quantities.

Suppose that T associates a particular value to two or more observations X. Then T generally compresses the original stock of data reflected in X to a smaller set of points. *Sufficiency* refers to the notion that the data reduction may be achieved without losing information—namely, our ability to draw inferences about the underlying phenomenon.[1]

To illustrate, consider a sampling situation. A manufacturer of mobile robots has received 5000 batteries for use in its current production run. Since it would be too costly to evaluate the entire shipment, the manufacturer decides to test 10 batteries at random. Let X_i be the indicator variable for defectiveness: $X_i = 1$ if the ith sample tested is defective, and $X_i = 0$ otherwise. Then $X = (X_1, X_2, \ldots, X_{10})$ denotes the complete set of observations.

In estimating the proportion of defective items from X, it is clear that only the total number of faulty batteries is of consequence, rather than their precise sequence. For instance, the sequence $X = (0, 0, 1, 0, 0, 1, 0, 0, 0, 1)$ leads to the

---

[1] The concept of sufficiency was first introduced by Fisher (1922). Other interpretations are due to Halmos and Savage (1949), and Kolmogorov (1942). But these interpretations can be shown to be equivalent.

same estimate for the proportion of defectives—namely, 3/10—as the observation $X' = (1, 0, 0, 1, 0, 0, 1, 0, 0, 0)$. More generally, the total number of defective items given by $T = \sum_{i=1}^{10} X_i$ is a sufficient statistic for estimating the proportion of faulty batteries. In this case, the data space has been reduced by almost 99%: from the 2^{10} possibilities for X to only 11 alternatives for T.

The notion of sufficiency can be defined more precisely as follows. A statistic $T(X)$ is said to be *sufficient* for a parameter θ if the conditional distribution of X given $T(X) = t$ is independent of θ. In other words, $X = (X_1, \ldots, X_n)$ embodies no further information about the parameter θ once the statistic T is specified.

Example

Let us consider the general inspection situation. Suppose that a production line generates n items, each of which passes inspection with probability θ and fails with probability $1 - \theta$. The value of θ is unknown and must be estimated. Suppose that the quality of each item has no impact on its successors. Let X_i be the indicator of acceptability: $X_i = 1$ if the ith item passes, and equals 0 otherwise. The vector $X = (X_1, \ldots, X_n)$ represents a sequence of independent binomial trials. Let $x = (x_1, x_2, \ldots, x_n)$ be the actual sequence of observations, where $x_i = 0$ or 1. Then the probability of a particular outcome is

$$P\{X = x\} = \prod_{i=1}^{n} \theta^i (1 - \theta)^{1-i}$$

$$= \theta^t (1 - \theta)^{n-t}$$

where $t \equiv \sum_i x_i$.

Now the conditional distribution of X given T is

$$P\{X = x \mid T = t\} = \frac{P\{X = x, T = t\}}{P\{T = t\}} = \frac{\theta^t (1 - \theta)^{n-t}}{\binom{n}{t} \theta^t (1 - \theta)^{n-t}} = \frac{1}{\binom{n}{t}}$$

The last expression is independent of θ. T is therefore a sufficient statistic for θ.   ◆

In general, sufficient statistics are not unique. This is reasonable since a sufficient statistic reduces the richness of information in a set of observations into a more sparse set of values.

For instance, knowing that 8 out of 10 items passed an inspection test is tantamount to the knowledge that the ratio of acceptance to rejection was 4 in a sample size of 10.

More generally, two statistics T and T' are equivalent if they can be written in terms of each other. In the inspection example, let X_i be the indicator of success: $X_i = 1$ if a product passes inspection, and 0 otherwise. Further, we take A to be the number of accepted items and R the ratio of acceptances to rejections.

The first statistic may be written as

$$T(X_1, X_2, \ldots, X_n) = \sum_{i=1}^{n} X_i \equiv A$$

and the second as

$$T'(X_1, X_2, \ldots, X_n) = \frac{\sum_{i=1}^{n} X_i}{n - \sum_{i=1}^{n} X_i} \equiv R.$$

These statistics can be written as functions of each other. In particular, the ratio is given by $R = A/(n - A)$, whereas the number of accepted items is $A = nR/(1 + R)$. As a result, the statistics T and T' are equivalent. If T is a sufficient statistic for, say, estimating the probability θ that any particular item is defective, then so is T', and vice versa.

Attempting to determine whether a statistic is sufficient by using the definition of sufficiency is cumbersome: the conditional distribution is often a nontrivial quantity to compute. An easier strategy is available in the following theorem, which presents necessary and sufficient conditions for the sufficiency of a statistic.

Theorem (Factorization for Sufficiency). Let $T(X)$ be a statistic taking values in a space I. Then $T(X)$ is sufficient for parameter θ if and only if there exist functions $g(t, \theta)$ and $h(x)$ such that the joint probability of x and θ is

$$p(x, \theta) = g(T(x), \theta)h(x) \tag{C-1}$$

for all $X \in \mathcal{R}^n$. The functions g and h are defined on some subset of $\mathcal{R}^n$.

Proof. We will prove the result for the case of discrete variables. [The full result is given in sources such as Lehman (1959, pp. 47–49).]

To validate the theorem in the forward direction, we assume that T is a sufficient statistic. We define functions g and h in the following way:

$$g(T(x), \theta) = P_\theta\{T = T(x)\}$$
$$h(x) = P\{X = x \mid T = T(x)\}$$

Here $P_\theta\{T = T(x)\}$ refers to the probability that T takes value $T(x)$ for a given value of θ. The joint probability of x and θ can be written as

$$p(x, \theta) = P_\theta\{X = x, T = T(x)\}$$

We can expand the right-hand side of the equation to obtain

$$p(x, \theta) = P_\theta\{T = T(x)\}P\{X = x|T = T(x)\}$$
$$= g(T(x), \theta)h(x)$$

which is the desired result.

To validate the theorem in the backward direction, we consider the set of values $x_1, x_2, \ldots$ that the discrete variable X can take. Let $T_i \equiv T(x_i)$ be the value of the statistic corresponding to observation x_i. Since X is discrete, so is T. Further, the t_is form a complete event space:

$$\sum_{i=1}^{\infty} P_\theta\{T = t_i\} = 1$$

for each value of state θ.

Our next task is to show that $P_\theta\{X - x_j|T = t_i\}$ is independent of θ for all values of t_i and x_j. This can be achieved by showing that $P_\theta\{x = x_j|T = t_j\}$ is independent of θ on each consequential value of θ: namely, each set $S_i = \{\theta: P_\theta\{T = t_i\} > 0\}$. The values of T taking positive probability can be written as follows:

$$P_\theta\{T = t_i\} = \sum_{\{x:\, T(x)=t_i\}} p(x, \theta)$$
$$= g(T(x) = t_i, \theta) \sum_{\{x:\, T(x)=t_i\}} h(x) \tag{C-2}$$

where the second equality relies on Eq. (C–1).

For the state $\theta \in S_i$, the conditional probability of x_j given t_i is

$$P_\theta\{X = x_j|T = t_i\} = \frac{P_\theta\{X = x_j, T = t_i\}}{P_\theta\{T = t_i\}}$$
$$= \frac{p(x_j, \theta)}{P_\theta\{T = t_i\}} \tag{C-3}$$

The first equation follows from the expansion of joint probabilities, and the second from the fact that the value of $T(X)$ is completely determined by that of X. Using Eq. (C–1) once more, we write Eq. (C–3) as

$$P_\theta\{X = x_j | T = t_i\} = \begin{cases} g(t_i, \theta)h(x_j)/P_\theta\{T = t_i\} & \text{if } T(x_j) = t_i \\ 0 & \text{otherwise} \end{cases}$$

By introducing Eq. (C–3), the last result can be expressed as

$$P_\theta\{X = x_j | T = t_i\} = \begin{cases} h(x_j)/ \displaystyle\sum_{\{x_j:\, T(x_j)=t_i\}} h(x_j) & \text{if } T(x_j) = t_i \\ 0 & \text{otherwise} \end{cases}$$

The right-hand side is free of θ. Since the conditional distribution of X given $T(X)$ is independent of θ, the function T is a sufficient statistic.   ∎

Example

Let X_1, X_2, ..., X_n be a set of random variables having an exponential distribution with parameter θ:

$$f(X_i = x_i, \theta) = \theta e^{-\theta x_i} I_{[0, \infty)}(x_i)$$

Here $I_{[0, \infty)}(x)$ denotes an indicator function: it takes the value 1 when x lies on the internal $[0, \infty)$, and 0 otherwise. The joint density of the random variables is given by

$$f(x_1 \cdots x_n, \theta) = \prod_{i=1}^{n} f(x_i, \theta)$$

$$= \theta^n \exp\left[-\theta \sum_i x_i \right]$$

Here each $x_i \geq 0$. We take the statistic $T(X_1, ..., X_n) \equiv \sum_i X_i$ as well as the functions $g(t, \theta) \equiv \theta^n e^{-\theta t}$ and $h(x_1, ..., x_n) \equiv 1$. Then,

$$f(x_1, ..., x_n, \theta) = g(t, \theta)h(x_1, ..., x_n)$$

We conclude that T is a sufficient statistic.   ◆

Example

Let X_1, X_2, ..., X_n be a collection of independent random variables, each having a normal density with mean μ and variance σ^2. Both parameters are assumed unknown, and the state is a two-dimensional vector: $\theta \equiv (\mu, \sigma^2)$. The joint density of the X_i is

$$f(x_1,\ldots,x_n,\theta) = (2\pi\sigma^2)^{-n/2} \exp\left(-\frac{1}{2\sigma^2} \sum_{i=1}^{n} (x_i - \mu)^2\right)$$

$$= (2\pi\sigma^2)^{-n/2} \exp\left(-\frac{n\mu^2}{2\sigma^2}\right) \exp\left[-\frac{1}{2\sigma^2}\left(\sum_{i=1}^{n} x_i^2 - 2\mu \sum_i x_i\right)\right]$$

We take the statistic as the two-dimensional quantity $T(X_1,\ldots,X_n) \equiv (\sum_i X_i, \sum_i X_i^2)$. Then f is seen to be free of the individual observations x_i, depending only on the statistic T and the state θ. Therefore, T is a sufficient statistic for θ.

A sufficient statistic need not be unique. An alternative statistic which suffices for this example is $T(X_1,\ldots,X_n) \equiv [\overline{X}, (1/n)\sum_i (X_i - \overline{X})^2]$, where $\overline{X} \equiv (1/n)\sum_i X_i$. The first component of this statistic is the sample mean, and the second component is the sample variance. These two quantities are sufficient condensations of the observations $X_1, X_2, \ldots, X_n$. The first component also happens to be a good estimator for μ, and the second a fair estimator for σ^2. ◆

FURTHER READING

Bickel, Peter J. and Kjell A. Doksum. 1977. *Mathematical Statistics: Basic Ideas and Selected Topics*. Oakland, CA: Holden-Day.

Fisher, R.A. 1950. "On the mathematical foundations of theoretical statistics," 1922. Reprinted in R.A. Fisher, *Contributions to Mathematical Statistics*. New York: Wiley.

Halmos, P.R. and L.J. Savage. 1949. Applications of the Radom–Nikodym theorem to sufficient statistics, *Annual Mathematics and Statistics*. 20: 225–241.

Kolmogorov, A. 1942. Sur l'estimation statistique des parametres de la loi de Gauss, *Izv. Akad. Nauk S.S.S.R. Ser. Mat.* 6: 3–32.

Lehman, E.L. 1959. *Testing Statistical Hypotheses*. New York: Wiley.

Mandel, J. 1964. *The Statistical Analysis of Experimental Data*. New York: Wiley.

Answers to Selected Problems

Chapter 2

2.1A (a) Drawing a single card. (b) Any of the cards. (c) Some subset of the cards, e.g., an ace. (d) All the cards in the deck. (e) None of the cards in the deck. (f) Two picture cards are drawn; or at least one "7"; etc.

2.1B (a) Stock price at the end of a day. (b) Any non-negative value, such as 0 or 24.5. (c) Stock price is within 30% of original purchase price, etc. (d) Trial is a list of 6 consecutive end-of-day prices; 6 are needed to make 5 comparisons. An outcome is "5 consecutive increases"; or "Alternating increases and decreases"; etc.

2.2A (a) Frequentist: based on observed data. (b) Objective assessment: equal chance for each outcome. (c) Subjective assessment: personal judgment.

2.3A (a) $P(\text{Both blue}) + P(\text{Both brown}) = (20/30)(19/29) + (10/30)(9/29) = 47/87$. (b) 1.

2.4A $P(\text{Success}) = 0.7\,(0.8) = .56$.

2.5A $P(\text{Err}|1) = 1/400{,}000$. $P(\text{Err}|2) = 1/800{,}000$.
$$P(\text{Err}) = P(1)P(\text{Err}|1) + P(2)P(\text{Err}|2)$$
$$= (2/3)(1/400{,}000) + (1/3)(1/800{,}000) = (5/3)/800{,}000.$$
$$P(1|\text{Err}) = P(1)P(\text{Err}|1)/P(\text{Err})$$
$$= [(2/3)/400{,}000]/[(5/3)/800{,}000] = 4/5.$$

2.6B $P(\text{Finish}) = P(X > 30) = \int_{30}^{\infty}(1/90)e^{-x/90}\,dx = e^{-1/3} = .7165$.

2.7A $E(\text{Revenue}) = 0.6(500) + 0.3(600) + 0.1(200) = \$500$.

2.7B $E(X + Y) = E(X) + E(Y) = E(X) + 2E(X) = 3E(X) = 3np$.

2.8A   (a) $f(x, y) = a^2 e^{-a(x+y)} = (ae^{-ax})(ae^{-ay}) = f_X(x)f_Y(y)$. Since the joint density is factorable, X and Y are independent.

(b) $P\{X > 2/a, Y < 1/a\} = P\{X > 2/a\}P\{Y < 1/a\}$
$= e^{-a(2/a)}(1 - e^{-a(1/a)})$
$= e^{-2}(1 - e^{-1}) = .08555$

2.9A $E(X^2) = \int_a^b x^2 \frac{1}{b-a} dx = \frac{1}{b-a}\left(\frac{b^3 - a^3}{3}\right) = \frac{a^2 + ab + b^2}{3}.$

$\mathrm{Var}(X) = E(X^2) - E^2(X) = \frac{a^2 + ab + b^2}{3} - \left(\frac{a+b}{2}\right)^2 = \frac{(b-a)^2}{12}.$

2.9B $\mathrm{Var}(X) = \frac{1}{3}(1-3)^2 + \frac{1}{3}(3-3)^2 + \frac{1}{3}(5-3)^2 = \frac{8}{3}.$
$\mathrm{Var}(X + Y) = \mathrm{Var}(X) + \mathrm{Var}(Y) = 2(\frac{8}{3}) = \frac{16}{3}.$

2.9C $E[\delta(X)] = E[X - E(X)] = E(X) - E(X) = 0$. The expected deviation is always 0, so we would learn nothing about the deviation of X overall. Further, for any actual samples $x_1, \ldots, x_n$, the deviations $\delta_i = x_i - \bar{x}$ will always sum to 0.

2.10A   (a) $p_X(3) = \binom{10}{3}(.05)^3(1 - .05)^{10-3} \cong .0105.$
(b) $P\{X > 1\} = 1 - p_X(0) = 1 - \binom{10}{0}(.05)^0(1 - .05)^{10} \cong .4013.$

2.10B   (a) $p_X(7) = e^{-3} 3^7/7! \cong .0216.$ (b) $p_X(0) = e^{-3} 3^0/0! \cong .0498.$

2.10C $P\{X > 167\} = P\left\{\frac{x - \mu}{\sigma} > \frac{167 - 163}{2}\right\} = P\{Z > 2\} = 1 - \Phi(2) = 1 - .9772 = .023.$

2.10E   Let $Z \equiv (X - \mu)/\sigma$. Then $dZ = dX/\sigma$. We have

$$E(Z) = \int_{-\infty}^{\infty} z\frac{1}{\sqrt{2\pi}}e^{-z^2/2}\,dz = \frac{1}{\sqrt{2\pi}}e^{-z^2/2}\Big|_{+\infty}^{-\infty} = 0.$$

Also, $E(Z) = E[(X - \mu)/\sigma] = [EX - \mu]/\sigma$. Since $E(Z) = 0$, we obtain $E(X) = \mu$. To determine the variance, first consider the quantity $A = \int_{-\infty}^{\infty} e^{-u^2/2}\,du$. Its square is $A^2 = \int_{-\infty}^{\infty} e^{-u^2/2}\,du \int_{-\infty}^{\infty} e^{-v^2/2}\,dv = \int_{-\infty}^{\infty}\int_{-\infty}^{\infty} e^{-(u^2+v^2)/2}\,du\,dv$. We change variables from rectangular to polar coordinates. Thus, $u^2 + v^2 = r^2$ and $du\,dv = r\,dr\,d\theta$, with appropriate changes in limits of integration: $A^2 = \int_0^{2\pi}\int_0^{\infty} e^{-r^2/2} r\,dr\,d\theta = 2\pi$. Thus $A = \sqrt{2\pi}$. Now

$$\mathrm{Var}(Z) = \int_{-\infty}^{\infty} z^2 \frac{e^{-z^2/2}}{\sqrt{2\pi}}\,dz = \frac{1}{\sqrt{2\pi}}\int_{-\infty}^{\infty} e^{-z^2/2}\,dz = 1.$$

The second equality follows from integration by parts. Finally, $\text{Var}(Z) = \text{Var}([X - \mu]/\sigma) = \text{Var}(X)/\sigma^2$. Since $\text{Var}(Z) = 1$, we conclude $\text{Var}(X) = \sigma^2$.

Chapter 3

3.1A (a) $\Omega = \{(x, y, z): x, y, z$ are integers, and $1 \leq x, y, z \leq 6\}$.
      (b) $E = \{(1, 1, 2), (1, 2, 1), (2, 1, 1)\}$.

3.1B The result holds for the base case $n = 2$:

$$P\left\{\bigcup_{i=1}^{2} E_i\right\} = P\{E_1 \cup E_2\} = P\{E_1\} + P\{E_2\} - P\{E_1 \cap E_2\}$$

$$\leq P\{E_1\} + P\{E_2\} = \sum_{i=1}^{2} P\{E_i\}$$

Assume result holds for n: the inductive hypothesis is

$$P\left\{\bigcup_{i=1}^{n} E_i\right\} \leq \sum_{i=1}^{n} P\{E_i\}.$$

Now let $\bigcup_{i=1}^{n} E_i = E$. Then

$$P\left\{\bigcup_{i=1}^{n+1} E_i\right\} = P\{E \cup E_{n+1}\} = P\{E\} + P\{E_{n+1}\} - P\{E \cup E_{n+1}\}$$

$$\leq P\{E\} + P\{E_{n+1}\}$$

$$\leq \sum_{i=1}^{n} P\{E_i\} + P\{E_{n+1}\} = \sum_{i=1}^{n+1} P\{E_i\}$$

The second inequality follows from the inductive hypothesis.

3.1C $P\{E \cup F \cup G\} = P\{E \cup (F \cup G)\} = P\{E\} + P\{F \cup G\} - P\{E \cap (F \cup G)\}$. From DeMorgan's laws, $E \cap (F \cup G) = (E \cap F) \cup (E \cap G)$. So $P\{E \cap (F \cup G)\} = P\{E \cap F\} + P\{E \cap G\} - P\{(E \cap F) \cap (E \cap G)\}$. Therefore, $P\{E \cup F \cup G\} = P\{E\} + [P\{F\} + P\{G\} - P\{F \cap G\}] - [P\{E \cap F\} + P\{E \cap G\} - P\{E \cap F \cap G\}]$, which yields the desired result.

3.3A (c) No. Independence would require $F(x, y) = F_X(x)F_Y(y)$.

3.3B (a) $F(x, y) = \int_0^x \int_0^y \lambda^2 e^{-\lambda(x+y)} \, dy \, dx$
          $= (1 - e^{-\lambda x})(1 - e^{-\lambda y}) = F_X(x)F_Y(y)$
      $F_X(x) = \int_0^x \int_0^\infty \lambda^2 e^{-\lambda(x+y)} \, dy \, dx = 1 - e^{-\lambda x}$.

By symmetry, $F_Y(y) = 1 - e^{-\lambda y}$.
(b) $f_x(x) = \lambda e^{-\lambda x}$ and $f_Y(y) = \lambda e^{-\lambda y}$
(c) Yes, since $f(x, y) = f_x(x)f_Y(y)$.

3.3C $F(x, y) = \sum_x \sum_y p(x, y) = \sum_x \sum_y p_X(x)p_Y(y).$
$$= \sum_x p_X(x) \sum_y p_Y(y) = F_X(x)F_Y(y).$$

3.4A $E(X, Y) = \sum_x \sum_y xy\, p(x, y) = \sum_x \sum_y xy p_x(x)p_Y(y)$
$$= \sum_x x p_X(x) \sum_y y P_Y(y) = E(X)E(Y).$$

3.5A The basic case for $n = 2$ is $\text{Var}(X_1 + X_2) = \text{Var}(X_1) + \text{Var}(X_2)$. The inductive step assumes that the relation holds for n variables:

$$\text{Var}\left(\sum_{i=1}^{n} X_i\right) = \sum_{i=1}^{n} \text{Var}(X_i).$$

Let $Y = \sum_{i=1}^n X_i$. For $n + 1$ variables, we have

$$\text{Var}\left(\sum_{i=1}^{n+1} X_i\right) = \text{Var}(Y + X_{n+1}) = \text{Var}(Y) + \text{Var}(X_{n+1})$$

$$= \text{Var}\left(\sum_{i=1}^{n} X_i\right) + \text{Var}(X_{n+1})$$

$$= \sum_{i=1}^{n} \text{Var}(X_i) + \text{Var}(X_{n+1})$$

$$= \sum_{i=1}^{n+1} \text{Var}(X_i),$$

as desired.

3.5B $\text{Cov}(X, Y) = E(XY) - E(X)E(Y) = [\frac{1}{3}(-1)(1) + \frac{1}{3}(0)(0) + \frac{1}{3}(1)(1)]$
$- [\frac{1}{3}(-1) + \frac{1}{3}(0) + \frac{1}{3}(1)][\frac{1}{3}(0) + \frac{2}{3}(1)] = 0.$ However, $p(0, 0) = \frac{1}{3} \neq p_X(0)p_Y(0) = \frac{1}{3}(\frac{1}{3}) = \frac{1}{9}.$ In fact, $p(x, y) \neq p_X(x)p_Y(y)$ for all (x, y) pairs.

3.5C $$\text{Var}\left(\sum_{i=1}^{n} X_i\right) = E\left[\sum_{i=1}^{n} X_i - E\left(\sum_{i=1}^{n} X_i\right)\right]^2$$

$$= E\left(\sum_{i=1}^{n} X_i - \sum_{i=1}^{n} EX_i\right)^2 = E\left(\sum_{i=1}^{n} (X_i - EX_i)\right)^2$$

$$= E\left[\sum_{i=1}^{n} (X_i - EX_i)^2 + 2\sum_{i<j}\sum (X_i - EX_i)X_j - EX_j\right]$$

$$= \sum_{i=1}^{n} \text{Var}(X_i) + 2\sum_{i<j}\sum \text{Cov}(X_i, X_j).$$

3.6A $M_X(t) = \sum_{x=0}^{n} (e^{tx}) \left[\binom{n}{x} p^x (1-p)^{n-x} \right] = \sum_{x=0}^{n} \binom{n}{x} (pe^t)^x (1-p)^{n-x} =$
$[(pe^t) + (1-p)]^n$. The last step follows from the fact that $(u+v)^n = \sum_{x=0}^{n} u^x v^{n-x}$ for any u and v. The mean and variance of X are straightforward calculations.

3.6B   (a) $M_X(t) = \int_{-\infty}^{\infty} \exp(tx)(1/\sigma\sqrt{2\pi})\exp[-(x-\mu)^2/2\sigma^2]\,dx$. Let $Z = (X-\mu)/\sigma$. Then $X = \sigma Z + \mu$ and $dX = \sigma\,dZ$.

$$M_X(t) = \frac{1}{\sqrt{2\pi}} \int_{-\infty}^{\infty} \exp\left(\sigma t Z + \mu t - \frac{Z^2}{2}\right) dZ$$

$$= \frac{\exp(\mu t)}{\sqrt{2\pi}} \int_{-\infty}^{\infty} \exp\left[\sigma t Z - \frac{Z^2}{2} + \left(\frac{\sigma^2 t^2}{2} - \frac{\sigma^2 t^2}{2}\right)\right] dZ$$

$$= \frac{\exp(\mu t + \sigma^2 t^2/2)}{\sqrt{2\pi}} \int_{-\infty}^{\infty} \exp\left(\frac{-Z^2 + 2\sigma t Z - \sigma^2 t^2}{2}\right) dZ$$

The second equality follows from adding and subtracting $\sigma^2 t^2/2$. Now let $Y = Z - \sigma t$. Then the integral, with the $\sqrt{2\pi}$ denominator, is a normal density function whose integral equals 1. Thus, $M_X(t) = \exp(\mu t + \sigma^2 t^2/2)$.

(b) $E(X) = M_X'(t)|_0 = [\mu + (\sigma^2)(0)]e^{(0)} = \mu$

$E(X^2) = M_X''(t)|_0 = \sigma^2 e^0 + [\mu + (\sigma^2)(0)]\{[\mu + \sigma^2(0)]e^0\}$

$= \sigma^2 + \mu^2$

$\text{Var}(X) = EX^2 - E^2 X = \sigma^2$

3.6C $M_{X_i}(t) = \exp[\lambda_i(e^t - 1)]$. $M_Y(t) = M_{X_1}(t) M_{X_2}(t) = \exp[(\lambda_1 + \lambda_2)(e^t - 1)]$. Since the moment generating function uniquely characterizes a distribution, we see that $(X_1 + X_2)$ is another Poisson random variable with distribution parameter $(\lambda_1 + \lambda_2)$.

3.7A $E(X|0) = E(Y|0) = 4/5$. $E(X|1) = E(Y|1) = 1/5$. $E(X) = E(Y) = 1/2 = E[E(X|Y)] = E[E(Y|X)]$.

3.7B $$E[E(X|Y)] = \sum_y p_Y(y) E(X|Y=y) = \sum_y p_Y(y)\left(\sum_x x \frac{p(x,y)}{p_Y(y)}\right)$$

$$= \sum_x x \sum_y p(x,y) = \sum_x x p_X(x) = E(X).$$

3.7C   Let N be the number of orders, X the amount per order, and $Y = NX$ the monthly revenue. The mean is $E(Y) = E(N)E(X) = 500(\$9{,}000) = \$4{,}5000{,}000$. Also, $\mathrm{Var}(Y) = E(N)\,\mathrm{Var}(X) + (EX)^2\,\mathrm{Var}(N) = 500(81 \times 10^6) + (9{,}000)^2(500) = \$81 \times 10^9$, while $\sigma(Y) = \$284{,}605$.

3.8A $E(\overline{X}_n) = 5$ for all n. $\mathrm{Var}(\overline{X}_n) = \sigma^2/n \to 0$ as $n \to \infty$. So the probability of $\overline{X}_n$ differing from the mean of 5 vanishes: $P\{|\overline{X}_n - 5| > \varepsilon\} \to 0$ for any $\varepsilon > 0$.

3.10A $P(\text{Like} = 3/4)$. $P(\text{Ask}|\text{Like}) = 2/3$. $P(\text{Accept}|\text{Ask, Like}) = 1/2$.
   (a) $P_{R_1}(3) = (3/4)(1 - 3/4)^2 = 3/64$.
   (b) Memoryless assumption implies that the required probability is the chance of meeting the second likeable boy at the fourth party.

$$P_{R_2}(4) = \binom{4-1}{2-1}\left(\frac{3}{4}\right)^2\left(1 - \frac{3}{4}\right)^2 = \frac{27}{256}$$

   (c) $P(\text{Accept}) = P(\text{Like})P(\text{Ask}|\text{Like})P(\text{Accept}|\text{Ask, Like}) = \frac{3}{4}\left(\frac{2}{3}\right)\frac{1}{2} = \frac{1}{4}$.
   Let $S_i = $ variable for the ith date.

$$P_{S_2}(5) = \binom{5-1}{2-1}\left(\frac{1}{4}\right)^2\left(1 - \frac{1}{4}\right)^3 = \frac{27}{256}$$

3.11A (a)
$$M_X(t) = E(e^k) = \sum_{k=0}^{\infty} e^{tk}\frac{e^{-\lambda}\lambda^k}{k!}$$

$$= e^{-\lambda}\sum_{k=0}^{\infty}\frac{(\lambda e^t)^k}{k!} = e^{-\lambda}e^{\lambda e^t}$$

$$= e^{\lambda(e^t - 1)}.$$

The fourth equality follows from $e^x = \sum_{k=0}^{\infty} x^k/k!$.

   (b)
$$E(X) = M_X'(t)|_0 = e^{\lambda(e^t - 1)}(\lambda e^t)|_0 = \lambda$$

$$E(X^2) = M_X''(t)|_0 = \lambda e^0(e^0) + \lambda e^0(\lambda e^0 e^0) = \lambda + \lambda^2$$

$$\mathrm{Var}(X) = EX^2 - E^2X = \lambda.$$

3.11B Let X be the time until he is discovered by a beast of prey, and Y the time to rescue. The chance of survival equals the probability that X

exceeds Y:

$$E(X > Y) = \int_{y=0}^{\infty} \int_{x=y}^{\infty} (\mu e^{-\mu y})(\lambda e^{-\lambda x}) \, dx \, dy$$

$$= \int_{0}^{\infty} \mu e^{-\mu y} \int_{x=y}^{\infty} \lambda e^{-\lambda x} \, dx \, dy = \int_{0}^{\infty} \mu e^{-(\lambda+\mu)y} \, dy$$

$$= \frac{\mu}{\lambda + \mu} = \frac{1/39}{1/13 + 1/39} = \frac{1}{4}$$

Note that if $\mu \ll \lambda$, then the probability of survival falls toward 0. On the other hand, if $\mu \gg \lambda$, then the chance of rescue approaches unity.

3.11C (a) Newborn rabbit lives if second-order interarrival time exceeds 10 days: $P\{R_2 > 10\} = \lambda r e^{-\lambda r} + e^{-\lambda r} = 0.2(10)e^{-0.2(10)} + e^{-0.2(10)} = 0.406$.

(b) Severely ill if R_2 lies between 10 and 20 days. Prob. = $P\{R_2 > 10\} - P\{R_2 > 20\} = [.2(10)e^{-.2(10)} + e^{-.2(10)}] - [.2(20)e^{-.2(20)} + e^{-.2(20)}] = .314$.

(c) Poisson process has no memory. Chance of being severely ill is chance of second-order interarrival time being under 10 days. From Part (a), we know this is $1 - .406 = 0.594$.

(d) Viral and bacterial processes are independent. Overall arrival rate is $\lambda'' = 0.2 + 0.3 = 0.5/\text{day}$. Thus, $P\{R_2 > 10\} = (.5)10e^{-.5(10)} + e^{-.5(10)} = .0404$.

Chapter 4

4.3A (a) $\bar{x} = 6.4$. $s^2 = 2.54$.

(b) $\bar{x} = 6.4$. $s^2 = 2.64$.

4.3B We have $y/(n-1)$ where y is given by

$$y = \sum_i (x_i - \bar{x})^2 = \sum_i (x_i^2 - 2x_i\bar{x} + \bar{x}^2)$$

$$= \sum_i x_i^2 - 2\bar{x} \sum_i x_i + n\bar{x}^2 = \sum_i x_i^2 - 2\bar{x}(n\bar{x}) + n\bar{x}^2$$

$$= \sum_i x_i^2 - n\bar{x}^2.$$

4.4A $\mathscr{L}(x|\lambda) = \prod_{i=1}^{n} (\lambda e^{-\lambda x_i}) = \lambda^n e^{-\lambda \sum_i x_i}$. Taking logarithms yields $\ln \mathscr{L} = n \ln \lambda - \lambda \sum_i x_i$. Differentiating $\ln \mathscr{L}$ and setting it to zero yields $\hat{\lambda} = n/\sum_i x_i = 1/\bar{x}$

4.5A   The sample average $\overline{X}$ is distributed normally with mean μ and standard deviation $\sigma/\sqrt{n}$. So $Z \equiv (\overline{X} - \mu)/(\sigma/\sqrt{n})$ is a standard normal variable. Since $\Phi(2.58) = .995$, the 99% interval is defined by $\mu = \overline{x} \pm z_c\sigma/\sqrt{n} = 60 \pm 2.58(20)/\sqrt{100}$. The interval is (54.84, 65.16).

4.6A   For $\Phi(z_c) = .99$, we have $z_c = 2.33$. The threshold is given by $\overline{x}_c = \mu + z_c\sigma/\sqrt{n} = 500 + 2.33(100)/\sqrt{400} = 511.65$. Since $\overline{x} = 510 < \overline{x}_c$, we must accept the null hypothesis.

4.6B   The sample average is $\overline{x} = 6.4$, whereas the sample standard deviation is $s = 1.62$. Since the variance is unknown, it must be estimated by s. Consequently, the procedure $t = (\overline{x} - \mu)/(s/\sqrt{n})$ follows a t-distribution. Since one parameter is being estimated in computing t, the degree of freedom is $v = n - 1 = 10 - 1 = 9$.
  (a) From Table B–2, 90% of the area falls within $\pm t_c = 1.8331$ for $v = 9$. Hence, $\mu = \overline{x} \pm t_c s/\sqrt{n} = 6.4 \pm 1.8331(1.62)/\sqrt{10} = 6.4 \pm 0.94$. That is, the 90% interval is (5.46, 7.34).
  (b) 99% of the area falls within $\pm t_c = 3.2498$ for $v = 9$. The thresholds are at $\overline{x} = 6.4 \pm 3.2498(1.62)/\sqrt{10} = 6.4 \pm 1.66$. So the interval is (4.74, 8.06).

Chapter 5

5.2A   (a)

Parameter θ_i	Procedure δ_j	Loss $L(\theta_i, \delta_j(X_k))$			
		$x_1 = 0$	$x_2 = 1$	$x_3 = 2$	$x_4 = 3$
$\theta_1 = .1$	δ_1	0	0	0	0
	δ_2	.1	.1	.1	.1
	δ_3	0	.1	.1	.1
$\theta_2 = .2$	δ_1	.1	.1	.1	.1
	δ_2	0	0	0	0
	δ_3	.1	0	0	0

(b) $$R(\theta_1, \delta_1) = 0. \quad R(\theta_1, \delta_2) = 0.1.$$

$$R(\theta_1, \delta_3) = 0(.9)^3 + .1[3(.9)^2(.1) + 3(.9)(.1)^2 + (.1)^3]$$

$$= 0.0271$$

$$R(\theta_2, \delta_1) = 0.1. \quad R(\theta_2, \delta_2) = 0.$$

$$R(\theta_2, \delta_3) = (.1)(.8)^3 = 0.0512.$$

(c) Maximum risk to δ_1 and δ_2 are both 0.1. Maximum risk for δ_3 is 0.0512; so δ_3 is minimax procedure.

(d)
$$r(\delta_1) = .6(0) + .4(.1) = .04$$
$$r(\delta_2) = .6(.1) + .4(0) = .06$$
$$r(\delta_3) = .6(.0271) + .4(.0512) = .0367$$

Decision rule δ_3 is the Bayes' procedure as well.

5.2B   (a) $R(\theta, \delta) = EL(\theta, \delta(X)) = E(cX - \theta)^2 = c^2 E[(X - \theta) + \theta - \theta/c)]^2$
$$= c^2[\text{Var}(X) + 0 + (\theta - \theta/c)^2]$$
$$= c^2[1 + 0 + \theta^2(c - 1)^2/c^2] = c^2 + \theta^2(c - 1)^2$$

(b) Risk R increases indefinitely with increasing θ. The only bounded value occurs for $c = 1$. So $\delta(X) = X$ is the minimax procedure.

(c) $r(\delta) = ER(\theta, \delta) = \int_0^\infty [c^2 + \theta^2(c - 1)^2]e^{-\theta} d\theta = 3c^2 - 4c + 2$. The minimum occurs at $c = 2/3$; the Bayes' procedure is $\delta(X) = 2X/3$, for which the mean risk is $2/3$.

Chapter 6

6.1F   Let $\gamma(X) \equiv n\delta/\theta$.

(a) $n = E(\gamma) = nE(\delta)/\theta$. Thus, $E(\delta) = \theta$.

(b) $2n = \text{Var}(\gamma) = n^2 \text{Var}(\delta)/\theta^2$. Thus, $\text{Var}(\delta) = 2\theta^2/n$.

(c) The Fisher information due to a single observation is $I_1(\theta) = E[(\gamma/\gamma\theta)\ln f(x|\theta)]^2 = E[-1/2\theta + Z/(2\theta)]^2$. Here $Z \equiv (X - \mu)^2/\theta$, which follows a χ^2-distribution with 1 degree of freedom. Completing the square leads to $I_1(\theta) = 1/4\theta^2 - E(Z)/2\theta^2 + E(Z^2)/4\theta^2$. Further, $E(Z) = 1$ and $E(Z^2) = \text{Var}(Z) + E^2(Z) = 2 + 1 = 3$. Thus, $I_1(\theta) = 1/2\theta^2$. The lower bound is $\mathcal{M}(\theta) = [nI_1(\theta)]^{-1} = 2\theta^2/n = \text{Var}(\delta)$. So efficiency is 1.

Chapter 7

7.6A   (a) $\Theta = (0, 7) \cup (7, \infty)$ and $\Theta_0 = \{7\}$.

(b) We have $\mathcal{L}(x|\Theta) = \prod_{i=1}^{10} e^{-\mu}\mu^{x_i}/x_i! = e^{-10\mu}\mu^{\Sigma_i x_i}/(\prod_i x_i!)$, whereas $\mathcal{L}(x|\Theta_0) = e^{-10(7)}(7)^{\Sigma_i x_i}/(\prod_i x_i)$. The maximum likelihood estimator is $\hat{\mu} = \bar{x}$. Also, $\sum_i x_i = 10\bar{x}$. So $\lambda = [e^{\bar{x}-7}(7)^{\bar{x}}/(\bar{x})^{\bar{x}}]^{10}$. We obtain $-2\ln \lambda = -20[\bar{x} - 7 + \bar{x}\ln(7/\bar{x})] = 4.745$, since $\bar{x}$ can be calculated as 8.9.

(c) There is one independent parameter in Θ, and none in Θ_0. So $-2\ln \lambda$ follows a χ^2-density with 1 degree of freedom. From Table

B–3, the threshold is $\chi_c^2 = 3.84$. The critical region is defined by $\lambda \le \lambda_c = \exp(-\chi_c^2/2)$, or the set of $\bar{x}$ for which $\bar{x}[\ln(\bar{x}/7) - 1] \ge \chi_c^2/20 - 7$.

Chapter 8

8.4A (a) $\gamma \equiv \bar{x} = 7.9$ kg. For the Bayes' procedure, we first find $\rho = \sigma^2/(n\tau^2) = 100^2/20(10^2) = 5$. Then

$$\delta(\mu) = E(\mu|x) = \frac{\rho}{1 + \rho} v + \frac{1}{1 + \rho} \bar{x}$$

$$= \frac{5}{1 + 5}(8) + \frac{1}{1 + 5}(7.9) = 7.983 \text{ kg}$$

(b) $r(\delta)/r(\gamma) = 1/(1 + \rho) = 1/(1 + 5) = 1/6$.

Chapter 9

9.3A Wendy's indifference between lottery and fruit implies $u([$ice cream, 3/4; cheesecake$]) = u($fruit$)$. The left-hand side equals $(3/4)u($ice cream$) + (1 - 3/4)u($cheesecake$) = (3/4)u($ice cream$) + (1 - 3/4)(-2)$. This equals $u($fruit$) \equiv 1$. Result is $u($icecream$) = 2$.

9.3B (a) Utility of lottery $= (3/4)v($ice cream$) + (1 - 3/4)v($cheesecake$) = (3/4)(8) + (1 - 3/4)0 = 6$; this equals $u($fruit$)$.
(b) $v = 2u + 4$.

9.4F We know the formula holds for $n = 2$: $u([C_1, c_1; C_2]) = c_1 u(C_1) + c_2 u(C_2)$. For the inductive hypothesis, assume $u([C_1, c_1; \ldots; C_n]) = \sum_1^n c_i u(c_i)$. Now consider the lottery $[C_1, c_1; \ldots; C_{n+1}]$. It is equivalent to $[B_n, b_n; C_{n+1}, c_{n+1}]$, where $b_n \equiv \sum_1^n c_i$ and B_n is the sublottery $[C_1, d_1; \ldots; C_n, d_n]$. In this sublottery, $\sum_1^n c_i = 1$; so $d_i = c_i/\sum_1^n c_i = c_i/b_n$. We have

$$u([C_1, c_1; \ldots; C_{n+1}, c_{n+1}]) = u([B_n, b_n; C_{n+1}, c_{n+1}])$$

$$= b_n u(B_n) + c_{n+1} u(C_{n+1})$$

Now $u(B_n) = \sum_1^n d_i u(C_i) = \sum_1^n [c_i/b_n] u(C_i) = \sum_1^n c_i u(C_i)/b_n$. Substituting this expression into the above equation leads to

$$u([C_1, c_1; \ldots; C_{n+1}]) = \sum_{i=1}^{n+1} c_i u(C_i).$$

9.5A
$$\text{EMV} = \sum_{n=1}^{\infty} p_{Mn} M_n = \sum_{n=1}^{14} \left(\frac{1}{2^n}\right)(2^n) + \sum_{n=15}^{\infty} \left(\frac{1}{2^n}\right)(2^{14})$$

$$= \sum_{1}^{14} 1 + \left(\frac{1}{2} + \frac{1}{4} + \cdots\right) = 15.$$

9.5B   (a) Chart is a ramp from $(0,0)$ to $(16,4)$, then a concave rising arc.
   (b) EMV = $(\$0 + \$1 + \$2)/3 = \$8.67$.
   (c) $u([\$0, 1/3; \$1, 1/3; \$25]) = [u(\$0) + u(\$1) + u(\$25)]/3 = 2$.
   Since $u(\$7) = 1.75$, we conclude CME = $7.

9.5D   (a) $-\$3,000$ to $+\$10,000$.
   (b) EMV = $(-3 - 1 + 0 + 2 + 7 + 10)/6 = \$15,000/6 = \$2,500$.
   (c) ———.

Appendix A

A.1A $\mathbf{A} = \{1, 3, 3, 5\} = \{1, 3, 5\}$; $\mathbf{B} = \{1, 3, 5\}$; $\mathbf{C} = \{1, 3, 5\}$. Therefore $\mathbf{A} = \mathbf{B} = \mathbf{C}$.

A.2A   (a) Yes. $\mathbf{B}_i = (0, 1/i)$, whereas $\mathbf{B}_{i+1} = (0, 1/(i + 1))$. Since $1/(i + 1) < 1/i$, $\mathbf{B}_{i+1} \subset \mathbf{B}_i$.
   (b) No, since $\mathbf{B}_i \subset \mathbf{B}_{i+1}$ is false.

A.2B   We assume $\mathbf{A} \subset \mathbf{B}$. Consequently, if $x \in \mathbf{A}$, then $x \in \mathbf{B}$.
   (a) *Necessity.* Suppose that $x \in \mathbf{A} \cup \mathbf{B}$. Then $x \in \mathbf{A}$ or $x \in \mathbf{B}$. By the assumption, $x \in \mathbf{B}$ if $x \in \mathbf{A}$. So $x \in \mathbf{B}$. Thus, $\mathbf{A} \cup \mathbf{B} \subset \mathbf{B}$.
   Sufficiency. Suppose $x \in \mathbf{B}$. Then $x \in \mathbf{A} \cup \mathbf{B}$. Thus, $\mathbf{B} \subset \mathbf{A} \cup \mathbf{B}$. These two results imply $\mathbf{A} \cup \mathbf{B} = \mathbf{B}$.
   (b) *Necessity.* Suppose $x \in \mathbf{A} \cap \mathbf{B}$. Then, $x \in \mathbf{A}$ and $x \in \mathbf{B}$. From $x \in \mathbf{A}$, we infer $\mathbf{A} \cap \mathbf{B} \subset \mathbf{A}$.
   Sufficiency. Suppose $x \in \mathbf{A}$. From the assumption, we know $x \in \mathbf{B}$ also. Thus $x \in \mathbf{A} \cap \mathbf{B}$. We infer $\mathbf{A} \subset \mathbf{A} \cap \mathbf{B}$. These two results imply $\mathbf{A} \cap \mathbf{B} = \mathbf{A}$.

A.2C $\mathbf{A} \supset \mathbf{B}$ is equivalent to $\mathbf{B} \subset \mathbf{A}$: if $x \in \mathbf{B}$, then $x \in \mathbf{A}$. $\mathbf{B} \supset \mathbf{C}$ is equivalent to $\mathbf{C} \subset \mathbf{B}$; so if $x \in \mathbf{C}$, then $x \in \mathbf{B}$. From these assumptions, we infer that whenever $x \in \mathbf{C}$, then $x \in \mathbf{B}$, and, in turn, $x \in \mathbf{A}$. In other words, $\mathbf{C} \subset \mathbf{A}$, which is equivalent to $\mathbf{A} \supset \mathbf{C}$.

A.3A   (a) $\mathbf{A} \cup \mathbf{B} = \{1, 2, 3, a, b, c, d\}$.
   (b) $\mathbf{A} \cap \mathbf{B} = \{2\}$.

A.3B   (a) $\bigcup_{i=1}^{M} \mathbf{A}_i = \{n : n \text{ is an integer } \& n \geq 1\} = \mathbf{A}_1$.
   (b) $\bigcap_{i=1}^{M} \mathbf{A}_i = \{n : n \text{ is an integer } \& n \geq M\} = \mathbf{A}_M$.

Further Reading and Selected Bibliography

FURTHER READING

This section offers some suggestions for further reading and reference. For convenience, the publications are classified into three areas: probability, statistics, and decision theory. Although many of the books could fit into more than one category, each is grouped according to its primary focus. All the publications are listed in full in the second part of this addendum.

Probability. Some advanced topics in probability can be found in *Introduction to Stochastic Processes* by Paul Hoel, Sidney Port, and Charles Stone, as well as *Stochastic Processes* by Sheldon Ross. A theoretical introduction to probability is available in *Real Analysis and Probability* by Robert Ash.

Statistics. A textbook of simplicity and elegance is *Introduction to Statistical Theory* by Hoel, Port, and Stone. Another excellent text is *Introduction to the Theory of Statistics* by Alexander Mood and Franklin Graybill.

Decision Theory. The work that engendered the field of decision theory in its modern form is *The Foundations of Statistics* by Leonard Savage; the serious student will want to read through this book at some point in his scholastic lifetime. A book that presents a systematic view of statistics and decision theory is *Mathematical Statistics: A Decision Theoretic Approach* by Thomas Ferguson. A standard textbook on decision analysis is *Decisions with Multiple Objectives* by Ralph Keeney and Howard Raiffa.

SELECTED BIBLIOGRAPHY

A selected list of references follows, including the publications mentioned in the previous section. Additional references are available at the end of each chapter in the body of this book.

Ash, Robert B. 1972. *Real Analysis and Probability*. New York: Academic Press.

Bickel, Peter J., and Kjell A. Doksum. 1977. *Mathematical Statistics: Basic Ideas and Selected Topics*. Oakland, CA: Holden-Day.

Blackwell, David, and M.A. Girshick. 1954. *Theory of Games and Statistical Decisions*. New York: Dover.

Breiman, L. 1968. *Probability*. Reading, MA: Addison-Wesley.

Chernoff, Herman, and Lincoln E. Moses. 1959. *Elementary Decision Theory*. New York: Wiley.

DeGroot, Morris H. 1970. *Optimal Statistical Decisions*. New York: McGraw-Hill.

Doob, J.L. 1953. *Stochastic Processes*. New York: Wiley.

Drake, Alvin W. 1967. *Fundamentals of Applied Probability Theory*. New York: McGraw-Hill.

Drymes, Phoebus J. 1989. *Topics in Advanced Econometrics*. New York: Springer-Verlag.

Feller, W. 1957. *An Introduction to Probability Theory and Its Applications*. Vols. 1 and 2. New York: Wiley.

Ferguson, Thomas S. 1967. *Mathematical Statistics: A Decision Theoretic Approach*. San Diego, CA: Academic.

Fisher, R.A. 1958. *Statistical Methods for Research Workers*. 13th ed. New York: Hafner.

Hoel, Paul G., Sidney C. Port, and Charles J. Stone. 1971. *Introduction to Statistical Theory*. Boston: Houghton Mifflin.

Hoel, Paul G., Sidney C. Port, and Charles J. Stone. 1971. *Introduction to Probability Theory*. Boston: Houghton Mifflin.

Hoel, Paul G., Sidney C. Port, and Charles J. Stone. 1971. *Introduction to Stochastic Processes*. Boston: Houghton Mifflin.

Intriligator, Michael D. 1971. *Mathematical Optimization and Economic Theory*. Englewood Cliffs, NJ: Prentice-Hall.

Keeney, Ralph L., and Howard Raiffa. 1976. *Decisions with Multiple Objectives*. New York: Wiley.

Kendall. M.G., and Stuart, A. 1967. *The Advanced Theory of Statistics*. Vol. II. 2nd ed. New York: Hafner.

Kim, Steven H. 1990. *Designing Intelligence: A Framework for Smart Systems*. New York: Oxford University Press.

Mood, Alexander M., and Franklin A. Graybill. 1963. *Introduction to the Theory of Statistics*. 2nd ed. New York: McGraw-Hill.

Moulin, Hervé. 1986. *Game Theory for the Social Sciences*. 2nd ed. New York: New York University Press.

Rich, Elaine. 1963. *Artificial Intelligence*. New York: McGraw-Hill.

Ross, Sheldon M. 1983. *Stochastic Processes*. New York: Wiley.

Ross, Sheldon M. 1985. *Introduction to Probability Models*. 3rd ed. New York: Academic.

Salvatore, Dominick. 1982. *Statistics and Econometrics*. New York: McGraw-Hill.

Samson, Danny. 1988. *Managerial Decision Analysis*. Homewood, IL: Irwin.
Savage, Leonard J. 1972. *The Foundations of Statistics*. 2nd ed. New York: Dover.
von Neumann, John, and Oskar Morgenstern. 1980. *Theory of Games and Economic Behavior*. Princeton, NJ: Princeton University Press.
Wald, Abraham. 1947. *Sequential Analysis*. New York: Wiley.
Wilks, Samuel S. 1962. *Mathematical Statistics*. New York: Wiley.

Index

Milton Keynes UK
Ingram Content Group UK Ltd.
UKHW020018071024
449327UK00032B/2832